Biological Properties of Peptidoglycan

Biological Properties of Peptidoglycan

Proceedings
Second International Workshop
Munich, Federal Republic of Germany
May 20–21, 1985

Editors
P. H. Seidl · K. H. Schleifer

W DE G

Walter de Gruyter · Berlin · New York 1986

Editors
Peter H. Seidl, Dr. rer. nat.
Karl Heinz Schleifer, Prof. Dr. rer. nat.
Lehrstuhl für Mikrobiologie
Technische Universität München
Arcisstraße 21
D-8000 München 2
Federal Republic of Germany

Library of Congress Cataloging in Publication Data

Biological properties of peptidoglycan.
„Second International Workshop on the Biological Properties of Peptidoglycan"--Pref.
Includes bibliographies and indexes.
1. Peptidoglycans--Congresses. I. Seidl, P. H. (Peter H.), 1948– .
II. Schleifer, Karl H. III. International Workshop on the Biological Properties of Peptidoglycan (2nd : 1985 : Munich, Germany)
[DNLM: 1. Peptidoglycan--congresses. QW 52 B6138 1985]
QP702.P47B56 1986 616.07'9 86-19791
ISBN 0-89925-262-1 (U.S.)

CIP-Kurztitelaufnahme der Deutschen Bibliothek

Biological properties of peptidoglycan : proceedings, 2. internat. workshop, Munich, Fed. Republic of Germany, May 20–21, 1985 / [2. Internat. Workshop on the Biolog. Properties of Peptidoglycan]. Ed. P. H. Seidl ; K. H. Schleifer. – Berlin ; New York : de Gruyter, 1986.
ISBN 3-11-010737-6

NE: Seidl, Peter H. [Hrsg.]; International Workshop on the Biological Properties of Peptidoglycan <02, 1985, München>

 Printing: Gerike GmbH, Berlin. – Binding: D. Mikolai, Berlin. – Printed in Germany.

PREFACE

This book is based on the lectures presented at the "Second International Workshop on the Biological Properties of Peptidoglycan" in Munich from May 20 to 21, 1985. The aim of this workshop was to summarize the current state of knowledge and to demonstrate the manifold progress made since the first workshop held on this topic in 1974. Included in this book are contributions from various scientific disciplines, covering such fields as immunology, pathology, microbiology, medicine, preparative organic chemistry, ultrastructure and tumor research.

Some of the data presented at the workshop will indeed result in diagnostic as well as therapeutic application. Determining antibodies to peptidoglycan may be useful in the diagnosis of bacterial infections and 'MDP drugs' could be helpful as adjuvants for veterinary and human vaccines and as anticancer, antitumor and antiparasitic agents.

Besides the applied aspect, numerous papers enable us to understand the biology of peptidoglycan in more detail. Such presentations as the description of soluble peptidoglycan as the natural immunogen may lead to a better understanding of the mechanism of immunogenicity of peptidoglycan. The elegant elucidation of the induction of slow wave sleep by muramyl peptides was one of the highlights of this workshop.

We would like to express our gratitude to the Deutsche Forschungsgemeinschaft for its financial support and to the Carl Friedrich von Siemens Stiftung for its hospitality. Furthermore, the success of our workshop is mainly due to the excellent presentations of the participants and we thank all authors for their contributions. To the publisher we convey our thanks for good co-operation in publishing these proceedings so rapidly. Finally we are grateful to all our colleagues, in particular to B. Heymer, for their help in organizing the workshop and/or in the realization of the proceedings.

Munich, June 1986

P.H. Seidl
K.H. Schleifer

Sponsors

Bayer AG, 5090 Leverkusen, FRG

Behringwerke AG, 3550 Marburg, FRG

Biotest Pharma GmbH, 6000 Frankfurt, FRG

Boehringer Ingelheim Diagnostica, 8046 Garching b. München, FRG

Boehringer Mannheim GmbH, 6800 Mannheim, FRG

Carl Friedrich von Siemens Stiftung, 8000 München, FRG

Ciba-Geigy GmbH, 7867 Wehr/Baden, FRG

Deutsche Forschungsgemeinschaft, 5300 Bonn, FRG

Deutsche Gesellschaft für Hygiene und Mikrobiologie e.V., FRG

Dr. Karl Thomae GmbH, 7950 Biberach an der Riss, FRG

Heinrich Mack Nachf., 7918 Illertissen, FRG

Luitpold-Werk München, 8000 München, FRG

CONTENTS

II. DEGRADATION OF PEPTIDOGLYCAN, PEPTIDOGLYCAN-LIKE PRODUCTS IN THE HOST

III. ACTION OF PEPTIDOGLYCAN ON CELLS

IV. DIVERSE BIOLOGICAL ACTIVITIES OF PEPTIDOGLYCAN

GUEST LECTURE

V. ADJUVANT ACTIVITY AND IMMUNOMODULATION BY PEPTIDOGLYCAN

STRUCTURE AND IMMUNOCHEMISTRY OF PEPTIDOGLYCAN

Peter H. Seidl, Karl Heinz Schleifer
Lehrstuhl für Mikrobiologie, Technische Universität München,
D-8000 München 2, FRG

I. Structure of Peptidoglycan

Peptidoglycan (murein, mucopeptide), the main cell wall polymer of eubacteria, is common to both Gram-positive and Gram-negative bacteria. Only a few prokaryotic organisms such as the Mycoplasmatales and archaebacteria lack peptidoglycan. Chemically, it is a heteropolymer consisting of glycan strands cross-linked through short peptides (Fig. 1). Investigations of the chemical structure of peptidoglycans from numerous bacterial strains demonstrated the

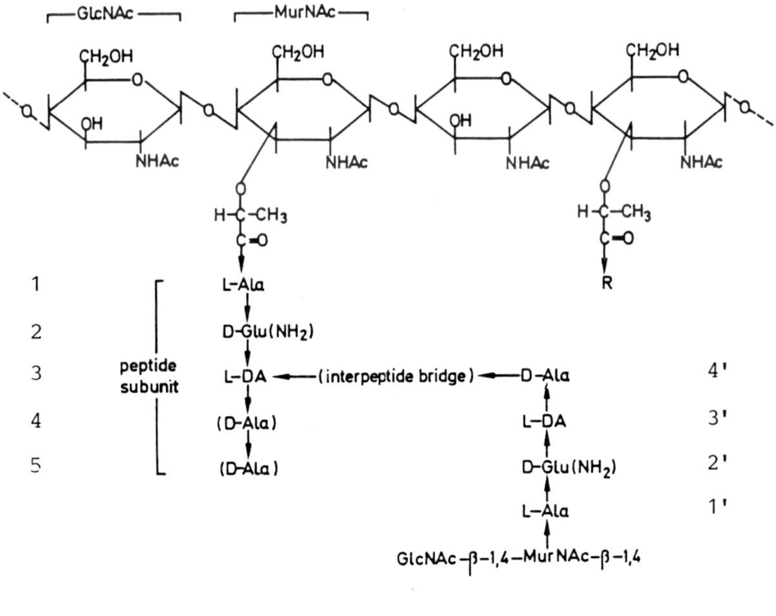

Fig. 1. Scheme of the primary structure of peptidoglycan. Unusual abbreviations: MurNAc, N-acetylmuramic acid; DA, diamino acid. Substituents in brackets may be missing.

Biological Properties of Peptidoglycan

existence of almost 100 different peptidoglycan types (1,2). The glycan moiety of the various peptidoglycans is rather uniform and reveals only a few variations such as acetylation or phosphorylation of the muramyl-6-hydroxyl groups, the occasional absence of N-acetyl substitution or lacking of part of the peptide substituents. In some mycobacteria and nocardiae, the N-acetyl group of muramic acid is oxidized to give an N-glycolyl-residue.

The number of chemically different amino acids occurring in a particular peptidoglycan is restricted to three to six. Peptidoglycan does not contain branched amino acids, aromatic amino acids, cysteine, methionine, arginine, histidine and proline. The stem peptide or peptide subunit linked to the carboxyl group of muramic acid always consists of alternating L- and D-amino acids. Depending on the mode of crosslinking, two main groups of peptidoglycan named A and B, have been distinguished (1). The crosslinkage of Group A peptidoglycans extends from the distal amino group of the diamino acid in position 3 of one stem peptide to the carboxyl group of D-alanine in position 4' of an adjacent stem peptide either directly or via an interpeptide bridge (Fig. 1). Peptidoglycans of Group B are characterized by a crosslinkage between the α-carboxyl group of D-glutamic acid in position 2 in one stem peptide and the C-terminal D-alanine in position 4' of an adjacent stem peptide. There is always a diamino acid necessary in the interpeptide bridge since two carboxyl groups must be crosslinked. Group B peptidoglycans are rather rare and are only found in some coryneform bacteria and in a few anaerobic organisms (2).

The peptidoglycan of Gram-negative bacteria is rather uniform. It usually forms a monomolecular layer and is directly crosslinked. In most cases meso-diaminopimelic acid is present as the diamino acid in position 3 of the stem peptide, but a few exceptions are known. The peptidoglycan of spirochaetes often contains ornithine instead of diaminopimelic acid (3,4) among some fusobacteria, meso-diaminopimelic acid is replaced by its sulphur analogue meso-lanthionine (5).

Gram-positive bacteria contain a multilayered peptidoglycan and reveal a great variation in the composition and primary structure of the constituent peptide moiety. Various peptidoglycan types can be distinguished, depending on the different amino acid sequence of the stem peptide and/or the interpeptide bridge (for review, see 1,2).

The amino acid sequence of the stem peptide shows some variation among the different peptidoglycans (Fig. 2). The amino acids in parantheses may replace the corresponding amino acids or substituents. The greatest variation is found in position 3. To date, 11 different amino acids have been detected in this position, meso-diaminopimelic acid and lysine are the most common ones (for review, see 1,2).

Position 4 is occupied by D-alanine. The carboxyl group of this D-alanine is often blocked by crosslinkage to an adjacent stem peptide. In non-crosslinked stem peptides the D-alanine in position 4 may be split off or substituted by another D-alanine; the latter D-alanine represents a remnant of the peptidoglycan precursor.

Regarding the biological properties of peptidoglycan, one should keep in mind that the structure of peptidoglycan is not as regularly as depicted in the scheme of Fig. 1. Several enzymes of the bacterial cell may split bonds in the peptidoglycan (6). Glucosaminidases and muraminidases split glycosidic linkages in the glycan strand; amidases may split the linkage between muramic acid and L-alanine in position 1 of the peptide subunit, thus leading to muramic acid residues with a free carboxyl group; endopeptidases may split linkages in the stem peptide or in the interpeptide

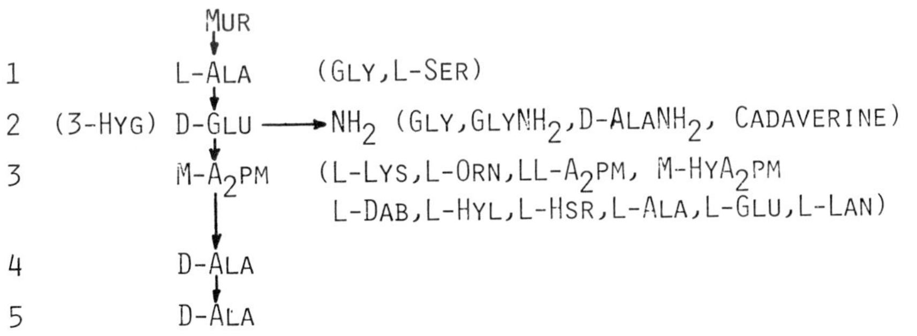

Fig. 2. Variations of the stem peptide (peptide subunit). Amino acids in parentheses may replace the corresponding amino acids or substituents. Unusual abbreviations: A_2pm, 2,6-diaminopimelic acid; Dab, diaminobutyric acid; HyA_2pm, 2,6-diamino-3-hydroxypimelic acid; Hyg, threo-ß-hydroxyglutamic acid; L-Hyl, L-hydroxylysine; Hsr, homoserine; Lan, lanthionine.

bridge, causing additional free amino groups and carboxyl groups; D,D-carboxypeptidases may split off the D-alanine in position 5 of the peptide subunit and L,D-carboxypeptidases may split off the D-alanine in position 4. Strains lacking carboxypeptidases, such as many staphylococci and streptococci, reveal up to 20 per cent of the stem peptides as non-crosslinked pentapeptide subunits with a C-terminal D-alanyl-D-alanine.

II. Immunochemistry of Peptidoglycan

Cell wall peptidoglycan is immunogenic. Up to date, at least five main antigenic epitopes (Fig. 3) of peptidoglycan have been discovered (7) that is a) the glycan strand, b) and c) N-terminal and

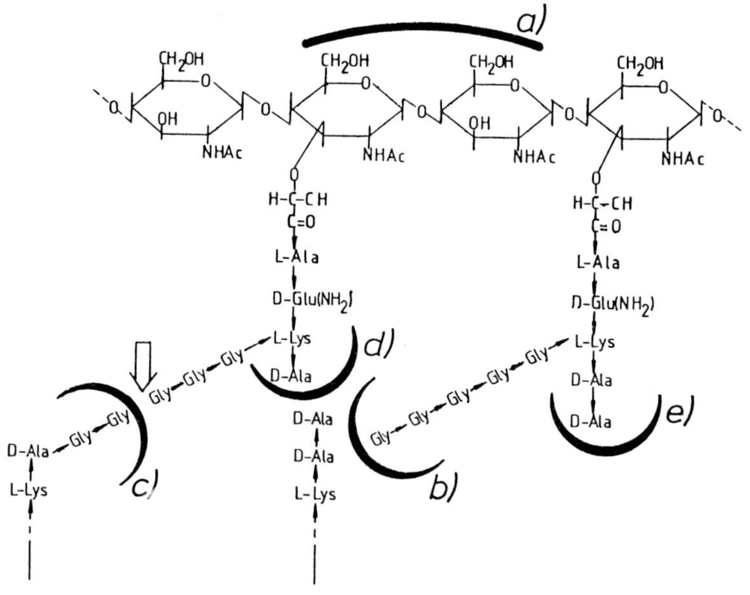

Fig. 3. Schematic representation of the five antigenic epitopes of peptidoglycan: a) the glycan strand; b) N-terminal and c) C-terminal sequences of the interpeptide bridge; d) the non-crosslinked peptide subunit tetrapeptide; e) the non-cross-linked peptide subunit pentapeptide. ⇩: endopeptidases splitting glycyl-glycine bindings.

C-terminal sequences of the interpeptide bridge, d) the non-crosslinked tetrapeptide subunit, and e) the non-crosslinked pentapeptide subunit. In man, specific antibodies to the glycan moiety, to the tetrapeptide subunit and to the pentapeptide subunit have been hitherto detected.

1. Antigenic properties of the glycan moiety

The findings on the immunological properties of the glycan strand are summarized in Table 1. Early studies by Karakawa and Krause (8) on the antigenic properties of peptidoglycan indicated that the hexosamine polymer of peptidoglycan was a main antigenic epitope. Rolicka and Park (9) demonstrated that antibodies against the glycan strand recognized acetylglucosamine and not acetylmuramic acid, whereas Wikler (10), investigating antisera to *Micrococcus luteus*, claimed that acetylmuramic acid and not acetylglucosamine was the immunodominant sugar. Schleifer and Seidl (11) were finally able to resolve this controversy. A glycan strand specific antibody population directed to acetylglucosamine and a po-

ANTIGENIC PROPERTIES OF THE GLYCAN STRAND

GlcNAc immunodominant; cross-reactivity between A-polysaccharide and Peptidoglycan	Karakawa et al., 1967	Specific antibodies to MDP: a) MDP conjugated to carriers via glutamic acid b) MDP conjugated to carriers via anomeric C-1 carbon atom of MurNAc	Reichert et al., 1980
GlcNAc and not MurNAc immunodominant	Rolicka and Park, 1969		
MurNAc and not GlcNAc immunodominant	Wikler, 1975		
Antibody populations to GlcNAc and MurNAc as determinant sugars	Schleifer and Seidl, 1977	Specific antibodies to the glycan moiety in human sera; antibody combining site at least $(GlcNAc\text{-}MurNAc)_2$	Seidl et al., Zauner et al., 1985
MDP does not bind to antibodies against Peptidoglycan	Audibert et al., 1978		

Table 1. Antigenic properties of the glycan strand

pulation directed to acetylmuramic acid were found in the same antiserum.

The occurrence of antibodies to the glycan strand led to the question whether the synthetic immunoadjuvant MDP would bind to such antibodies or not. An answer to this question might indicate whether reactivity of MDP with natural occurring peptidoglycan antibodies could induce clinical hazards, preventing the application of MDP or its derivatives in man. No binding of radiolabelled MDP to naturally occurring or experimentally induced antibodies was detected (12). This lack of reactivity was surprising since MDP represents part of the monomeric subunit of the peptidoglycan backbone. Although MDP in isolated form was found not to elicit an immune response, it became immunogenic when it was coupled to an appropriate carrier. Conjugates, carrying MDP substituents via the γ-carboxyl group of glutamic acid of MDP, or via the anomeric C-1 carbon atom of the N-acetylmuramyl portion of MDP, were prepared (13). Both conjugates were immunogenic and the antibodies reacted with free synthetic MDP.

Recently, we developed a Farr-type radioactive hapten binding assay for detecting glycan moiety specific antibodies. As an antigen, we employed low-molecular weight glycopeptides from _Bacillus subtilis_ peptidoglycan. The structural features of the glycopeptides are presented in detail by Zauner et al., this book. It is important that the glycopeptides isolated did not reveal any pentapeptide subunits with C-terminal D-alanyl-D-alanine, and did not contain non-crosslinked tetrapeptide subunits. Thus, it was excluded that antibodies to the tetra- or pentapeptide subunit were detected employing the glycopeptide as an antigen. The glycopeptides were radiolabelled with iodine and binding of the radioactive glycopeptide hapten to animal hyperimmune sera, to human normal sera and to sera from patients was measured. High binding (70-80%) of radioactively labelled glycopeptides was obtained with rabbit reference hyperimmune sera (11) containing high levels of antibodies to the glycan moiety. No binding was achieved, however, to specific antibodies against the peptide moiety of peptidoglycan (14-18), and to antisera against the A-carbohydrate of streptococci, known to be specifically directed against N-acetylglucosamine (19). Specificity of the test system was further examined in appropriate hapten binding inhibition studies (Fig. 4). Besides the homologous glycopeptide, soluble peptidoglycans from various bacterial strains revealing variations in the glycan moiety as O-acetylation (staphylococci), partial lack of N-acetylation (bacilli) or peptide substitution (Micrococcus luteus), gave a strong inhibitory effect. The size of the antibody combining site was at least

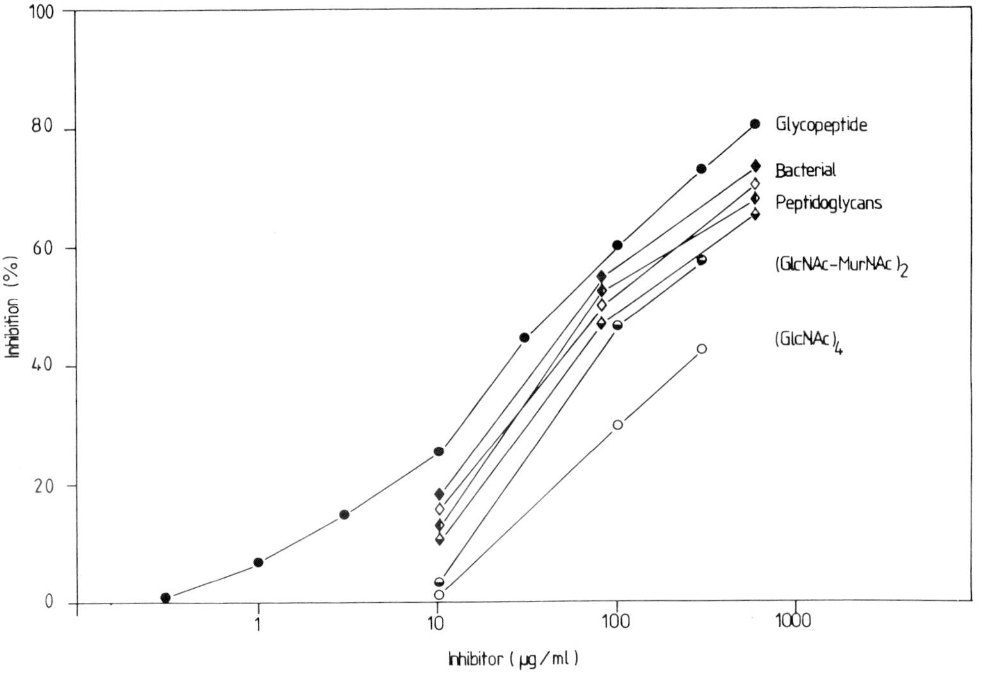

Fig. 4. Binding inhibition reaction of radio-iodinated glycopeptides to peptidoglycan antiserum (A-variant streptococcal antiserum. (●) Bacillus subtilis W23 glycopeptide; (o) acetylchitotetraose; (◒) glycan oligosaccharide $(GlcNAc\text{-}MurNAc)_2$. Soluble peptidoglycans (MW > 10 000) from (◆) Bacillus subtilis, (◇) Bacillus licheniformis, (◆) Micrococcus luteus, (◆) Staphylococcus epidermidis. The following compounds gave no significant inhibition (<10%) at a concentration of 0.3 μmoles/ml: L-rhamnose, N-acetylglucosamine, N-acetylmuramic acid, H-L-Ala-D-Glu (L-Lys-D-Ala-OH)-OH, H-L-Ala-D-Glu(L-Lys-D-Ala-D-Ala-OH)-OH, D-lactyl-L-Ala-D-Glu(meso-A_2pm-OH)-OH, MurNAc-L-Ala-D-Glu(OH)-NH_2.

complementary to that of a tetrasaccharide (compare Fig. 4). No inhibition was achieved in the radioactive glycopeptide binding employing as an inhibitor L-rhamnose, N-acetylglucosamine, N-acetylmuramic acid, the tetrapeptide subunit, the pentapeptide subunit, the lactyltripeptide lactyl-L-Ala-D-Glu(meso-A_2pm-OH)-OH and MDP. The specificity of the test system and the independence of antibodies directed against the glycan moiety and those directed against the peptide part were thus clearly demonstrated. It was further established that the antibody population detected by the iodinated glycopeptides did not recognize a single N-acetyl-amino sugar, thus excluding an interaction of terminal N-acetylglucosamine from bacterial polysaccharides.

At a first glance, specificity of glycan moiety antibodies reported (8-13, Zauner et al., this book) seems to be somewhat contradictory. However, the results only reflect great immunochemical complexity contributed by particular antibody populations.

2. Antigenic properties of the interpeptide bridge

The interpeptide bridge of peptidoglycan contains independent antigenic sites (compare Fig. 2). This was proved applying synthetic peptidyl-protein immunogens carrying as substituents peptides with structural similarity to amino-terminal (17,20,21, Fig. 5a-c) or to

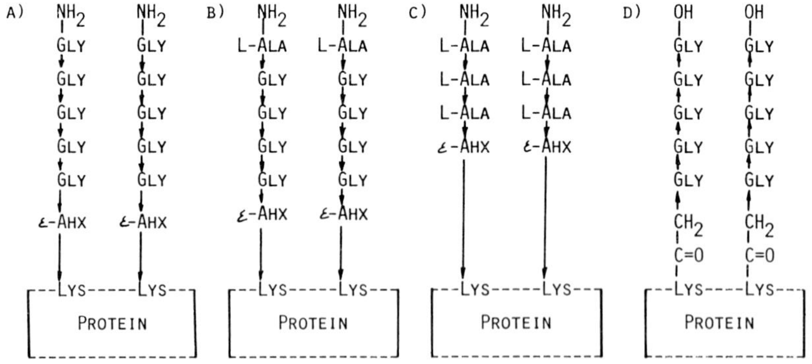

Fig. 5. Scheme of synthetic immunogens with structural similarity a-c) to amino terminal and d) to carboxyl terminal sequences of peptidoglycan interpeptide bridges.

carboxyl-terminal sequences (15, Fig. 5d) of interpeptide bridges. Antisera to (pentaglycyl-ε-aminocaproyl)$_{20}$-albumin (Fig. 5a) gave without exception strong specific precipitin reactions with peptidoglycans from staphylococci, revealing pentaglycine interpeptide bridges. No reaction was obtained, however, with micrococcal peptidoglycans containing three to four L-alanine residues in the interpeptide bridge or with other bacterial peptidoglycans lacking oligoglycine peptides (Fig. 6a (17)). It was noteworthy that even the difference in a single N-terminal amino acid (glycine/L-alanine (compare Fig. 2a,b) was discriminated (20). Antisera to (tri-L-alanine-ε-aminocaproyl)$_{22}$-albumin (Fig. 5c) strongly precipitated micrococcal peptidoglycans with interpeptide bridges composed of three to four L-alanine molecules, but gave a negligible reaction with peptidoglycans lacking such oligoglycine peptides, Fig. 6b, (17). The specific antisera can be used for the serological sepa-

Fig. 6a

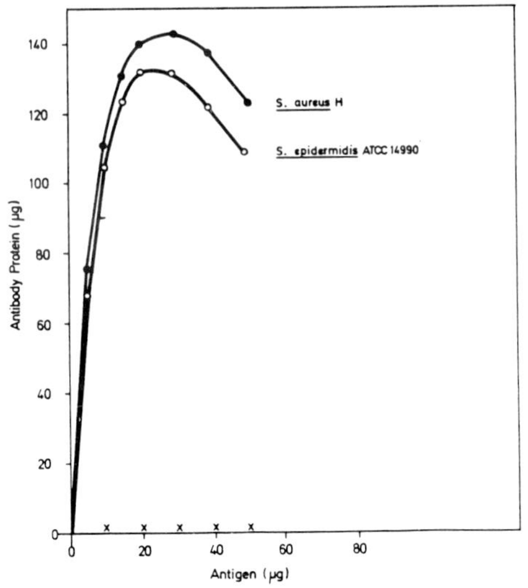

Fig. 6b

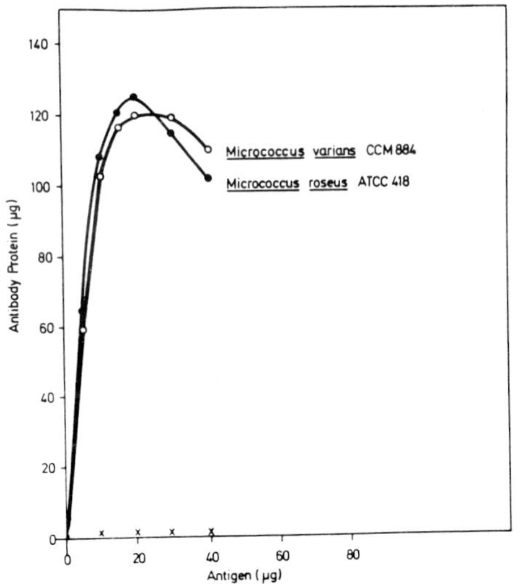

Fig. 6a. Quantitative precipitin reaction of antiserum to (Gly$_5$-ε-Ahx)$_{20}$-albumin with peptidoglycans from (●) *Staphylococcus aureus* H, (o) *S. epidermidis* ATCC 14990, and (x) *Micrococcus varians* CCM 884 or *M. roseus* ATCC 418.

Fig. 6b. Quantitative precipitin reaction of antiserum to (L-Ala$_3$-ε-Ahx)$_{22}$-albumin with peptidoglycans from (●) *Micrococcus roseus* ATCC 418, (o) *M. varians* CCM 883, and (x) *S. aureus* H or *S. epidermidis* ATCC 14990.

ration of staphylococci from micrococci (21).

Carboxyl terminal oligoglycine sequences of the interpeptide bridges from staphylococci further reveal an independent antigenic epitope of peptidoglycan as was proved using appropriate synthetic protein-peptide conjugates (Fig. 5d). Oligoglycine peptides revealing free carboxyl groups are the product of endopeptidases splitting glycylglycine linkages (22).

3. Antigenic properties of the pentapeptide subunit

a) Characterization of animal hyperimmune sera

The stem peptide was presumed to be an antigenic epitope of peptidoglycan already in 1968 by Karakawa and Krause (23). Employing a series of synthetic peptides as inhibitors in the quantitative peptidoglycan precipitin reaction, Schleifer and Krause (1971) demonstrated an antibody population with specificity for the pentapeptide subunit in A-variant streptococcal antisera (24). The conclusions drawn from their results are summarized in Table 2. The pentapeptide subunit being an antigenic epitope was consequently corroborated by several approaches using synthetic antigens. Zeiger and Maurer (1973) coupled synthetic pentapeptide subunits to a random alanine/glutamic acid copolymer (25). Schleifer and Seidl

1. THE PENTAPEPTIDE AND NOT THE PREDOMINANTLY OCCURRING TETRAPEPTIDE IS THE ANTIGENIC DETERMINANT OF THE PEPTIDE SUBUNIT
2. THE ANTIBODIES ARE DIRECTED AGAINST THE C-TERMINAL PORTION OF THE PEPTIDE
3. THE C-TERMINAL D-ALANYL-D-ALANINE IS THE IMMUNODOMINANT SITE OF THE ANTIGEN
4. THE CONTRIBUTION OF THE OTHER AMINO ACIDS TO THE ANTIGENIC SITE IS OF LESSER IMPORTANCE (LYSINE AND GLUTAMIC ACID) OR IS INSIGNIFICANT (L-ALANINE)

Table 2. Characterization of the pentapeptide subunit as an antigenic epitope of peptidoglycan (24)

(1974) synthesized the protein peptide conjugate albumin-(CH_2CO-Gly-L-Ala_2-D-Ala_2-OH)$_{39}$, carrying peptide substituents with C-terminal D-alanyl-D-alanine; the amino acids L-alanine, D-glutamic acid and L-lysine from the stem peptide were replaced (14). Employing the conjugates synthesized by Zeiger and Maurer (1973) and by Schleifer and Seidl (1974) as an immunogen specific antibodies directed against the peptide substituents were obtained. Monospecific antibodies to the C-terminal D-alanyl-D-alanine moiety of the pentapeptide subunit (14) allowed demonstrating the distribution of pentapeptide subunits in the cell walls of several Gram-positive bacteria (26,27, Franken et al., this book).

b) Characterization of human sera

Antibodies directed to the pentapeptide subunit of peptidoglycan are frequent in sera from normal blood donors or from patients (28,29). Recently, Franken et al. (1984, 1985) linked tri-D-alanine covalently to albumin via the amino group (18,30) and employed this protein peptide conjugate as an antigen in an ELISA. Specific IgG (18) and IgA class antibodies (30) against the immunodominant D-alanyl-D-alanine moiety of the pentapeptide subunit were detected in human sera. It was furthermore clear from appropriate inhibition studies in the ELISA (Figure 7) or from inhibition of radiolabelled H-L-Ala-D-Glu(L-Lys-D-Ala_2-OH)-OH hapten binding (18, 31, Figures 8,9) that human sera also may contain antibodies with exclusive or predominant specificity for the carboxyl-terminal D-alanyl-D-alanine sequence of the pentapeptide subunit, similar as shown for streptococcal hyperimmune sera (24, compare Table 2).

The results from the radioactive hapten binding inhibition studies with specific antibodies in human serum 004, depicted in Fig. 9, are of interest for several reasons:

1. In contrast to the strong inhibitory effect of Ac_2-L-Lys-D-Ala-D-Ala-OH or of tri-D-alanine, oligoglycine peptides or D-Ala-Gly-OH did not inhibit, thus excluding cross-reactivity between the C-terminal sequences of the interpeptide bridge (16) and the pentapeptide subunit.

2. Ac_2-L-Lys-D-Ala-D-Ala-OH and Ac_2-L-Lys-D-Ala-D-lactic acid are both substrates for the penicillin-sensitive D-alanine carboxypeptidases of Bacillus subtilis, Escherichia coli and Staphylo-

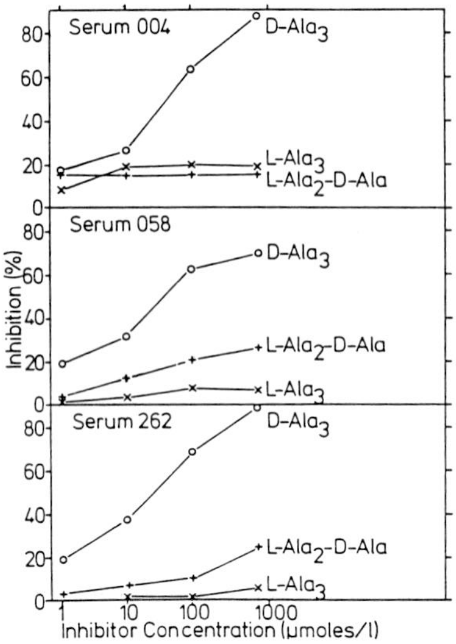

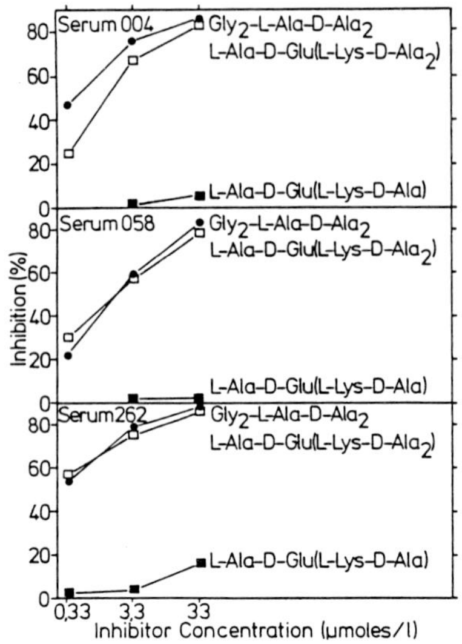

Fig. 7. Inhibition of binding of specific IgG in human sera nos. 004, 058 and 262 to albumin-(D-Ala-D-Ala-D-Ala-OH)$_9$ in the ELISA technique, using as inhibitors several peptides: (o) H-D-Ala-D-Ala-D-Ala-OH, (x) H-L-Ala-L-Ala-L-Ala-OH, (+) H-L-Ala-L-Ala-D-Ala-OH (18).

Fig. 8. Inhibition of radiolabelled H-L-Ala-D-Glu(L-Lys-D-Ala-D-Ala-OH)-OH hapten binding (human sera nos. 004, 058 and 262) by various synthetic inhibitors: (●) H-Gly-Gly-L-Ala-D-Ala-D-Ala-OH), (□) H-L-Ala-D-Glu(L-Lys-D-Ala-D-Ala-OH)-OH, (■) H-L-Ala-D-Glu(L-Lys-D-Ala-OH)-OH (18,31).

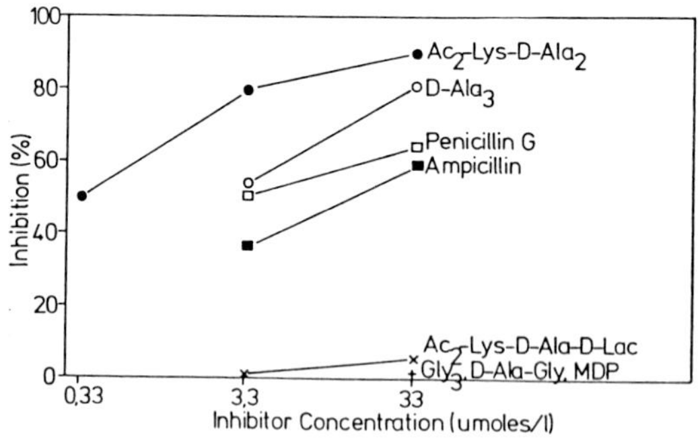

Fig. 9. Inhibition of radiolabelled H-L-Ala-D-Glu(L-Lys-D-Ala-D-Ala-OH)-OH hapten binding (human serum no. 004) by various inhibitors.

coccus aureus (32). The ester substrate was hydrolysed faster than the peptide analogue, diacetyl-L-Lys-D-Ala-D-Ala-OH, by the B. subtilis (15-fold) and E. coli (4-fold) carboxypeptidases, no rate acceleration was observed for the S. aureus carboxypeptidase (32). However, in contrast to the strong inhibition found with Ac_2-L-Lys-D-Ala-D-Ala-OH, Ac_2-L-Lys-D-Ala-D-lactic acid did not react as hapten in the radioimmuno-assay.

3. Due to proposed structural analogy between acyl-D-alanyl-D-alanine and penicillin (32), we investigated a possible crossreactivity between ß-lactam antibiotics and R-D-alanyl-D-alanine and came to the following conclusions:
   - All 22 penicillins and cephalosporins investigated (Penicillin G and Ampicillin depicted in Fig. 9) bound to antibodies with specificity against D-alanyl-D-alanine.
   - In general, cephalosporins bound less than penicillins.
   - An intact ß-lactam ring was absolutely necessary for the interaction of the antibiotic with the antibody.

A failing or very weak inhibitory effect of penicillin or tri-D-alanine on the binding of radioactive H-L-Ala-D-Glu(L-Lys-D-Ala_2-OH)-OH to several human sera led us to studying in detail specificity of human antibodies to the pentapeptide subunit and investigating binding of several iodine labelled synthetic peptides to human sera. Fig. 10 gives some typical examples. As depicted in Fig. 10a, all radioactively labelled peptides employed bound quite similarly to human serum 004. As was already shown from the previous studies (compare Figs. 7-9), this human serum revealed specific antibodies exclusively directed against D-alanyl-D-alanine, similar as described for A-variant streptococcal antisera by Schleifer and Krause, 1971 (24). In contrast, it is evident from Fig. 10b that serum 426 strongly bound the pentapeptide subunit H-L-Ala-D-Glu(L-Lys-D-Ala_2-OH)-OH and the peptide H-Gly-Gly-L-Lys-D-Ala_2-OH; the peptide H-L-Ala-D-Glu(L-Ala-D-Ala_2-OH)-OH revealing L-alanine instead of lysine, or tri-D-alanine inhibited less, thus obviously indicating a dominant role of lysine.
On the other hand, serum 019 bound radioactively labelled H-L-Ala-D-Glu(L-Lys-D-Ala_2-OH)-OH or the peptide L-Ala-Glu(L-Ala-D-Ala_2-OH)-OH to the same extent, peptides lacking D-glutamic acid inhibited less. The results regarding specificity of antibodies direc-

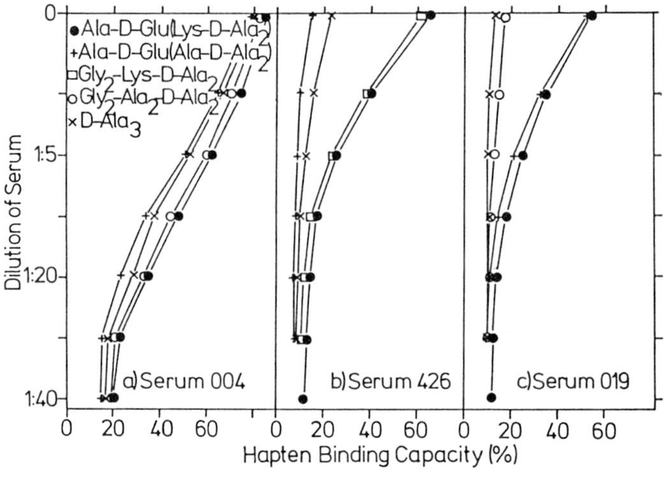

Fig. 10. Binding of radioactively labelled synthetic peptides with structural similarity to the pentapeptide subunit to human sera nos. 004, 426 and 019.

ted against the pentapeptide subunit in human sera are summarized in Table 3.

1. FOR SOME HUMAN SERA, ANTIBODIES DIRECTED AGAINST THE PENTAPEPTIDE SUBUNIT REVEAL PREDOMINANT SPECIFICITY FOR D-ALANINE AS THE IMMUNODOMINANT GROUP, L-LYSINE AND/OR D-GLUTAMIC ACID PLAYING A MINOR ROLE, SIMILAR AS DEMONSTRATED FOR ANIMAL HYPERIMMUNE SERA BY SCHLEIFER AND KRAUSE, 1971 (24).
2. FOR SOME HUMAN SERA, L-LYSINE AND/OR D-GLUTAMIC ACID SIGNIFICANTLY CONTRIBUTE TO THE ANTIGENIC SITE, THE SEQUENCE D-ALANYL-D-ALANINE PLAYING A MINOR ROLE.

Table 3. Specificity of human antibodies directed against the pentapeptide subunit.

4. Antigenic properties of the tetrapeptide subunit

The tetrapeptide subunit further represents an independent antigenic epitope of peptidoglycan. This was clear from serial radioactive hapten binding studies employing as haptens the radioactively labelled tetrapeptide subunit or the pentapeptide subunit. It is obvious from Fig. 11 that serum 033 strongly bound the radiolabelled pentapeptide subunit and lacked specific antibodies to the tetrapeptide subunit. Serum 327, on the contrary, strongly bound the radioactively labelled tetrapeptide subunit whereas the pentapeptide subunit was bound less. The final proof for the occurrence of specific antibodies to the tetrapeptide subunit in serum 327 was furnished by appropriate inhibition studies (Fig. 12).
The tetrapeptide subunit and C-terminal sequences of the tetrapeptide inhibited strongly, whereas the sequences H-L-Lys-L-Ala or C-terminal sequences of the pentapeptide subunit did not inhibit or only insignificantly. The tetrapeptide from the pseudomurein, H-L-Ala-L-Lys-L-Glu-L-Ala-OH did also not inhibit.

Fig. 11

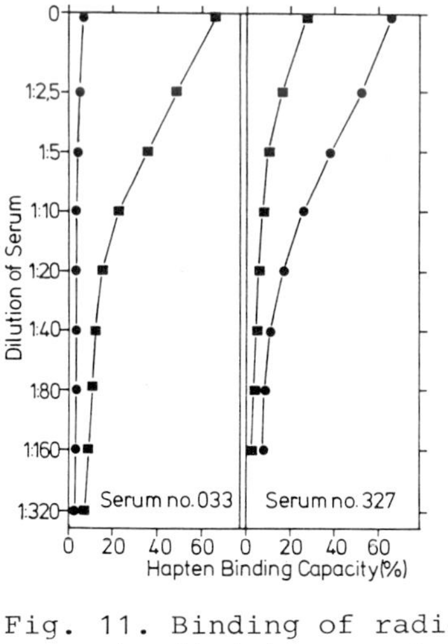

Fig. 12

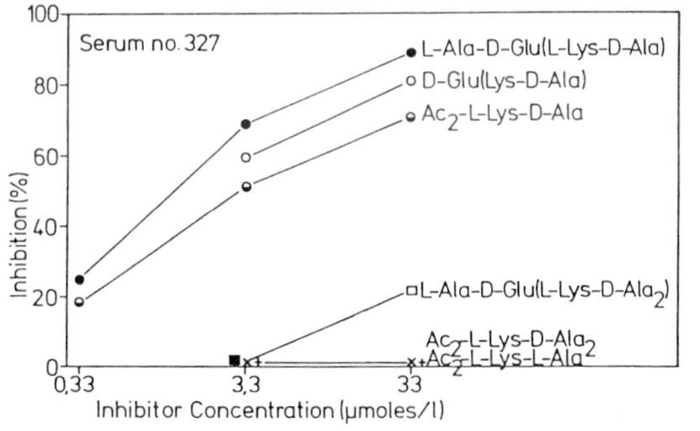

Fig. 11. Binding of radioactively labelled synthetic tetrapeptide subunit (●) or pentapeptide subunit (■) to human sera nos. 033 and 327.

Fig. 12. Inhibition of radiolabelled tetrapeptide subunit H-L-Ala-D-Glu(L-Lys-D-Ala-OH)-OH hapten binding (human serum no. 327) with several synthetic inhibitors.

In conclusion, the stem peptide of peptidoglycan reveals at least two main antigenic epitopes, i.e. the pentapeptide subunit and the tetrapeptide subunit. With few exceptions, no cross-reactivity was observed between the tetrapeptide and the pentapeptide of the stem peptide.

5. Immunoglobulin classes involved in the immune response to peptidoglycan

Specific antibodies to peptidoglycan have been reported of the IgG

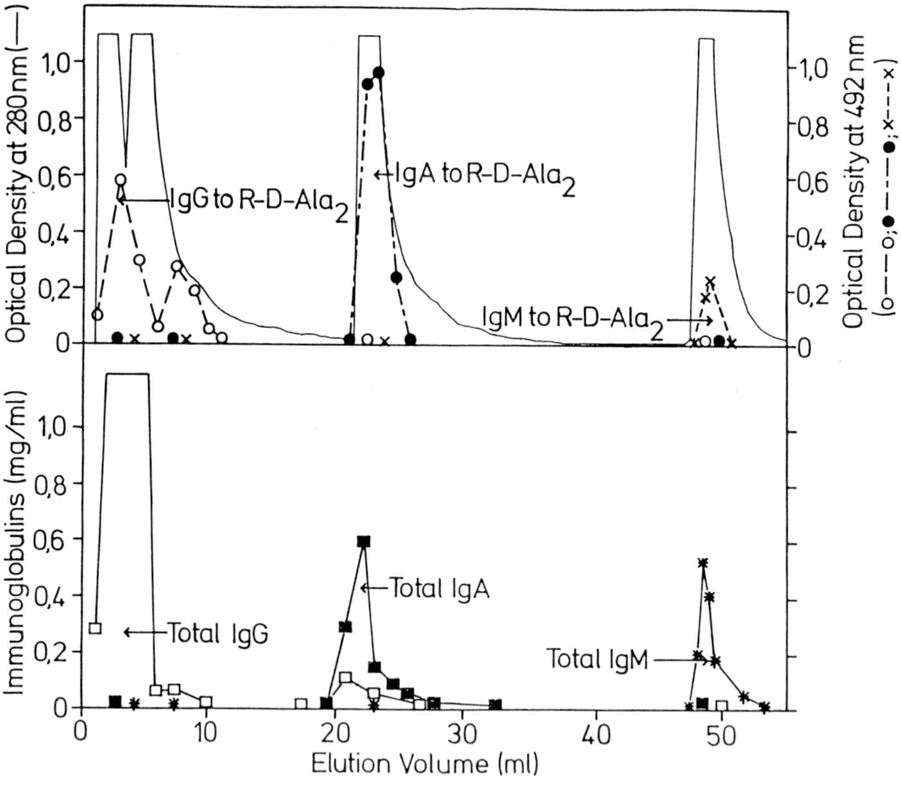

Fig. 13. Separation of IgG, IgM and IgA antibodies by the Pharmacia FPLC-system. Two 1-ml samples of serum no. 449, revealing specific immunoglobulins to peptidoglycan sequence R-D-Ala-D-Ala-OH of the IgG, IgA and IgM classes, respectively, were separated on a Mono Q anion-exchange column (for detail, see (30)). Specific antibodies of the particular immunoglobulin classes were detected by ELISA (18,30).

class (18,29,33,37), of the IgA class (29,30,33,36), of the IgM class (29,35-37) and of the IgE class (38). However, only Zeiger et al. (29) and Franken et al. (18,30) employed chemically well-defined antigens thus clearly proving that the immunoglobulin class identified was directed to a sequence of peptidoglycan. Recently, separating the particular immunoglobulin classes by the Pharmacia FPLC chromatography system we could demonstrate in human sera the occurrence of specific IgG, IgM and IgA class antibodies to the same epitope of peptidoglycan (Fig. 13, (30)). Regarding specific immunoglobulin M antibodies reported (35-37), one should be aware of the fact that unless carefully excluded, all indirect immunoassays may be severely disturbed by the presence of rheumatoid factor. This holds particularly true for sera exhibiting both specific IgG and IgM antibodies to the same antigen. Similarly, reported IgE antibodies with specificity for peptidoglycan may not yet be accepted (Seidl et al., this book).

Conclusion

The immunochemistry of peptidoglycan in man is rather complex. At least 3 main classes of immunoglobulins may be directed to at least five main antigenic epitopes, with a broad spectrum in particular antigenic specificity. Studying the immunochemistry of peptidoglycan may, however, be useful for understanding the immune response to peptidoglycan in man.

References

1. Schleifer, K.H. and O. Kandler. 1972. Peptidoglycan types of bacterial cell walls and their taxonomic implications. Bacteriol. Rev. 36, 407-477.

2. Schleifer, K.H. and P.H. Seidl. 1985. In: Chemical Methods in Bacterial Systematics (M. Goodfellow and D.E. Minnikin, eds). Academic Press, London, pp. 201-219.

3. Schleifer, K.H. and R. Joseph. 1973. A directly cross-linked L-ornithine containing peptidoglycan in cell walls of *Spirochaeta stenostrepta*. FEBS Letters 36, 83-86.

4. Yanagihara, Y., K. Kamisango, S. Yasuda, S. Kobayashi, I. Mifuchi, I. Azuma, Y. Yamamura and R.C. Johnson. 1984. Chemical compositions of cell walls and polysaccharide fractions of spirochetes. Microbiol. Immunol. 28, 535-544.

5. Vasstrand, E.N., T. Hofstad, C. Endresen and H.B. Jensen. 1979. Demonstration of lanthionine as a natural constituent of the peptidoglycan of Fusobacterium nucleatum Fevl. Infection and Immunity 25, 775-780.

6. Rogers, H.J., H.R. Perkins and J.B. Ward. 1980. Microbial cell walls and membranes. Chapman and Hall, London, U.K.

7. Heymer, B., P.H. Seidl and K.H. Schleifer. 1985. Immunochemistry and biological activity of peptidoglycan. In: Immunology of the bacterial cell envelope (Stewart-Tull DES, ed). John Wiley, Chichester, U.K.

8. Karakawa, W.W., H. Lackland and R.M. Krause. 1967. Antigenic properties of the hexosamine polymer of streptococcal mucopeptides. J. Immunol. 99, 1179-1182.

9. Rolicka, M. and J.D. Park. 1969. Antimucopeptide antibodies and their specificity. J. Immunol. 103, 196-203.

10. Wikler, M. 1975. Isolation and characterization of homogenous rabbit antibodies to Micrococcus lysodeikticus with specificity to the peptidoglycan and to the glucose-N-acetylaminomannuronic polymer. Z. Immun.-Forsch. 148, 193-200.

11. Schleifer, K.H. and P.H. Seidl. 1977. Structure and immunological aspects of peptidoglycans. In: Microbiology (D. Schlesinger, ed). American Society for Microbiology, Washington, U.S.A.

12. Audibert, F., B. Heymer, C. Gross, K.H. Schleifer, P.H. Seidl and L. Chedid. 1978. Absence of binding of MDP, a synthetic immunoadjuvant, to antipeptidoglycan antibodies. J. Immunol. 121, 1219-1222.

13. Reichert, C.M., C. Carelli, M. Jolivet, F. Audibert, P. Lefrancier and L. Chedid. 1980. Synthesis of conjugates containing N-acetylmuramyl-L-alanyl-D-isoglutaminyl (MDP). Their use as hapten-carrier systems. Mol. Immunol. 17, 357-363.

14. Schleifer, K.H. and P.H. Seidl. 1974. The immunochemistry of peptidoglycan. Antibodies against a synthetic immunogen cross-reacting with peptidoglycan. Eur. J. Biochem. 43, 509-519.

15. Seidl, P.H. and K.H. Schleifer. 1975. Immunochemical studies with synthetic immunogens chemically related to peptidoglycan. Z. Immun.-Forsch. 149, 157-164.

16. Seidl, P.H. and K.H. Schleifer. 1977. The immunochemistry of peptidoglycan. Antibodies against a synthetic immunogen cross-reacting with an interpeptide bridge of peptidoglycan. Eur. J. Biochem. 74, 353-363.

17. Seidl, P.H. and K.H. Schleifer. 1978a. Specific antibodies to the N-termini of the interpeptide bridges of peptidoglycan. Arch. Microbiol. 118, 185-192.

18. Franken, N., P.H. Seidl, E. Zauner, H.J. Kolb, K.H. Schleifer, L. Weiss. 1985. Quantitative determination of human IgG antibodies to the peptide subunit determinant of peptidoglycan by an enzyme-linked immunosorbent assay. Mol. Immunol. 22, 573-579.

19. Eichmann, K., D. G. Braun, T. Feizi and R.M. Krause. 1970. The emergence of antibodies with either identical or unrelated individual antigenic specificity during repeated immunizations with streptococcal vaccines. J. Exp. Med. 131, 1169-1189.

20. Seidl, P.H. and K.H. Schleifer. 1979. The immunochemistry of peptidoglycan. Serological detection of a difference in a single N-terminal amino acid. Mol. Immunol. 16, 385-388.

21. Seidl, P.H. and K.H. Schleifer. 1978b. Rapid test for the serological separation of staphylococci from micrococci. Appl. Environ. Microbiol. 35, 479-482.

22. Wadström, T. 1973. Bacteriolytic enzymes from Staphylococcus aureus. Contrib. Microbiol. Immunol. 1, 397-405.

23. Karakawa, W.W., D.G. Braun, H. Lackland and R.M. Krause. 1968. Immunochemical studies on the cross-reactivity between streptococcal and staphylococcal mucopeptide. J. Exp. Med. 128, 325-340.

24. Schleifer, K.H. and R.M. Krause. 1971. The immunochemistry of peptidoglycan. The immunodominant site of the peptide subunit and the contribution of each of the amino acids to the binding properties of the peptides. J. Biol. Chem. 246, 986-993.

25. Zeiger, A.R. and P.H. Maurer. 1973. Immunochemistry of a synthetic peptidoglycan-precursor pentapeptide. Biochemistry 12, 3387-3394.

26. Seidl, P.H., J.R. Golecki, N. Franken and K.H. Schleifer. 1985. Immunoelectron microscopic studies on the localization of peptidoglycan peptide subunit pentapeptides in bacterial cell walls. Arch. Microbiol. 142, 121-127.

27. Seidl, P.H., P. Zwerenz, J.R. Golecki and K.H. Schleifer. 1985. Streptococcus pyogenes grown under sublethal concentrations of penicillin G accumulates close to the septum pentapeptide subunits of peptidoglycan. FEMS Microbiol. Lett. 30, 325-329.

28. Heymer, B., K.H. Schleifer, S. Read, J.D. Zabriskie and R.M. Krause. 1976. Detection of antibodies to bacterial cell wall peptidoglycan in human sera. J. Immunol. 117, 23-26.

29. Zeiger, A.R., C.U. Tuazon and J.N. Sheagren. 1981. Antibody levels to bacterial peptidoglycan in human sera during the time course of endocarditis and bacteremic infections caused by Staphylococcus aureus. Infect. Immun. 33, 795-800.

30. Franken, N., P.H. Seidl, T. Kuchenbauer, H.J. Kolb, K.H. Schleifer, L. Weiss and K.-D. Tympner. 1984. Specific immunoglobulin A antibodies to a peptide subunit sequence of bacterial cell wall peptidoglycan. Infect. Immun. 44, 182-187.

31. Heymer, B., D. Bernstein, K.H. Schleifer and R.M. Krause. 1975. A radioactive hapten-binding assay for measuring antibodies to the pentapeptide determinant of peptidoglycan. J. Immunol. 114, 1191-1196.

32. Rasmussen, J.R. and J.L. Strominger. 1978. Utilization of a depsipeptide substrate for trapping acyl-enzyme intermediates of penicillin-sensitive D-alanine carboxypeptidases. Proc. Natl. Acad. Sci. USA 75, 84-88.

33. Helgeland, S.M. and A. Grov. 1971. Immunochemical characterization of staphylococcal and micrococcal mucopeptides. Acta Pathol. Microbiol. Scand. Sect. B, 79, 819-826.

34. Heymer, B., W. Schachenmayr, B. Bültmann, R. Spanel, O. Haferkamp and W.C. Schmidt. 1973. A latex agglutination test for measuring antibodies to streptococcal mucopeptides. J. Immunol. 111, 478-484.

35. Verbrugh, H.H., R. Peters, M. Rozenberg-Arska, P.K. Peterson and J. Verhoef. 1981. Antibodies to cell wall peptidoglycan of Staphylococcus aureus in patients with serious staphylococcal infections. J. Infect. Dis. 144, 1-9.

36. Wilhelm, J.A., L. Matter and K. Schopfer. 1982. IgG, IgA and IgM antibodies to S. aureus purified cell walls (PCW) in normal and infected individuals. Experientia 38, 1375-1376.

37. Wheat, L.J., B.J. Wilkinson, R.B. Kohler and A.C. White. 1983. Antibody response to peptidoglycan during staphylococcal infections. J. Infect. Dis. 147, 16-22.

38. Schopfer, K., S.D. Douglas and B. Wilkinson. 1980. Immunoglobulin E antibodies against Staphylococcus aureus cell walls in the sera of patients with hyperimmunoglobulinemia E and recurrent staphylococcal infection. Infect. Immun. 27, 563-568.

THE DIAGNOSTIC VALUE OF ANTIBODIES TO PEPTIDOGLYCAN AND OTHER STAPHYLOCOCCUS AUREUS ANTIGENS IN SERIOUS STAPHYLOCOCCAL INFECTIONS

B. Christensson

Department of Infectious Diseases, University Hospital, S-221 85 Lund, Sweden

Introduction

Staphylococcus aureus infections constitute an increasing problem in modern medicine. Intravenous devices, often used in seriously ill patients, are well-known portals of entry for staphylococcal infections. The use of valvular, vascular and joint prostheses as well as intravenous drug abuse are other predisposing factors for achieving serious bacteremic staphylococcal infections (1-4). However, there has been an increasing number of community-acquired S.aureus septicemia. These patients often lack obvious primary foci, and the origin of these infections may be difficult to establish (2).

The diagnosis of S.aureus septicemia is certainly still mainly based upon bacteriological isolation of the bacteria, but S.aureus serology can be a valuable complement in many instances. When blood cultures are negative, which can be due to previous antibiotic treatment, staphylococcal serology can be of diagnostic value, as well as when deep seated infections, unavailable to bacteriological isolation, are suspected. It has also been suggested, that the response to treatment could be reflected by a fall in antibody levels against S.aureus teichoic acid (5), but this has been opposed by others (6-8).

The present investigations have focused upon the question whether a combination of different serological assays can help to distinguish between an uncomplicated course of S.aureus septicemia and a septicemia complicated by endocarditis or metastatic abscesses.

Biological Properties of Peptidoglycan

This distinction is important, since patients with these complications require a more aggressive and long-standing antibiotic treatment and often also surgical intervention like heart valve replacement and drainage of deep abscesses.

Materials and Methods

Patients and controls

S.aureus endocarditis. Twenty-two to 25 patients with S.aureus endocarditis were investigated in the various assays. The diagnostic criteria were elevated body temperature (>38.5°C) for at least two days together with two or more positive blood cultures. Additionally, the patients developed changes in heart murmurs (n=13), typical ultrasound echocardiographic signs (n=13), secondary manifestations (n=16) or diagnosis was made at surgery (n=2) or autopsy (n=6). Six of 10 drug addicts had tricuspid valve involvement; no patients had infected prosthetic valves.

S.aureus complicated septicemia. The criteria of elevated body temperature and positive blood cultures among the 10-23 patients investigated were the same as for the endocarditis patients, but there were otherwise no signs of endocarditis. These patients generally developed secondary infectious foci in the skeleton (n=7) but also deep seated abscesses in internal organs (n=5) and meningitis (n=4). Patients with persistently infected joint prostheses (n=7) that for some reason were not removed or drained within a few days were considered complicated even if metastatic infections did not develop.

S.aureus uncomplicated septicemia. This group comprised 14-23 patients and criteria of body temperature and blood cultures were as mentioned above. No infectious complications developed. When primary foci were present (n=19) they were easily eradicated within a few days. These foci were skin infections (n=10), intravenous devices (n=6) and osteosynthetic material (n=3).

Non-S.aureus endocarditis and septicemia. The diagnostic criteria of these infections were the same as for S.aureus endocarditis and septicemia. Most of the 27 endocarditis and 39 septicemia patients investigated were infected with gram-positive bacteria, mainly streptococci and S.epidermidis. In the anti-alpha toxin assay (see below) only four and 18 patients, respectively, were tested.

Febrile controls. When these 63 patients were admitted to the hospital, septicemia of unknown origin was initially suspected. They were eventually considered as non-septicemic when at least two blood cultures were negative. Their subsequent diagnoses were pneumonia (n=23), viral infection (n=18), urinary tract infection (n=8), fever of unknown origin (n=5), skin abscess (n=3) and one patient each proved to have toxoplasmosis, mastitis, cholangitis, pleural empyema, brain abscess and gas gangrene.

Healthy controls I. This group consisted of 160 individuals which were divided into different age groups. They were all healthy at the time of serum sampling and had no previous history of staphylococcal infection.

Healthy controls II. These 11 individuals were followed with monthly serum sampling for six months.

Antigens

Alpha toxin. S.aureus alpha toxin from strain Wood 46 was purified by ion exchange chromatography, isoelectric focusing and gel chromatography according to Thelestam et al (9) and Wadström and Möllby (10). The purity of the preparation was verified by means of polyacrylamide gel electrophoresis and immunoelectrophoresis.

Peptidoglycan. Peptidoglycan (PG) from two different strains, E2371, rich in protein A (11), and E1369, protein A deficient (12), was prepared by a modification of the method of Park and Hancock (13) using trichloracetic acid and trypsin treatment of whole S.aureus cells. This insoluble preparation was solubilized in three different ways; by ultrasonication (SO-PG), lysozyme treatment (LZ-PG) or lysostaphin treatment (LS-PG). All PG preparations were investigated in crossed immunoelectrophoresis against polyspecific rabbit hyperimmune serum against S.aureus cell walls. Moreover, the different PG preparations were compared by crossed-line immunoelectrophoresis.

Crude staphylococcal antigen. Ultrasonication of the protein A negative strain Wood 46 was used to prepare the crude staphylococcal antigen (SA) (14). This preparation was investigated in crossed immunoelectrophoresis against both rabbit hyperimmune serum and human serum and more than 50 different precipitates were obtained. A similar preparation has also been proposed to contain teichoic acid, lipoteichoic acid as well as peptidoglycan (15).

Lipase. Purified lipase from S.aureus strain TEN 5 (16) was prepared according to the methods described by Jürgens et al (17) and Jürgens and Huser (18) using ammonium sulphate precipitation of the culture supernatant, ultrafiltration, adsorption on Octyl-Sepharose CL-4B and finally elution with a linear Triton X-100 gradient. The preparation contained both the 44 kD and the 43 kD lipase, where the latter probably represents the nicked 44 kD enzyme. The purity of the lipase preparation was verified by SDS-polyacrylamide gel electrophoresis, determination of enzyme activity and lack of hemolytic activity, inhibition of catalytic activity by corresponding antiserum and immunodiffusion (17-19).

Serological methods

Radioimmunoassay (RIA). Purified alpha toxin was iodinelabelled. After incubating diluted serum and labelled antigen in plastic tubes the antigen-antibody complexes were precipitated by adding an excess of a S.aureus Cowan I suspension, thus utilizing the non-specific ability of protein A to bind to the Fc portion of

IgG. After centrifugation, the radioactivity of the bacterial pellet was calculated as corresponding to the amount of IgG antibodies against alpha toxin (20).

Solid-phase radioimmunoasay (SPRIA). The SA- and the various PG-preparations were coated on plastic tubes, and remaining binding sites were blocked by bovine serum albumin. After incubating diluted serum, bound serum IgG antibodies were detected by adding iodinated protein A (14,21,22).

Enzyme-linked immunosorbent assay (ELISA). The same preparations as in SPRIA together with the lipase preparation were placed into the wells of microtiter plates. The following steps were according to a modification of the ELISA of Engvall and Perlmann (23) with the addition of bovine serum albumin, diluted patient serum, alkaline-phosphatase conjugated anti-human IgG and p-nitrophenyl-phosphate substrate. The amount of specific serum IgG antibody bound corresponded to the optical density value measured spectrophotometrically after the enzyme-substrate reaction (22,24,25).

Statistical analysis

The reproducibility of an assay was expressed as the analytical error of the method calculated as the standard deviation of the single determination (22).

Upper normal antibody levels were set at the 99 % confidence limit of the values in the febrile control group.

A significant change in antibody level was based upon two standard deviations of both the analytical error of the method and the mean variation of antibody levels in the healthy control group II (24,25).

Predictive values for positive (PV pos) and negative (PV neg) test results were calculated according to Vecchio (26) and Kronvall (27).

Results

There was no single serological assay capable of diagnosing all patients with S.aureus endocarditis/septicemia or excluding those patients with an uncomplicated course of infection, when peak antibody levels during one to four weeks after onset of infection were determined (Table 1).

Table 1. Number and Percentage of Patients and Controls with Positive Peak Antibody Levels in Staphylococcus aureus Serology

	S. AUREUS			NON-S. AUREUS			
	ENDO	COMPL SEPT	UNCOMPL SEPT	ENDO	SEPT	FEBRILE NON-SEPT	HEALTHY CONTROLS
Alpha-toxin RIA	14/22 64 %	4/10 40 %	1/14 7 %	0/4 0%	0/18 0 %	N.D.	4/160 2.5 %
PG-ELISA	20/25 80 %	7/23 30 %	6/23 26 %	10/27 37 %	1/39 3 %	0/63 0 %	N.D.
SA-ELISA	21/25 84 %	17/23 74 %	6/23 26 %	3/27 11 %	3/39 8 %	1/63 1.6 %	N.D.
Lipase-ELISA	19/24 79 %	14/23 61 %	3/23 13 %	0/27 0 %	0/39 0 %	0/63 0 %	0/157 0 %

Alpha toxin antibody

The alpha toxin RIA was highly specific, and no cross-reactions were seen in patients with non-S.aureus endocarditis or septicemia (Table 1). However, the sensitivity in diagnosing S.aureus endocarditis was only 64 %, and only 40 % and 7 % in the complicated and uncomplicated septicemia groups, respectively.

PG-antibody

When the solubilized PG-preparations SO-PG, LZ-PG and LS-PG were investigated in crossed immunoelectrophoresis they all showed two precipitates but with some variations in electrophoretic mobility (22). No traces of teichoic acid were seen. Moreover, all PG-preparations cross-reacted serologically when investigated in crossed-line immunoelectrophoresis. The time for enzyme treatment influenzed the serological activity of the preparations. The optimal time was two hours with lysostaphin and four hours with lysozyme while the sonication continued for 30 minutes in preparing SO-PG.

No differences were seen whether the protein A rich or protein A negative S.aureus strain were used for preparing PG. The most striking differences were seen between the SO-, LZ- and LS-PG preparations when they were used as antigens in SPRIA or ELISA. The sensitivity of the LS-PG was much too low, and using SPRIA it was impossible to distinguish positive from negative sera. The results obtained with LZ-PG and SO-PG were satisfactory with a slightly higher sensitivity using the SO-PG.

When comparing the PG-ELISA and the PG-SPRIA as routine laboratory assays the PG-ELISA had some advantages. The reproducibility was better (+/- 12.0 % vs. +/- 18.1 %) and higher serum dilutions could be used with improved specificity (1/3000 vs. 1/40). Furthermore, antibodies of all IgG subclasses were determined in ELISA, whereas possible IgG3 subclass antibodies remained undetected when using protein A in the SPRIA.

The ELISA with sonicated PG as antigen was chosen for testing all the patient sera and the results of peak antibody levels are shown in Table 1. The sensitivity in diagnosing S.aureus endocarditis and complicated septicemia could be raised to 94 % and 56 %, respectively, when not only a positive peak antibody level but also a significant rise in titer was considered as a positive test result (Table 2). A 50 % rise in optical density values was considered significant. However, the specificity was low, because cross-reacting antibodies were not uncommon in, mainly gram-positive, non-S.aureus endocarditis and septicemia.

SA-antibody

The results of the SA-assays were in many ways similar to those of the PG-assays. Again, the SA-ELISA was superior to the SA-SPRIA regarding reproducibility and convenience for the routine laboratory. The sensitivity in diagnosing S.aureus endocarditis using peak antibody levels was comparable to that of the PG-ELISA but complicated S.aureus septicemia patients were more often positive in the SA-ELISA than in the PG-ELISA (Table 1). Similar to the PG-ELISA, serial serum sampling raised the sensitivity, because significant changes in antibody levels (> 50 % rise in optical density values) could be detected even if a positive level was not reached (Table 2). The specificity was high, but some cross-reactions were seen in non-S.aureus infected patients (Table 1,2).

Lipase antibody

Antibodies to S.aureus lipase were only investigated by ELISA. This was the most sensitive assay diagnosing S.aureus endocarditis and complicated septicemia in 100 % and 89 %, respectively with repeated serum sampling (Table 2). Moreover, the major advantage with this assay was the extremely high specificity reaching 100 % (Table 1,2).

Table 2. Number and Percentage of Patients and Controls with Positive Peak Antibody Levels or Significant Titer Rises in Staphylococcus aureus Serology

	S. AUREUS			NON-S. AUREUS		
	ENDO	COMPL SEPT	UNCOMPL SEPT	ENDO	SEPT	FEBRILE NON-SEPT
PG-ELISA	15/16 94 %	10/18 56 %	7/18 39 %	3/11 27 %	4/20 20 %	3/63 5 %
SA-ELISA	14/16 88 %	15/18 83 %	8/18 44 %	0/11 0 %	2/20 10%	2/63 3 %
Lipase-ELISA	16/16 100 %	16/18 89 %	3/18 17 %	0/11 0 %	1/25 4 %	0/63 0 %
Lipase- + either PG- or SA-ELISA	16/16 100 %	14/18 78 %	0/18 0 %	0/11 0 %	0/25 0 %	0/63 0 %

Combined serology

The problem with cross-reacting antibodies could not be overcome with the PG- and SA-ELISAs. Instead, the three very sensitive assays (PG, SA and lipase), one of them (lipase) also highly specific, were evaluated in combination. These three assays could well be used in parallel due to the lack of absolute correlation between the individual assays. The criteria for positivity were positive peak value or significant titer rise in the lipase ELISA as well as in either the PG- or SA-ELISA. This proved to be a highly sensitive and highly specific combination (Table 2) where S.aureus endocarditis and complicated septicemia patients could be differentiated from those with an uncomplicated course or non-S.aureus infection.

Predictive values of combined serology

The clinical significance of a laboratory test or test combination should be evaluated in terms of predictive values for positive (PV pos) and negative (PV neg) test results. PV pos is defined as per cent positive patients among all positive individuals and PV neg as per cent negative non-diseased individuals among all negative individuals (26,27). The PV will be influenzed by the sensitivity and specificity of the test as well as by the prevalence of the infection. The sensitivity in the S.aureus endocarditis/complicated septicemia group will be 30/34 = 88 % (Table 2). The specificity may seem to be 100 % (Table 2), but the upper normal levels of each test was based upon the 99 % confidence limit of the results from the febrile controls corresponding to a specificity of 99.5 %. The true specificity of the combination of the three assays would then be at least 99.5 % but hardly as much as 100 %, which in reality must be rather unlikely. The prevalence of S.aureus endocarditis/complicated septicemia can only be estimated, but was 10 % among all diagnosed septicemia at Lund University Hospital in 1982. These values for the three different variables will result in PV higher than 95 % for positive as well as negative test results (Table 3).

Table 3. Predictive Values for Positive (PV pos) and Negative (PV neg) Results of the Combined Use of the PG-, SA- and Lipase ELISAs in Diagnosing S.aureus Endocarditis/Complicated Septicemia

Sensitivity	88 %	PV pos	95.1 %
Specificity	99.5 %	PV neg	98.7 %
Prevalence	10 %		

Discussion

When estimating the diagnostic value of a serological assay proposed to be used in routine serology, there are many aspects which need to be thoroughly investigated. The various results often obtained in different laboratories between what is supposed to be similar assays are probably often due to variations in methodology and patient characteristics.

Clinical characteristics

When defining a patient with S.aureus bacteremia it is important to distinguish between those with a complicated and uncomplicated course of infection. The higher antibody levels seen in patients with endocarditis or metastatic abscesses are probably due to the long persistence of the infectious foci. These patients also require a more aggressive treatment and it is natural to separate them from those with an uncomplicated course of infection which often have a primary focus that can be early and easily eradicated. The clinical diagnosis of e.g. S.aureus endocarditis is, however, often difficult, which underlines the importance of using the same diagnostic criteria between different investigators.

Positive IgG antibody levels can be expected 7-14 days after the the onset of bacteremic infections. Therefore, it is important to relate the time for serum sampling to the day for onset of symptoms (like fever, chills, general disability) and not to the day for the first positive blood culture or admission to hospital.

Adequate control groups

All patients as well as healthy controls generally have detectable antibody levels against the different S.aureus antigens when extremely sensitive methods as SPRIA and ELISA are used. Therefore, the choice of control group is important when defining the upper normal antibody levels. The control group should be as similar to the patient group as possible regarding age, sex and initial clinical symptoms. The febrile controls used in the present investigations represent patients where septicemia initially was suspected, and where routine staphylococcal serology may well have been used by the clinician. Moreover, when evaluating the clinical significance of a change in antibody level, this change must also be evaluated in the control group. This is certainly not possible if single serum samples from healthy controls are used.

Septicemia due to other bacteria than S.aureus should also be evaluated. The cell walls of S.aureus and other, mainly gram-positive, bacteria share antigenic determinants in e.g. the peptidoglycan/teichoic acid complex. It is therefore not surprising, that cross-reacting antibodies can be found in S.epidermidis and streptococcal infections. However, cross-reactions against exoproteins unique for S.aureus like alpha toxin and lipase are not expected, and consequently assays measuring alpha toxin and lipase antibodies show the highest specificity.

Methodological aspects of antigen preparation and assay technique

The need for standardization of antigen preparations used in S.aureus serology has previously been emphasized regarding teichoic acid (15). The method of antigen preparation, the lack of contaminating antigens when using purified preparations,the antigen concentrations used in the assay and the reproducibility of the assay are important data when comparing the results from different laboratories.

These aspects are illustrated in the comparisons between the different PG-preparations. Although all PG-preparations cross-reacted in crossed immunoelectrophoresis, the LS-PG could not be recommended as antigen in SPRIA or ELISA measuring serum anti-PG antibodies due to its poor sensitivity. Moreover, the time for enzyme treatment with both lysostaphin and lysozyme was crucial in reaching an optimal sensitivity. Variations in enzyme activity of different batches may also have a disturbing influenze. The solubilization of PG by ultra-sonication is probably more easy to standardize, and the SO-PG was chosen as antigen in the present PG-ELISA.

Although the alpha toxin preparations were highly purified and lacked contaminations with delta toxin, which previously has been a great problem (9,10), the anti-alpha toxin RIA had a low sensitivity. One possible reason could be the recent findings of an unexpected low prevalence of alpha toxin producing S.aureus septicemia strains. It is previously generally maintained, that alpha toxin production is found in about 95 % of all S.aureus strains (28,29). However, in a recent study Christensson and Hedström found only 58 % (51/88) of S.aureus septicemia strains to be alpha toxin producers (30).

The use of single serum dilutions in serology compared with serial serum titrations is much more convenient for the laboratory. Single dilutions generally show a good correlation to end-point titrations (31,32). However, not only the absolute antibody antibody level but also the antibody affinity influence the amount of bound antibody (32-34). Therefore, also the serum dilution needs to be standardized between different laboratories.

The normal variation of antibody levels against S.aureus antigens is often very wide. Therefore, a change in antibody level during infection may be significant although the positive cut off level is not reached. The advantage of evaluating titer rises as well as positive peak values is illustrated by the higher values for sensitivity reached in Table 2 as compared to Table 1.

Predictive values

When a new serological assay or assay combination is presented, the clinical value is often expressed in terms of statistically significant differences between patient and control groups. Even if these differences are highly significant ($p<0.001$) no conclusions can be drawn regarding its diagnostic value as a routine laboratory assay. Instead, the predictive values (PV) should be calculated. Often too much emphasis is placed on reaching a high level of sensitivity, while in terms of PV, high specificity is far more important. A high specificity is especially important if the prevalence of the disease is low (27). Therefore, specificity levels very close to 100 % are required if the test is to be used for screening purposes of large populations with prevalence rates often far below 1 %. However, in the clinical situation, when standing in front of a febrile patient where septicemia is suspected, the expected prevalence of the disease is much higher. In the present studies calculations were made with a prevalence of 10 %. From Table 1 and 2 it can be seen, that no single serological assay will reach a specificity close to 100 %. However, the combined use of PG-, SA- and lipase ELISAs will give a specificity of 100 % in the present patient and control material and a true specificity of at least 99.5 %. Both PV pos and PV neg of this combination in diagnosing S.aureus endocarditis and complicated septicemia will then be higher than 95 % (Table 3) which is quite satisfactory for routine serological assays.

Conclusions

The S.aureus PG-, SA- and lipase ELISAs are highly sensitive serological assays in diagnosing S.aureus endocarditis and complicated septicemia. Together with the alpha toxin RIA the lipase ELISA is also highly specific for S.aureus infections. This is in contrast to the common occurrence of cross-reacting antibodies to PG, and to a lesser extent to SA, seen in patients with streptococcal and S.epidermidis endocarditis or septicemia. Although there is no single serological assay capable of diagnosing all patients with S.aureus bacteremia, predictive values higher than

95 % for diagnosing S.aureus endocarditis and complicated septicemia can be reached with a combination of assays. Positivity in the lipase ELISA as well as in either the PG- or SA-ELISA strongly suggests a complicated course of S.aureus septicemia or S.aureus endocarditis.

References

1. Chambers, H.F., M.A. Korzeniowski and M.A. Sande with the National Collaborative Endocarditis Study Group. 1983. Staphylococcus aureus endocarditis: Clinical manifestations in addicts and nonaddicts. Medicine 62, 170-177.

2. Hedström, S.Å. and B. Christensson. 1983. Staphylococcus aureus septicaemia and endocarditis at the University Hospital in Lund 1976-1980. Scand. J. Infect. Dis. Suppl. 41, 38-46.

3. Johnson Jr, W.D. 1976. Prosthetic valve endocarditis. In: Infective endocarditis (D. Kaye, ed.). Baltimore: University Park Press, pp. 129-142.

4. Watanakunakorn, C. 1977. Changing epidemiology and newer aspects of infective endocarditis. Adv. Intern. Med. 22, 21-47.

5. Sheagren, J.N. 1984. Guidelines for the use of the teichoic acid antibody assay. Arch. Intern. Med. 144, 250-251.

6. Kaplan, J.E., D.L. Palmer and K.S.K. Tung. 1981. Teichoic acid antibody and circulating immune complexes in the management of Staphylococcus aureus bacteremia. Am. J. Med. 70, 769-774.

7. Wheat, L.J., R.B. Kohler, Z.A. Tabbarah and A. White. 1981. IgM antibody response to staphylococcal infection. J. Infect. Dis. 144, 307-311.

8. Wheat, L.J., R.B. Kohler, A. White and M. Garten. 1983. IgM and IgG antibody response to teichoic acid in infections due to Staphylococcus aureus. J. Infect. Dis. 147, 1101.

9. Thelestam, M., R. Möllby and T. Wadström. 1973. Effects of staphylococcal alpha-, beta-, delta- and gammahemolysins on human diploid fibroblasts and HeLa cells: evaluation of a new quantitative assay for measuring cell damage. Infect. Immun. 8, 938-946.

10. Wadström, T. and R. Möllby. 1972. Some biological properties of purified staphylococcal hemolysins. Toxicon. 10, 511-519.

11. Espersen, F. and I. Clemmensen. 1982. Isolation of a fibronectin-binding protein from Staphylococcus aureus. Infect. Immun. 37, 526-531.

12. Schiötz, P.O., N. Höiby and J.B. Hertz. 1979. Cross-reactions between Staphylococcus aureus and fifteen other bacterial species. Acta Pathol. Microbiol. Immunol. Scand. Sect. B. 87, 329-336.

13. Park, J.T. and R. Hancock. 1960. A fractionation procedure for studies of the synthesis of cell-wall mucopeptide and of other polymers in cells of Staphylococcus aureus. J. Gen. Microbiol. 22, 249-258.

14. Christensson, B., S.Å. Hedström and G. Kronvall. 1982. Detection of Staphylococcus aureus antibodies in patients with S.aureus infections and in normals using solid-phase radioimmunoassay. Acta Pathol. Microbiol. Immunol. Scand. Sect. B. 90, 251-255.

15. White, A., L.J. Wheat and R.B. Kohler. 1983. Diagnostic and therapeutic significance of staphylococcal teichoic acid antibodies. Scand. J. Infect. Dis. Suppl. 41, 105-116.

16. Arbuthnott, J.P., C.G. Gemmell, J. Kent and A. Lyell. 1969. Haemolysis and enzyme patterns of coagulase-positive staphylococci isolated from toxic epidermal necrolysis, RITTER's disease and impetigo contagiosa. J. Med. Microbiol. 2, 479-487.

17. Jürgens, D., H. Huser, H. Brunner and F.J. Fehrenbach. 1981. Purification and characterization of Staphylococcus aureus lipase. FEMS Microbiol. Letters 12, 195-199.

18. Jürgens, D. and H. Huser. 1981. Large-scale purification of staphylococcal lipase by hydrophobic interaction chromatography. J. Chromat. 216, 295-301.

19. Kötting, J., D. Jürgens and H. Huser. 1983. Separation and characterization of two isolated lipases from Staphylococcus aureus (TEN 5). J. Chromat. 281, 253-261.

20. Christensson, B., S.Å Hedström and G. Kronvall. 1983. Antibody response to alpha- and betahemolysin from Staphylococcus aureus in patients with staphylococcal infections and in normals. Acta Pathol. Microbiol. Immunol. Scand. Sect. B. 91, 351-356.

21. Christensson, B., F. Espersen, S.Å. Hedström and G. Kronvall. 1983. Solid-phase radioimmunoassay of immunoglobulin G antibodies to Staphylococcus aureus peptidoglycan in patients with staphylococcal infections. Acta Pathol. Microbiol. Immunol. Scand. Sect. B. 91, 401-406.

22. Christensson, B., F. Espersen, S.Å Hedström and G. Kronvall. 1984. Methodological aspects of Staphylococcus aureus peptidoglycan serology: Comparisons between solid-phase radioimmunoassay and enzyme-linked immunosorbent assay. J. Clin. Microbiol. 19, 680-686.

23. Engvall, E. and P. Perlmann. 1972. Enzyme-linked immunosorbent assay, ELISA. III. Quantitation of specific antibodies by enzyme-labelled anti-immunoglobulin in antigen-coated tubes. J. Immunol. 109, 129-135.

24. Christensson, B., F. Espersen, S.Å Hedström and G. Kronvall. 1985. Serological assays against Staphylococcus aureus peptidoglycan, crude staphylococcal antigen and staphylolysin in the diagnosis of serious S.aureus infections. Scand. J. Infect. Dis. 17, 47-53.

25. Christensson, B., F.J. Fehrenbach and S.Å. Hedström. 1985. A new serological assay for Staphylococcus aureus infections: IgG antibodies to S.aureus lipase using enzyme-linked immunosorbent assay. J. Infect. Dis. 152, 286-292.

26. Vecchio, T.J. 1966. Predictive value of a single diagnostic test in unselected populations. N. Engl. J. Med. 274, 1171-1173.

27. Kronvall, G. 1981. The potential of immunological tests as tools in the epidemiology of leprosy. Lepr. Rev. Suppl. 1. 52, 207-219.

28. Elek, S.D. and E. Levy. 1950. Distribution of hemolysins in pathogenic and non-pathogenic staphylococci. J. Path. Bact. 62, 541-554.

29. Jeljaszewicz, J. 1972. Toxins (hemolysins). In: The staphylococci. (J.O. Cohen, ed.). Wiley-Interscience, New York. pp. 249-250.

30. Christensson, B. and S.Å. Hedström. 1985. Biochemical and biological properties of Staphylococcus aureus septicemia strains in relation to clinical characteristics. Scand. J. Infect. Dis. in press.

31. Granström, M., I.G. Julander, S.Å. Hedström and R. Möllby. 1983. Enzyme-linked immunosorbent assay for antibodies against teichoic acid in patients with staphylococcal infections. J. Clin. Microbiol. 17, 640-646.

32. Shekarchi, I.C., J.L. Sever, Y.J. Lee, G. Castellano and D.L. Madden. 1984. Evaluation of various plastic microtiter plates with measles, toxoplasma, and gamma globulin antigens in enzyme-linked immunosorbent assays. J. Clin. Microbiol. 19, 89-96.

33. Butler, J.E., T.L. Feldbush, P.L. McGivern and N. Stewart. 1978. The enzyme-linked immunosorbent assay (ELISA): a measure of antibody concentration or affinity? Immunochem. 15, 131-136.

34. Hill, H.R. and J.M. Matsen. 1983. Enzyme-linked immunosorbent assay and radioimmunoassay in the serologic diagnosis of infectious diseases. J. Infect. Dis. 147, 258-263.

ON THE RELATIONSHIPS BETWEEN CONFORMATIONAL AND BIOLOGICAL PROPERTIES OF MUREIN

H. Labischinski, L. Johannsen
Robert Koch-Institute of the Federal Health Office
D-1000 Berlin 65

Introduction

In contrast to earlier considerations a notion of the bacterial peptidoglycan (synonym: murein) as a flexible, dynamic and essentially non-crystalline network becomes more and more accomplished (cf. 1, 2, 3, 4, 5, see also 6). Nevertheless, the state of order of the peptidoglycan network cannot be properly understood in terms of a completely non-ordered, amorphous structure (3). We will try to describe in the following how the molecular architecture of the peptidoglycan strands - being dynamic and rigid simultaneously - might relate to the biological functions of the bacterial cell wall network by the ability to form distinct, ordered types of packing, which, however, are all of the non-crystalline type and allow for the construction of murein networks with different biological properties.

Experimental basis of the peptidoglycan model

From experimental data published earlier (1-8) four basic postulates can be taken as a prerequisite for any realistic murein model:

1) Murein proved to exist as an essentially non-crystalline structure (X-ray data, IR- and NMR-spectroscopy, electron microscopy (2, 3, 5, 7, 8)).
2) Murein cannot only be regarded as a rigid network but shows a remarkable flexibility (osmotic swelling/ shrinkage experiments (2, 3, 4, 5)).
3) Both, the peptide and the sugar moiety of peptidoglycan

Biological Properties of Peptidoglycan

reveal considerable conformational flexibility. However, the ability of the murein network to change its dimensions under the influence of stress is mediated exclusively by its peptide part. The sugar chains always behave as relatively straight rods (conformational energy calculations and X-ray data (1, 2, 3)).

4) The bulky D-lactyl-groups at the peptide attachment sites of the murein sugar strands force a twist between N-acetyl-glucosamine and N-acetyl-muramic acid residues along the sugar strands. This in turn results in a helical organisation not only of the sugar strands themselves, but also of their peptide substituents mediating the crosslinking of neighboured peptido-glycan chains (conformational energy calculations, X-ray data, infrared measurements, requirement for rapid lysozyme digestability (1, 2, 3, 7)).

Our atomic model (1) consistent with these requirements is shown in Fig. 1.

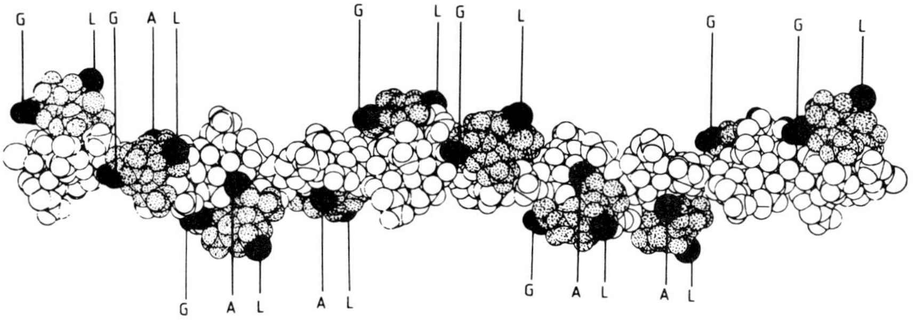

Fig. 1. Contours of a space-filling model of a peptidoglycan strand composed of ten repeating units. The sites of possible crosslinkage are denoted with A (last D-alanine), L (starting point for the side-chain of lysine), and G (carboxyl-group of glutamic acid). The peptides are dotted (from (1)).

Murein packing types

Based on the atomic model shown in Fig. 1 a simplified model can be constructed more suitable for packing considerations in which the glycan chains are approximated by cylindrical rods around which a continuous helix of the possible crosslinkage sites of the peptides is arranged (4). This model is not only easier to handle for the purpose of constructing space-filling and crosslinked peptidoglycan networks because possible crosslinking pattern can be recognized by the simple criterion that the helically arranged cross-linkage sites have to come in close proximity for crosslinking to occur; this model also involves possible variations in sugar helicity likely to occur in bacterial peptidoglycan since the model does not bear an atomic description of the crosslinkage sites, but a continuous description. The possible packing types of bacterial peptidoglycan based on this model have recently been analyzed (4). It seems, therefore, sufficient to summarize the results obtained in the following table (Table 1).

Table 1. Compilation of Murein Packing Pattern

Packing type	Cross-linking	Peptide-Oligomer-content	Swelling behaviour
I 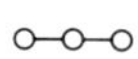	low (~25%)	extremely low, if any	non-isotropic
II 	medium (~50%)	low	non-isotropic
III 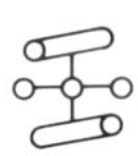	high (>60%)	high	isotropic

In the column "packing type" each circle represents a complete peptidoglycan chain projected down to the chain axis. The degree of crosslinking was estimated from the amount of crosslinkage regions of neighboured chains in close contact to each other in respect to total crosslinkage sites. The peptide oligomer content could be deduced from the amount of crosslinkage sites in close contact to each other which belong to more than two independent sugar chains. The swelling behaviour could be deduced as either isotropic or anisotropic, since swelling can occur only in the direction of crosslinked peptides (3, 4).

Three main murein packing types I, II, and III have to be taken into account (Table 1). Type I represents the only packing pattern possible in a single-layered murein, in which all strands are arranged more or less parallel and crosslinked mainly via peptide-dimers. Type II represents one of the possible arangements in a multi-layered murein. It can be thought of as being generated by successive layers of Type I, crosslinked via those parts of the crosslinkage regions above and below each layer, not involved in single layer crosslinking. Therefore, its degree of crosslinking is somewhat higher; since, however, peptide oligomer formation beyond peptide trimer-formation is rather unlikely in the packing type II, crosslinking indices higher than ~50 % are not expected: due to the helical nature of the peptidoglycan strands crosslinkage regions of neighbouring sugar strands within each layer are shifted against each other so that close contacts between the crosslinkage regions of different layers occur only between strands in close proximity to each other (Fig. 2). Thus for both types I and II a clear spatial separation of the crosslinking contacts is characteristic.

Fig. 2. Schematic representation of murein packing type I (cf. Table 1). The sugar chains of the peptidoglycan are replaced by more or less straight rods (light) around which the helical crosslinkage region (dark) is arranged. Crosslinking can occur only at those parts, where the "dark" regions of different peptidoglycan strands are in close contact. Therefore, well separated peptide dimers are characteristic for this packing type. Note, that the crosslinkage regions not involved in single layer crosslinking are shifted against each other in neighbouring peptidoglycan strands, such that new, quasi-continuous crosslinkage regions are formed with an angle of + 30° (above the layer) or - 30° (below the layer) relative to the sugar chain direction. Therefore, only a rotation of 60° between consecutive layers (packing type III, Table 1) can lead to a network with a high degree of crosslinking via the formation of high amounts of peptide-oligomers.

As described in Fig. 2, only a rotation of consecutive layers in a multi-layered peptidoglycan arrangement (type III) of 60° leads to a highly crosslinked network via peptide-oligomer formation. In contrast to packing types I and II, type III displays isotropic swelling behaviour (Table 1). It can only be mentioned here that two subgroups of type III are possible (rotation around + or - 60°) which might be differentiated by their different degrees of crosslinking (cf. (4)).

Biological implications of the model

The most important implication of the new model seems to be that different bacteria or even a given single bacterium might have a selection of different murein architectures at their disposal, capable of resulting in distinct murein qualities with different functional properties.
Since the stereochemical properties of the crosslinks in the different packing types are quite distinct, different enzymes would almost certainly be needed for the construction of the various murein types. No single lytic enzyme should be capable of degrading the various packing types with similar efficiency.

The porosity of the different packing types should also differ. One can define a lower limit of the "mesh size" of the murein as given by the distance between neighbouring peptidoglycan chains (~1-3 nm, depending on osmotic conditions) times the smallest distance between two crosslinked peptides connecting the same two chains (~4-5 nm) (not considering other pores resulting e.g. from finite sugar chain length). Although this "mesh size" in each of the individual murein layers would be similar, the effective mesh size would depend on the packing type, being smaller in type III murein as compared to type II. A change from a type II murein to type III murein could result in a change in spatial direction of the newly synthesized material. This might be important for the morphogenesis of septal wall material. Since in a single layered murein only one packing type (Type I) is possible which is not very suitable for the construction of strongly curved wall material (e.g. at the polar caps) and since its maximal crosslinking seems to be lower than actually found in some gram-negative bacteria, a

"more-than-single-layered" wall seems to be appropriate for these bacteria, too (cf. 4, 5).
These considerations remain, however to be sustained by direct experimental investigations. In the meantime we could already gather quite a few experimental indications that, in different bacteria and even within a single bacterium, different packing patterns could be realized (to be published elsewhere).

Acknowledgements

We thank Prof. Bradaczek and Prof. Giesbrecht for critical discussions on this subject. The crucial impetus of P. Keller during the determination of possible murein packing patterns is gratefully acknowledged. This investigation was supported by the Deutsche Forschungsgemeinschaft.

References

1. Barnickel, G., D. Naumann, H. Bradaczek, H. Labischinski and P. Giesbrecht. 1983. In: The Target of Penicillin (R. Hakenbeck, J.-V. Höltje and H. Labischinski, eds.). Walter de Gruyter, Berlin. New York, p. 61.

2. Burge, R.E., R. Adams, H.H.M. Balyuzi and D.A. Reaveley. 1977. J. Mol. Biol. 117, 955.

3. Labischinski, H., G. Barnickel and D. Naumann. 1983. In: The Target of Penicillin (R. Hakenbeck, J.-V. Höltje and H. Labischinski, eds.). Walter de Gruyter, Berlin . New York, p.49.

4. Labischinski, H., G. Barnickel, D. Naumann and P. Keller. 1985. Ann. Inst. Pasteur/Microbiol. 136 A, 45.

5. Hobot, J.A., E. Carlemalm, W. Villinger and E. Kellenberger. 1984. J. Bacteriol. 160, 143

6. Formanek, H. 1983. In: The Target of Penicillin (R. Hakenbeck, J.-V. Höltje and H. Labischinski, eds.). Walter de Gruyter, Berlin . New York, p. 55

7. Naumann, D., G. Barnickel, H. Bradaczek, H. Labischinski and P. Giesbrecht. 1982. Eur. J. Biochem. 125, 505

8. Lapidot, A. and L. Inbar. 1983. In: The Target of Penicillin (R. Hakenbeck, J.-V. Höltje and H. Labischinski, eds.). Walter de Gruyter, Berlin . New York, p. 79

A THREE-DIMENSIONAL MODEL OF PEPTIDOGLYCAN CORRESPONDING WITH THE RESULTS OF X-RAY- AND ELECTRON DIFFRACTION, HIGH RESOLUTION ELECTRON MICROSCOPY AND THE PROPERTIES OF PEPTIDOGLYCAN

H. Formanek

Botanisches Institut der Universität München
Menzinger Straße 67, D-8000 München 19, FRG

Experimental results

Foils have been prepared from the sacculi of murein of Gram-negative and Gram-positive bacteria. The planes of the collapsed sacculi are nearly parallel with the planes of the foils.

With X-ray diffraction two diffuse Debye-Scherrer rings have been obtained, corresponding to distances of about 4.5 and 10 Å within the planes of the foils (1, 2, 3). Vertical to the plane of the foils of sacculi of Gram-negative bacteria a reflex corresponding to a periodicity of 43 Å has been measured (4). A similar periodicity was found for the multilayered murein of Gram-positive bacteria (3).

The most intense Debye-Scherrer ring obtained by X-ray diffraction corresponding to a distance of 4.5 Å in the plane of foils composed of many murein sacculi could also be obtained in the planes of single murein sacculi by electron diffraction (5).

High resolution micrographs taken at 4 K with an electron microscope containing a superconducting lens system provide a fringe structure with a periodicity of about 4.5 Å in the planes of the murein sacculi of the Gram-negative bacterium Spirillum serpens (6). At 4 K the radiation damage of the object was considerably reduced.

Similar high density values of about 1.4 g/cm³ have been obtained for murein of different Gram-negative and Gram-positive bacteria. (1, 2, 3)

Biological Properties of Peptidoglycan

Interpretation of the structure of murein.

The distance of about 4.5 Å corresponds to the strongest Debye-Scherrer ring of X-ray diffraction, the Debye-Scherrer ring of electron diffraction and the lattice periodicity visualized directly by high resolution electron microscopy. It can be interpreted as function of the packing periodicity of the peptidoglycan strands (Fig. 1). This interpretation is supported by calculations of the two-dimensional Fourier transform for different network structures of murein (7). The calculations show, that the packing periodicity provides always the strongest reflex.

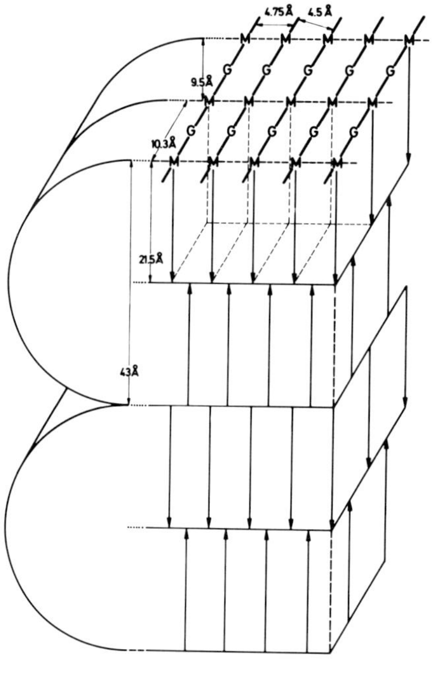

Fig. 1

Section through a foil prepared from murein sacculi of a Gram-negative bacterium. M = N-acetyl-muramic acid; G = N-acetylglu-cosamine

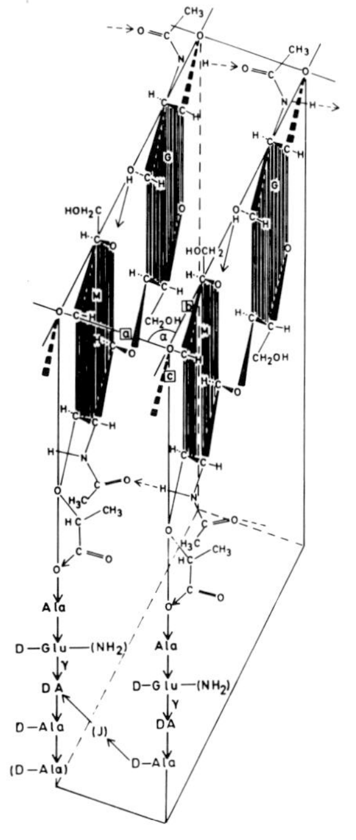

Fig. 2

Elementary cell of the murein layer containing one disaccharide peptide unit. a, b, c = dimensions of the elementary cell. α = angle between a and b.

The distance of about 10 Å can be interpreted as the repeating periodicity of the dissacharide peptide units along the peptidoglycan chain (Fig. 1).

The distance of 43 Å measured in the direction vertical to the planes of the murein sacculi can be interpreted as periodicity of a stack of superposed murein sacculi. In this stack the single head to head or tail to tail superposed murein layers are 21.5 Å thick (Fig. 1).

The density value of about 1.5 g/cm³ calculated with these data for an ideally crystalline structure of murein is only slightly higher, than the measured values for the paracrystalline and partially disordered real structure of murein. The distances of about 4.5 Å x 10 Å x 20 Å based on physical measurements can therefore be accepted as dimensions of an elementary cell of murein, containing one di-saccharide peptide unit (Fig. 2).

The dimensions of this elementary cell are in accordance with two closely related models of murein (1, 8) which propose, that its three dimensional structure can be derived from the well known three dimensional structure of chitin (9), where the 0_3 *atoms of every second sugar residue is substituted by a lactyl residue, to which a peptide chain is linked (Fig. 2).*

The sterical possibilities of the polysaccharide chains of murein have been determined, dependent on the rotation angles around the bonds to the glycosidic oxygen atoms (Fig. 2) using Ramachandran type calculations (7). Within the sterically allowed regions of these rotation angles, two to threefold screw axes are possible. Only a twofold screw axis or a screw axis oszillating around the twofold symmetry is however in agreement with the dimensions of the elementary cell. An oszillation of the screw axis around the twofold symmetry can explain both the stiffness and the gel like swelling and shrinksage of murein. Both functions are necessary for the rigidity of the murein layer and for the transport of molecules through the dynamic micellar structure of the murein layer.

As consequence of the twofold screw axis in murein all peptide chains point into identical or nearly identical directions, so that a separation of the carbohydrate and peptide moiety is obtained (Fig. 1).

Biosynthesis of murein

With this model the biosynthesis of murein can be explained by self association of single peptidoglycan strands to still existing oriented parts of murein in its micellar structure (Fig. 3). Self assembly can be supported by the possible performance of hydrogen bridges between the N-acetylgroups of adjacent peptidoglycan strands. Additional interchain hydrogen bonds for a further support of self assembly may be performed if the α-carboxylgroups of the D-glutamic acid residues are admidated, which is the case in the murein of many kinds of bacteria. This self assoziation brings the peptide chains in a position, that crosslinkage can easily be performed by an enzyme.

Fig. 3 *Scematic presentation of the biosynthesis of murein*

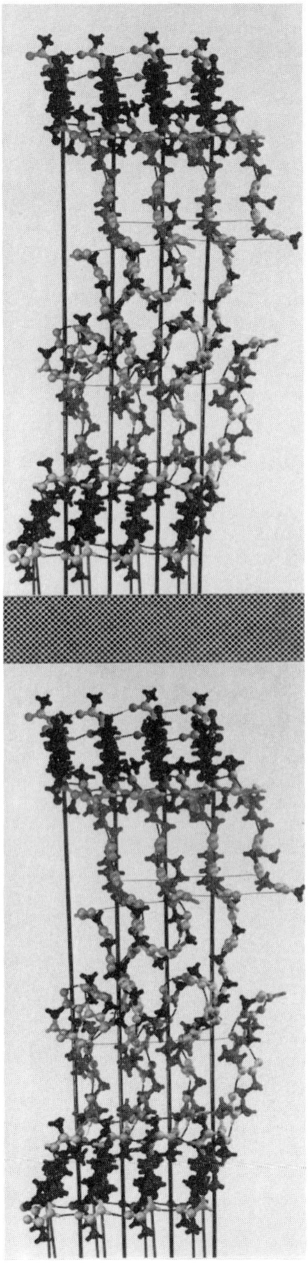

Murein of Gram positive bacteria

The multilayered murein of Gram positive bacteria may be explained by a head head or tail tail superposition of single murein layers (Fig. 4). In this case additional interpeptide bridges can be performed between the peptide moieties of two superposed murein layers. The polysaccharide moieties of two superposed murein layers may however be crosslinked by the phosphodiester linkage of teichoic acids between the C_6 OH-groups of muramic acid residues.

The twofold or approximately twofold screw axis in the polysaccharide chains of murein can explain, that on sterical reason, only the C_6OH-groups of the muramic acid residues are preferred for phosphorylation or acetylation.

Fig. 4 Scematic presentation of a part from one sacculus of a Gram-positive bacterium. Superposed layers consisting of teichoic acid, carbohydrate, peptide and interpeptide.

Digestion of murein by lysozyme

The digestion of murein by lysozyme (10) can be understood with the enzymatic mechanism described in atomic resolution by Phillips (11). Because of the separation of the carbohydrate and the peptide moiety in our model, the peptide chains can not induce any sterical hindrance to the attachment of lysozyme to a peptidoglycan chain. O-acetylation at the C_6OH-groups of N-acetyl muramic acid residues however hinders this attachment and therefore the digestion of murein by lysozyme.

Acknowledgement

This research was supported by grant no Fo 103/4 of the Deutsche Forschungsgemeinschaft.

References

1. Formanek, H., S. Formanek, H. Wawra, 1974. Eur. J. Biochem. 46, 279.
2. Burge, R.E., A.G. Fowler, D.A. Reavely, 1977. J. Mol. Biol. 117, 927.
3. Labischinski, H., G. Barnickel, H. Bradaczec, P. Giesbrecht, 1979. Eur. J. Biochem. 95, 147.
4. Wawra, H., 1976, Z. Naturf. 31c, 635.
5. Formanek, H., R. Rauscher, 1979. Ultramicroscopy 4, 37.
6. Dietrich, J., H. Formanek, F. Fox, E. Knapek, R. Weyl, 1979. Nature 277, 380.
7. Formanek, H., 1982. Z. Naturf. 37c, 226.
8. Kelemen, M.V., H.J. Rogers, 1971. Proc. Nat. Acad. Sci USA 68, 922.
9. Carlström, D., 1957. J. Biophys. Biochem. Cytol. 3, 669
10. Formanek, H., 1978. Biophys. Struct. Mechanism 4, 1.
11. Phillips, D.C., 1966. Sci. Amer. 215, 78.

STRUCTURE DETERMINATION OF PEPTIDOGLYCANS BY FAST ATOM BOMBARDMENT MASS SPECTROMETRY

S.A. Martin, K. Biemann
Department of Chemistry, Massachusetts Institute of Technology, Cambridge, MA 02139 U.S.A.

R.S. Rosenthal
Department of Microbiology and Immunology, Indiana University School of Medicine, Indianapolis, IN U.S.A.

Introduction

Since its introduction as an analytical technique mass spectrometry (MS) has been used to characterize a wide range of compound classes. The standard mode of operation involves electron ionization (EI) of the sample in the gas phase and the subsequent detection of the positive or negative ions formed due to the interaction of the electron beam with the sample. The major limitation of this method is the need to vaporize the sample prior to ionization.

Recently several techniques have been introduced which eliminate the need to vaporize the sample prior to ionization thereby greatly expanding the range of compounds amenable to analysis. These techniques, collectively referred to as particle desorption methods, provide molecular weight and structure information for a wide range of large, even thermally labile compounds. Fast atom bombardment mass spectrometry (FABMS) introduced by Barber et al. in 1981 (1) is one of these techniques. In FABMS a primary beam of neutral particles (produced by neutralization of an accelerated beam of charged species) with translational energies between 5 keV and 10 keV impinge upon a sample in a liquid matrix. The ions (positive or negative) produced from the interaction of the energetic primary beam with the sample or matrix are focused, accelerated, mass selected and detected.

Biological Properties of Peptidoglycan

The relative ease with which FABMS can be carried out with existing mass spectrometers has lead to the characterization of a variety of biological molecules. In the course of the determination of the structures of the somnogen Factor S and related peptidoglycans (2,3) the sample preparation, detection limits and fragmentation mechanisms of these compounds in FABMS have been studied in detail. The following provides an overview of the experimental conditions and extent of information which may be acquired by FABMS.

Methods

Instrumentation. A double focusing (Varian MAT 731) mass spectrometer of Mattauch-Herzog geometry with a mass range of 1-2000 daltons at 8 kV accelerating potential was employed for most of this work (4). The primary particle beam consisted of xenon atoms with 8 keV translational energy and beam currents of approximately 20 uA.

Sample Preparation. The liquid matrix is the most important parameter in FABMS providing a renewable surface from which the sample is desorbed. The matrix may be any material which is a liquid at room temperature, has a low vapor pressure so that it will not evaporate rapidly at $1x10^{6}$torr, dissolves the compound of interest and does not suppress the ionization of that compound. Several matrices, including glycerol, thioglycerol, dithiothreitol/dithioerythritol 5:1 and diethanolamine have been used successfully in a variety of applications. Glycerol is the most widely used matrix and is also the most effective one for the study of peptidoglycans. A typical sample preparation scheme (4) results in 2-3 µl of matrix containing a total of 0.01 nmol to 10 nmol of material. A small portion of this, approximately 0.5 µl, is applied to the probe tip, inserted into the high vaccum ion source region where it intersects the primary particle beam yielding ions which are mass analyzed and detected. At concentrations below 1 nmol/µl only molecular weight information is obtained, however, with concentrations in the range of 10 nmol/µl for single components fragment ions which are related to the structure may be observed. These are due to the cleavage of certain bonds which is sometimes accompanied by the transfer of one or two hydrogen atoms as shown in Figure 2.

Results

The predominant ions in the molecular weight region of the sample of interest in FABMS correspond to the protonated molecular species, $(M+H)^+$. In addition to this molecular weight information, which is the only reliable information obtained from mixtures, structurally significant fragment ions may also be observed in the spectra of single components. FAB mass spectra also exhibit a continuous background attributable to the liquid matrix (Figure 1,2). The FAB mass spectrum (Figure 1) of a mixture of anhydro-disaccharide peptidoglycan monomers obtained by digestion of gram negative bacteria, Neisseria gonorrhoeae,

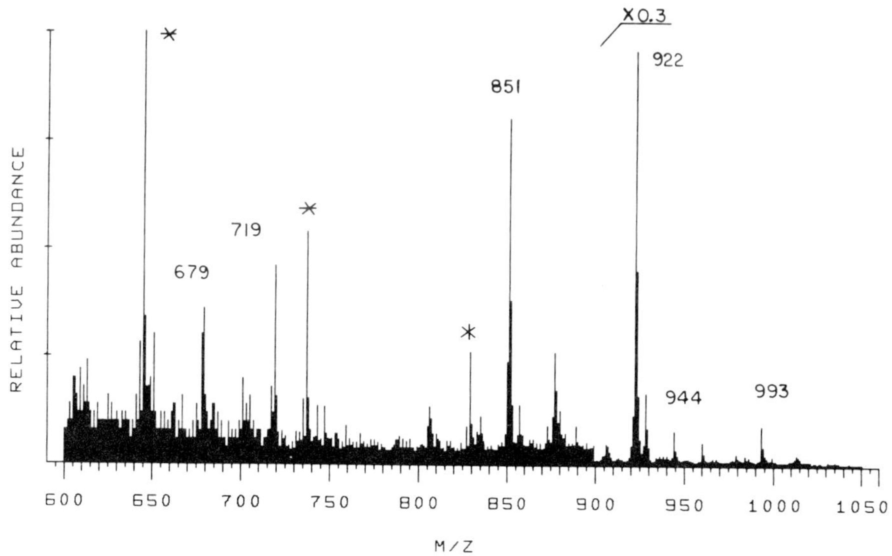

Figure 1. FAB mass spectrum of the molecular ion region of anhydro-disaccharide peptidoglycan monomer obtained from Neisseria gonorrhoeae. The concentration of this mixtue is 5 nmol/ul in acidified glycerol. The ions at m/z 679, 719, 851, 922 and 993 correspond to protonated molecular ions, MH^+. The ions at m/z 928, 944 and 960 are the lithium, sodium and potassium adducts, respectively, of m/z 922. Glycerol cluster ions are labeled * in this mass spectrum. The peaks plotted above m/z 900 are attenuated by a factor of 3.3 (x0.3) for comparison.

with E. coli transglycosylase/endopeptidase (5) illustrates the information obtained from mixtures. The abundant ions observed correspond to the protonated molecular species, $(M+H)^+$, of the peptidoglycans present in the mixture. This mixture was then separated into single components by HPLC (5) and each fraction examined by FABMS. The mass spectrum of the compound of M_r 921 (MH^+=922) exhibited fragment ions corresponding to the structure, NAG-(1-6 anhydro)NAM-Ala-Glu-Dap-Ala. The assignments were confirmed by esterifying the free carboxyl groups with excess 3N HCl/CH_3OH for 20 min at room temperature. The FAB mass spectrum (Figure 2) of the esterified product indicated a shift in molecular weight of 42 daltons corresponding to the formation of three methyl esters. In addition, the fragment ions observed in the mass spectrum of the esterified compound (Figure 2) support the structure assigned to the compound of M_r921 (MH^+= 922).

Conclusion

Recent advances in mass spectrometry permit the ionization and detection of large, underivatized thermally labile compounds such as peptidoglycans. Molecular weight information is obtained from complex mixtures whereas the spectra of individual components, after separation by HPLC, often provide additional structural information.

Acknowledgement

Peptidoglycans employed in the initial work were supplied by Drs. Manfred L. Karnovsky and John R. Pappenheimer of the Departments of Biological Chemistry and of Physiology, Harvard Medical School, Boston MA, Dr. James M. Krueger, Department of Physiology and Biophysics, University of Tennessee, Memphis, TN and Dr. Jelka Tomasic, Institute of Immunology, Zagreb, Yugoslavia. This work was supported by a grant from the National Institutes of Health (RR00317 to K. Biemann) and Public Health Service grants (R01AI-14826 and P01AI-20110 to R. Rosenthal) from the NIAID.

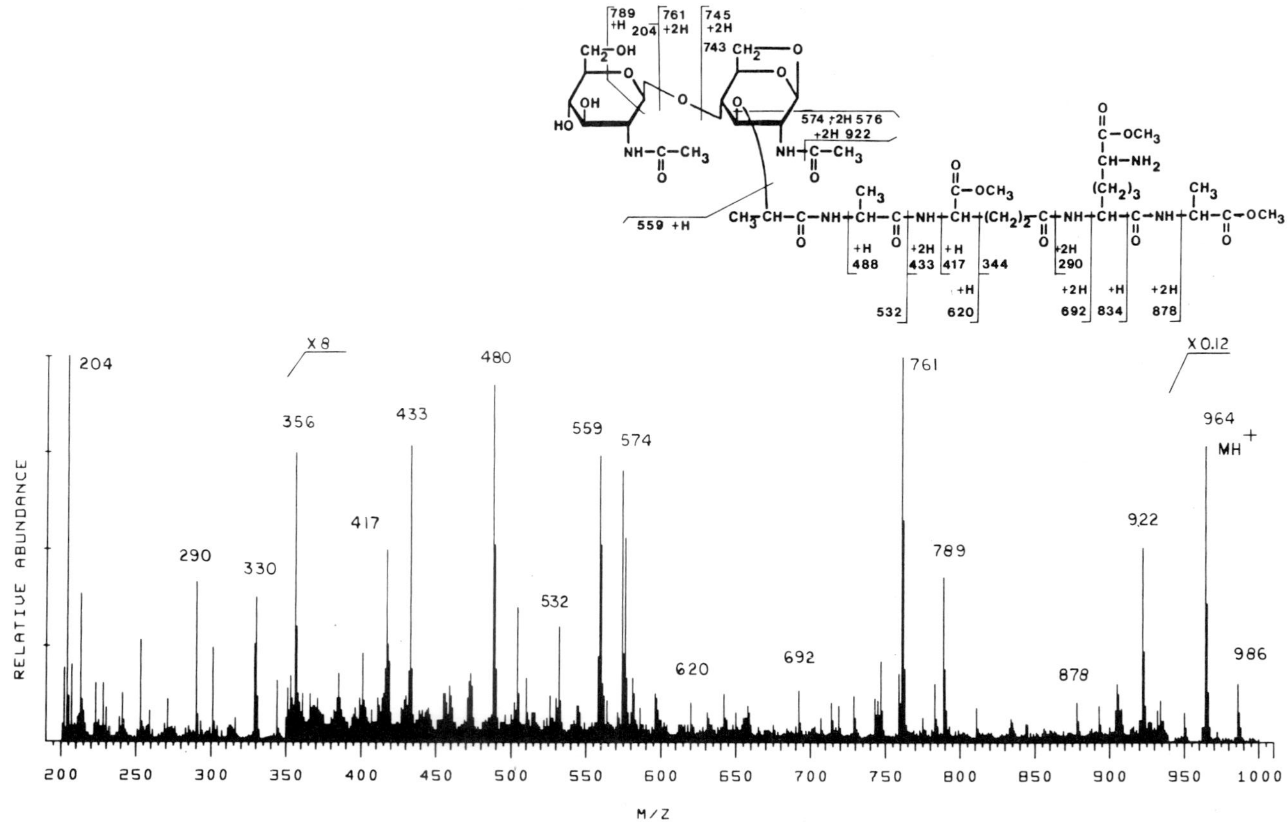

Figure 2. FAB mass spectrum and structure of methylated m/z 922, MH^+ 964, after separation of m/z 922 from the mixture (Figure 1). Sample concentration is 10 nmol/ul in acidified glycerol. Several ion which were employed in the assignment of the structure are listed in the mass spectrum and in the corresponding structure. Ion at m/z 986 is the sodium adduct of MH^+ 964.

References

1. Barber, M., R.S. Bordoli, R.D. Sedgwick, A.N. Tyler. 1981. J. Chem. Soc., Chem. Commun. 7, 325-327.

2. Martin, S.A., M.L. Karnovsky, J.M. Krueger, J.R. Pappenheimer, K. Biemann. 1984. J. Biol. Chem. 259, 12652-12658.

3. Krueger, J.M., M.L. Karnovsky, S.A. Martin, J.R. Pappenheimer, J. Walter, K. Biemann. 1984. J. Biol. Chem. 259, 12659-12662.

4. Martin, S.A., C.E. Costello, K. Biemann. 1982. Anal. Chem. 54, 2362-2368.

5. Melly, M.A., Z.A. McGee, R.S. Rosenthal. 1984. J. Infect. Dis. 149, 378-386.

peptidoglycan subunits presumably being linked to the polypeptide. Fatty acids were absent from the rigid layer fraction and the purified protein fraction. Thus, we are not dealing with a lipoprotein as described for some enterobacteriaceae (7).

Peptidoglycan/polysaccharide-complex

The peptidoglycan/polysaccharide-complex of Synechocystis PCC 6714 was prepared from rigid layer fractions by digestion with pronase (from Streptomyces griseus) followed by hot SDS-treatment (4 %, 100°C, 15 min) and ultracentrifugation (176,600 x g, 2 h). Non-peptidoglycan amino acids were completely removed as a soluble fraction (supernatant after ultracentrifugation) by this purification step. The pellet contained the peptidoglycan/polysaccharide-complex being composed of the peptidoglycan components (MurN, GlcN, L-Ala, D-Glu, m-A_2pm, D-Ala), the polysaccharide components (GlcN, ManN, GalN, Man, Glc), and phosphate. The peptidoglycan/polysaccharide-complex could be dissociated by hydrofluoric acid treatment in the cold (48 % HF, 0°C, 48 h) under conditions where only phosphodiester bonds were cleaved, yielding a de-phosphorylated, HF-insoluble peptidoglycan fraction and a HF-soluble polysaccharide fraction after centrifugation (45,000 x g, 1 h). The polysaccharide fraction was found to be free of phosphate after dialysis, demonstrating that phosphate is not a component of the polysaccharide moiety.

Enzymatic degradation of the peptidoglycan/polysaccharide-complex with N-acetylmuramidase (from hen egg white) released the glycopeptide-linked polysaccharide which was purified from peptidoglycan subunits not substituted with polysaccharide molecules by means of dialysis and gel filtration on Sephadex G-200. The glycopeptide-linked polysaccharide fraction had an apparent molecular weight of 25,000-30,000 d as determined by gel filtration on Sephadex G-75 in parallel to dextrans of known molecular weights. Peptidoglycan components as well as polysaccharide components, and organic phosphate were found to be present in the glycopeptide-linked polysaccharide fraction of Synechocystis PCC 6714.

Fragments of the peptidoglycan/polysaccharide-complex

Muramic acid-6-phosphate (MurN-6-P) was released from the peptidoglycan/polysaccharide-complex of *Synechocystis* PCC 6714 by partial acid hydrolysis (6 M HCl, 110°C, 1 h) (8) and separated by cation exchange chromatography on Dowex 50 WX-8 (H^+-form, 0.9 by 20 cm column, gradient: 0-2 M HCl) from the peptides, disaccharides, and monomers formed on hydrolysis. MurN-6-P co-chromatographed on the amino acid analyzer with MurN-6-P from the peptidoglycans of *Bacillus cereus* AHU 1356 and *Bacillus subtilis* 168, respectively, and yielded MurN and phosphate in equal amounts on total acid hydrolysis (4 M HCl, 105°C, 18 h). Glycolaldehyde-phosphate was formed on periodate oxidation of MurN-6-P from *Synechocystis* PCC 6714, which is produced only when MurN has a phosphate group in the C6-position.

Three MurN-6-P containing fragments were detected after partial acid hydrolysis (1 M HCl, 100°C, 1 h) of the peptidoglycan/polysaccharide-complex and separation by high voltage paper electrophoresis. The fragments were isolated from fluorescamine-stained electropherograms and were analyzed for their chemical composition. Amino acid analysis and gas-liquid chromatography revealed that peptidoglycan subunits, polysaccharide components, and phosphate were present in the isolated fragments. These fragments represent the linkage region of the peptidoglycan/polysaccharide-complex from *Synechocystis* PCC 6714 and demonstrate the covalent linkage of the polysaccharide to MurN-6-P of the peptidoglycan via phosphodiester bonds.

Discussion

From the presented and previous data (3-5, 10-12) we can conclude that the cell wall organization of the unicellular cyanobacterium *Synechocystis* PCC 6714 is neither typically Gram-positive nor Gram-negative, but instead combines structural elements of both these cell wall types. The presence of a peptidoglycan-linked polysaccharide with MurN-6-P as the binding site and the specific linkage type (phosphodiester bridges) are known only for Gram-positive bacteria (9). In addition, the thickness (12 nm) of the peptidoglycan (eight-layered network), the high degree of crosslinkage (56 %), and the positive Gram-reaction are all properties of a typical Gram-positive peptidoglycan. However, the presence of an outer membrane, composed of lipopolysaccharide, proteins, lipids (10, 11), and a rigid layer protein being anchored in the outer membrane, are common properties of Gram-negative cell walls.

According to the data of phylogenetical taxonomy, cyanobacteria have developed as a separate line, which is distinct from the lines of Gram-positive or Gram-negative bacteria, respectively, but have their origin in a common anchestor (13). Thus, it may be possible that the cyanobacteria represent intermediate forms between the Gram-positive and the Gram-negative cell wall types.

Acknowledments

We thank Y. Araki and H. Labischinski for supplying us with peptidoglycan fractions from Bacillus cereus AHU 1356 and Bacillus subtilis 168, respectively, and J.V. Höltje for the kindly given N,O-diacetylmuramidase (from Chalaropsis sp.). This research was supported by a grant from the Deutsche Forschungsgemeinschaft (We 670/8-1).

References

1. Drews, G. and J. Weckesser. 1982. In: The biology of cyanobacteria (N.G. Carr and B.A. Whitton, eds.). Blackwell Scientific Publications, Oxford, pp. 333-357.

2. Drews, G. 1973. In: The biology of blue-green algae (N.G. Carr and B.A. Whitton, eds.). Blackwell Scientific Publications, Oxford, pp. 96-116.

3. Jürgens, U.J., J.R. Golecki and J. Weckesser. 1985b. Characterization of the cell wall of the unicellular cyanobacterium Synechocystis PCC 6714. Arch. Microbiol. 142, 168-174.

4. Jürgens, U.J., G. Drews and J. Weckesser. 1983a. Primary structure of the peptidoglycan from the unicellular cyanobacterium Synechocystis sp. strain PCC 6714. J. Bacteriol. 154, 471-478.

5. Jürgens, U.J., G. Drews, J. Weckesser and D. Naumann. 1983b. In: The target of penicillin (R. Hakenbeck, H. Labischinski and J.V. Höltje, eds.). Walter de Gruyter, Berlin. pp. 19-24.

6. Rippka, R., J. Deruelles, J.B. Waterbury, M. Herdman and R.Y. Stanier. 1979. The cyanobacteria: generic assignments, strain histories and properties. J. Gen. Microbiol. 111, 1-61.

7. Lugtenberg, B. and L. van Alphen. 1983. Molecular architecture and functioning of the outer membrane of Escherichia coli and other gram-negative bacteria. Biochim. Biophys. Acta 737, 51-115.

8. Liu, T.Y. and E.C. Gotschlich. 1967. Muramic acid phosphate as a component of the mucopeptide of gram-positive bacteria. J. Biol. Chem. 242, 471-476.

9. Rogers, H.J., H.R. Perkins and J.W. Ward. 1980. Microbial cell walls and membranes. Chapman and Hall Ltd., New York.

10. Jürgens, U.J. and J. Weckesser. 1985a. The fine structure and chemical composition of the cell wall and sheath layers of cyanobacteria. Ann. Microbiol. (Inst. Pasteur) 136 A, 41-44.

11. Jürgens, U.J. and J. Weckesser. 1985c. Carotenoid-containing outer membrane of Synechocystis sp. strain PCC 6714. J. Bacteriol. (in press).

12. Jürgens, U.J. 1984. Charakterisierung der äußeren Membran und des Peptidoglycans des einzelligen Cyanobakteriums Synechocystis PCC 6714. Thesis, Freiburg, FRG.

13. Fox, G.E., E. Stackebrandt, R.B. Hespell, J. Gibson, J. Maniloff, T.A. Dyer, R.S. Wolfe, W.E. Balch, R.S. Tanner, L.B. Zablen, R. Blakemore, R. Gupta, L. Bonen and B.J. Lewis. 1980. The phylogeny of procaryotes. Science 209, 457-463.

CHEMICAL AND IMMUNOCHEMICAL STUDIES ON THE *IN VIVO* AND *IN VITRO* DEGRADATION OF PEPTIDOGLYCAN SUBUNITS AND CELL WALL COMPLEXES IN RELATION TO INFLAMMATORY DISEASES

A. Fox, J. Gilbart, J. Harrison, G. Pararajasegaram, A. Wells, M. Hammer, C-H Yang

Departments of Microbiology and Immunology, Ophthalmology, and Pathology, University of South Carolina School of Medicine, Columbia, S.C. 29208

Y. Ishikawa, R. Whiton, S. L. Morgan

Department of Chemistry, University of South Carolina, Columbia, S.C. 29208

The dissemination, biodegradation, and excretion of peptidoglycan (PG) containing complexes are each important factors in the perpetuation and termination of inflammation elicited during bacterial infection (1). Although certain peptidoglycans are readily degraded by lysozyme, others are highly resistant (2). Thus, other enzymes may be involved in the termination of PG mediated inflammatory events. Recently, it has been demonstrated that mammalian serum displays N-acetylmuramyl-L-alanine amidase activity (3, 4). Muramyl dipeptide (MDP) is one potential degradation product of PG and has only limited inflammatory properties *in vivo* (5). However, it has been suggested that N-acetylmuramyl L-alanine amidase does not digest MDP (3). Thus, urinary excretion of MDP would represent the only means of its elimination *in vivo* (6).

We have demonstrated using thin layer chromatography that when MDP is incubated in mammalian serum in a 0.05M Tris-0.02M $MgCl_2$ pH (7.9) buffer it is hydrolysed into its constituents N-acetylmuramic acid and the dipeptide L-alanine-D-isoglutamine.

Biological Properties of Peptidoglycan

Furthermore, the dipeptide is hydrolysed producing L-alanine and D-isoglutamine (7).

Table 1 summarizes some more recent quantitative data obtained using a Beckman 119C amino acid analyzer. Some breakdown of MDP could be detected within 6 h of incubation of MDP with serum, as determined by production of the dipeptide. Appreciable breakdown was noted within 24 h, with increasing amounts at 48 and 72 h.

Table 1 Degradation of muramyl dipeptide (2003 n mole/ml) by mammalian serum.

Incubation Time (h)	L-alanine (n mole/ml)	D-isoglutamine (n mole/ml)	L-alanine-D-isoglutamine (n mole/ml)
0	0	47*	0
1	0	31*	0
6	72	78	50
24	477	414	90
48	974	1044	116
72	1296	1299	0

*Serum incubated for 72 hours in a Tris buffer did contain 103 n mole/ml of L-alanine and 39 n mole/ml of a peak that had the same retention time as D-isoglutamine. MDP incubated for 72 hours in a Tris buffer in the absence of serum did not contain detectable L-alanine, D-isoglutamine or L-alanine-D-isoglutamine.

A more formidable task is to determine how high molecular weight peptidoglycan-polysaccharide complexes are degraded *in vivo*, where they may be present at extremely low levels. We have employed a model system in which concurrent bilateral uveitis and arthritis are elicited in the female Lewis rat by the systemic administration of an aqueous suspension of group A streptococcal peptidoglycan-polysaccharide complexes (PG-PS) and have studied the levels of PG-PS in tissues by enzyme immunoassay and gas chromatography-mass spectrometry (8-10).

Clinically, a uveitis was first observable within 24 to 72 h after injection of PG-PS and characterized by iritis, fibrin deposition, and hypopyon. Histologically, there was an acute inflammation with a polymorphonuclear cell infiltrate of the ciliary body, iris, and anterior chamber (Figure 1). Within one to two weeks, the uveitis had completely subsided both clinically and histologically with all eyes appearing normal at close of the experiments 58-62 days post-injection. In contrast, the arthritis was perpetuating with bone and cartilage erosion and pannus formation. The differences noted (acute and chronic inflammation in the eyes and joints respectively) may be the result of a number of factors which may include differing amounts of antigen localizing to different tissues (8).

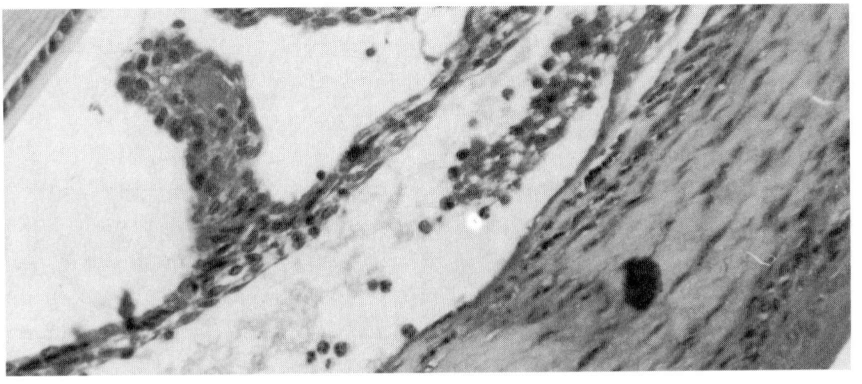

Figure 1 Hematoxylin and eosin stained section of an eye from an animal injected 3 days previously with PG-PS. A polymorphonuclear cell infiltrate is readily apparent in the anterior chamber.

An enzyme immunoassay was used to determine levels of PG-PS in tissue. Wheat germ lectin coated plates were used to bind PG-PS present in tissue homogenates followed by their detection with an affinity purified antibody specific for the group A streptococcal polysaccharide, biotin labelled anti-immunoglobulin and an avidin-peroxidase complex. The relative amounts of antigen in eye:limb:liver of PG-PS injected animals were 1:9:170 in the first several days after injection. The differences in the amounts of antigen detected in the eye and limb may help explain the development of the acute uveitis in contrast to the perpetuating polyarthritis observed on PG-PS administration (8).

More detailed biochemical studies defining the fate of PG-PS _in vivo_ need to be performed. Muramic acid is a component of bacterial peptidoglycan including group A streptococci (about 4% of the dry weight of PG-PS) which is not found elsewhere in nature. Rhamnose is a component of the polysaccharide of streptococcal cell wall complexes (about 20% of the dry weight) and is also not thought to be synthesized by mammalian enzyme systems. These two sugars can be used independently as chemical markers to measure trace levels of PG-PS in tissues. This has been accomplished using our modification of the alditol acetate procedure using selective ion monitoring gas chromatography-mass spectrometry (9, 10). Gas chromatography alone is sufficient for the analysis of high levels of sugars in isolated cell walls. (11).

Analysis of 48 experimental and control liver samples revealed mean levels of 154 ng rhamnose/mg wet weight of tissue and 60 ng muramic acid/mg at 4 days post-injection. At 60 days livers contained 27 ng rhamnose/mg and 6 ng muramic acid/mg. These chemical results confirm other immunochemical studies that streptococcal cell walls persist _in vivo_, however, they are slowly metabolized. Figure 2A illustrates the levels of rhamnose (peak 1) derived from cell walls in comparison to fucose (peak 2) derived from mammalian glycoproteins in the liver of an animal 4 days post-injection of PG-PS (i.p.). Figure 2B illustrates the considerably lower levels of rhamnose 60 days post-injection of PG-PS. By comparison normal mammalian tissues contain background levels of rhamnose (Figure 2C).

The size of PG-PS fragments is one factor in determining the severity and duration of inflammation they elicit which may be related to their dissemination and persistence in tissues (1, 12). Information is needed on the relative distribution of different particle sizes _in vivo_ and their relative degradation and elimination rates. The analysis of mixtures of such ultra

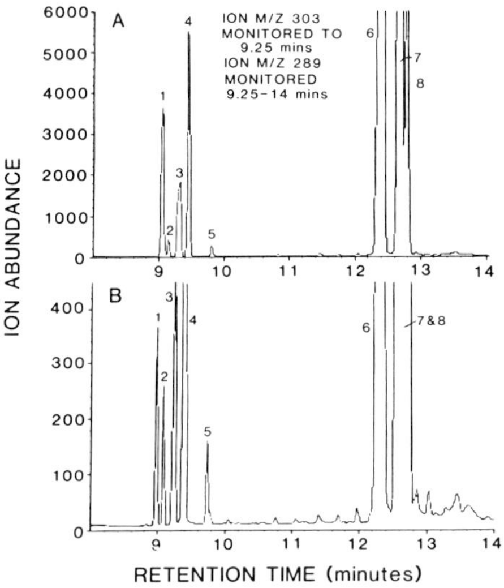

Figure 2 Selected ion monitoring gas chromatogram of liver samples from rats injected intraperitoneally: (A) 4 days previously with PG-PS, (B) 60 days previously with PG-PS, (C) 4 days previously with buffer alone. Peak identification: 1 = rhamnose, 2 = fucose, 3 = ribose, 4 = arabinose, 5 = xylose, 6 = mannose, 7 = galactose, 8 = glucose.

high molecular weight particles present in a tissue matrix at trace levels has proved extremely difficult to achieve with conventional procedures. A new approach (sedimentation field flow fractionation) has been demonstrated to be capable of determining the molecular weight distributions of heterogeneous mixtures of PG-PS ($\sim 10^6$-10^9) and is potentially capable of being applied to studies of PG degradation in vivo (13).

In conclusion, much remains to be done in determining the details of peptidoglycan catabolism in relationship to its inflammatory properties. We suggest new analytical approaches to the physical and chemical analysis of trace levels of PG-PS polymers may prove helpful in delineating some of these remaining questions.

Acknowledgements

This work was supported by grants from the U. S. Army Research Office, from the National Institutes of Health (Grant No. EY04715), the American Heart Association and from the National Science Foundation Biological Instrumentation Program.

References

1. Eisenberg, R., A. Fox, J. J. Greenblatt, S. K. Anderle, W. J. Cromartie, J. H. Schwab. 1982. Infect. Immun. 38, 127.

2. Rosenthal, R. S., J. K. Blundell, H. R. Perkins. 1982. Infect. Immun. 37, 826.

3. Ladešić, B., J. Tomašić, S. Kveder, I. Hršak. 1981. Biochim. Biophys. Acta. 678, 12.

4. Mollner, S., V. Braun. 1984. Arch. Microbiol. 140, 171.

5. Fox, A., M. E. Hammer, P. Lill, T. G. Burch, G. Burrish. 1984. Arch. Ophthalmol. 102, 1063.

6. Parant, M., F. Parant, L. Chedid, A. Yapo, J. F. Petit, E. Lederer. 1979. Int. J. Immunopharmac. 1, 35.

7. Harrison, J., A. Fox. 1985. Infect. and Immun. 50, 320.

8. Wells, A., G. Pararajasegaram, M. Baldwin, C-H Yang, M. Hammer, A. Fox. 1985. Invest. Ophthalmol. Vis. Sci. (accepted for publication).

9. Fox, A., J. H. Schwab, T. Cochran. 1980. Infect. Immun. 29, 526.

10. Whiton, R. S., P. Lau, S. L. Morgan, J. Gilbart, A. Fox. 1985. J. Chromatog. 347, 109.

11. Fox, A., S. L. Morgan, J. R. Hudson, Z. T. Zhu, and P. Y. Lau. 1983. J. Chromatog. 256, 429.

12. Fox, A., R. R. Brown, S. K. Anderle, C. Chetty, W. J. Cromartie, H. Gooder, J. H. Schwab. 1982. Infect. Immun. 35, 1003.

13. Fox, A., L. E. Schallinger, J. J. Kirkland. 1985. J. Microbiol. Meth. 3, 273.

SPECIFIC IMMUNOGLOBULIN E ANTIBODIES TO PEPTIDOGLYCAN?

Josef Seidl, Peter H. Seidl, Karl Heinz Schleifer

Lehrstuhl für Mikrobiologie, Technische Universität München, D-8000 München 2, FRG

Introduction

Specific immunoglobulin E to cells of *Staphylococcus aureus* or to purified cell walls was described by Schopfer et al. (1,2) using a solid phase radioimmunoassay (SPRIA). Employing the identical patient's serum as Schopfer et al. (1,2) we reinvestigated the reported data. The binding of radiolabelled anti-human IgE in the SPRIA was corroborated. However, from our experiments we draw the following conclusions:

1. Peptidoglycan was not involved as an antigen.
2. Binding of radioactivity reported in the SPRIA was most probably not due to IgE antibodies.

Material and methods

Solid phase radioimmunoassay (SPRIA). The methodology of Schopfer et al. (1,2) for performing the SPRIA was closely followed. Serum F.H. reported to reveal specific IgE antibodies against cells or cell walls of *S. aureus* was obtained from K. Schopfer, Children's Hospital, St. Gallen, Switzerland, further sera were collected at the City Hospital, Munich-Harlaching, FRG.

Bacterial strains and cell wall preparations. *Staphylococcus aureus* 52A5 a strain poor in cell wall teichoic acid was a gift from D. Mirelmann, The Weizman Institute, Rehovot, Israel; *S. aureus* Wood 46, poor in protein A., was obtained from the Deutsche Sammlung für Mikroorganismen, Göttingen, FRG. Crude cell walls were prepared by disruption of cells with glass beads in a Bühler cell desintegrator according to Schleifer and Kandler (3). Crude cell walls were trypsinized for 16 hours as previously described (3), the protein content of cell walls was additionally minimized by treatment of cell walls with subtilisin (4). Protein was quantified in cell wall preparations by a modified Lowry method (5).

Inactivation of IgE. IgE antibodies were inactivated by heating sera at 56°C for 4 hours (6,7).

Biological Properties of Peptidoglycan

Affinity chromatography. IgE antibodies were removed from sera by batch affinity chromatography at paper bound anti-human IgE (Pharmacia, Uppsala, Sweden).

HPLC. Serum F.H. was fractionated according to molecular weight by HPLC (LKB, Bromma, Sweden) on a TSK 4000 column (0.025 M KH_2PO buffer pH 6.8 containing 0.16 M NaCl; flow rate 0.15 ml/min). The amount of IgG and IgA eluted in the fractions was determined by kinetic nephelometry (8).

Quantitative determination of total IgE and of allergen specific IgE. Total IgE was determined in sera and in fractions by commercial test kit (PRIST®, Pharmacia, Uppsala, Sweden). Specific IgE to mite or to wheat was determined by RAST® (Pharmacia, Uppsala, Sweden).

Results and Discussion

The scheme of the SPRIA described by Schopfer et al. (1,2) is given in Fig. 1. Cells or trypsinized cell walls of *Staphylococcus aureus* were employed as an antigen. As the antibody source sera from patients with hyperimmunoglobulineamia E (Job's syndrome) were used. After incubation with ^{125}iodine-labelled anti-human IgE the radioactivity retained in the solid phase was measured.

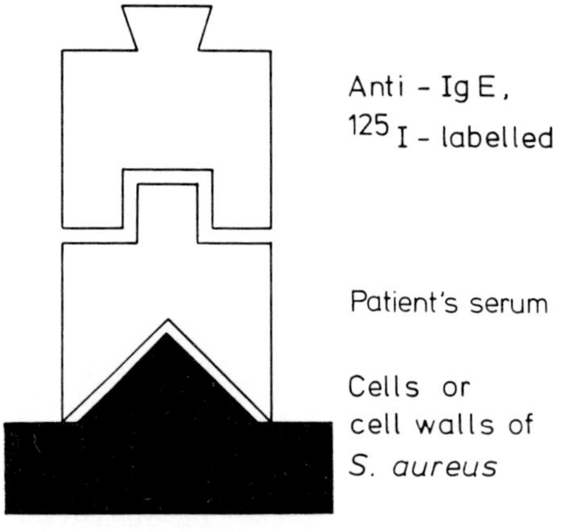

Fig. 1. Scheme of the SPRIA described by Schopfer et al. (1,2).

Employing the identical patient's serum as Schopfer et al. (1,2) we corroborated retainment of radioactivity by whole cells or crude cell walls (Table 1). It is, however, evident from the Table that binding to thoroughly trypsinized cell walls was markedly reduced.

Antigen	Percentage Binding in the SPRIA	
	S. aureus 52A5	*S. aureus* Wood 46
Whole Cells	40	38
Crude Cell Walls	38	37
Trypsinized Cell Walls	5	9
Trypsinized Cell Walls, Treated with Subtilisin	2	2

Table 1. Retainment of radioactivity in the SPRIA described by Schopfer et al. (1,2) by staphylococcal cells or cell wall preparations.

It was clear from chemically determining the protein content in cell walls that 4 hours' incubation with trypsin applied by Schopfer et al. (1,2) was not sufficient for complete removal of protein impurities. Additional treatment with subtilisin, a bacterial protease with broad specificity, further reduced protein content and binding of radioactivity (Table 1).

The assumption of Schopfer et al. (2) that binding was due to specific antibodies against the interpeptide bridge of peptidoglycan was thus obviously wrong. Our conclusion that peptidoglycan was not involved as an antigen was further corroborated from the reduction in binding after incubation of cells with 0.5 M sodium hydroxide for 30 minutes, a procedure known to remove protein.

In a second approach of the present work it was shown that binding of radioactivity reported (2) was probably not due to IgE antibodies.

It is well known that IgE antibodies are destroyed by heating at 56°C for 4 hours. This heat lability is one of the principle characteristics for distinguishing IgE from other anaphylactic antibodies (6,7). Consequently heat inactivation markedly reduced total IgE in sera nos. 3478, 2184 and F.H. as determined by PRIST ®

TOTAL IGE (IU/ML) AS DETERMINED BY PRIST

SERUM	UNTREATED	INCUBATED AT 56°C FOR 4 HOURS
No. 3478	186	19
No. 2184	3348	244
No. F.H.	7350	324

Table 2. Total IgE concentrations (IU/ml) in untreated sera and in sera incubated at 56°C for 4 hours.

(Table 2). Serum no. F.H. was reported by Schopfer et al. (2) to contain specific IgE antibodies against purified cell walls of <u>Staphylococcus aureus</u>, sera nos. 3478 and 2184 used as controls revealed no binding of radiolabelled anti-human IgE to staphylococcal cells. A drastic reduction after heat inactivation was also true for <u>allergen specific</u> IgE antibodies (Table 3) in serum no. 3478

SPECIFIC IGE (CPM) AS DETERMINED BY RAST OR SPRIA

SERUM	UNTREATED	INCUBATED AT 56°C FOR 4 HOURS
No. 3478 (SPECIFIC IGE TO MITE)	9406	986
No. F.H. (SPECIFIC IGE TO WHEAT)	4458	1621
No. F.H. ("SPECIFIC IGE TO S. AUREUS CELL WALLS")	9503	9578

Table 3. Allergen specific IgE concentrations (cpm) in untreated sera and in sera incubated at 56°C for 4 hours.

(specific IgE to mite) or serum no. F.H. (specific IgE to wheat) as determined by RAST® (Pharmacia, Uppsala, Sweden). It was, however, most surprising that in contrast specific IgE to S. aureus cells or cell walls in serum F.H. reported by Schopfer et al. (1,2), was not affected at all.

Removal of IgE antibodies from serum F.H. by affinity chromatography at solid-phase bound anti-human IgE further demonstrated that reported binding of radioactivity in the SPRIA (2) was not due to IgE antibodies. It is obvious from Table 4 that the concentration of total IgE antibodies in serum F.H. was reduced to zero level after affinity chromatography of this serum to matrix-bound anti-IgE. The same was true for allergen (wheat)-specific IgE in this serum (internal control). However, "specific IgE to S. aureus cell walls" was not affected thus demonstrating that the binding reported (2) was not due to IgE antibodies.

Fractionation of serum no. F.H. according to molecular weight by HPLC (TSK 4000 column, Pharmacia, Uppsala, Sweden) revealed, that the elution patterns of total IgE and of "specific IgE to S. aureus cell walls" were not identical with respect to molecular weight, thus again supporting the finding that binding obtained (2) was not due to IgE antibodies (Fig. 2). It was of interest that heat inactivation of IgE antibodies in the serum fraction revealing maximal binding in the SPRIA (2) again did not affect binding.

	Before	After
Total IgE (IU/ml)	7350	40
Specific IgE to Wheat (CPM)	4470	80
"Specific IgE to S. aureus Cell Walls" (CPM)	9701	9630

Table 4. Affinity chromatography of serum F.H. to matrix-bound anti-human IgE.

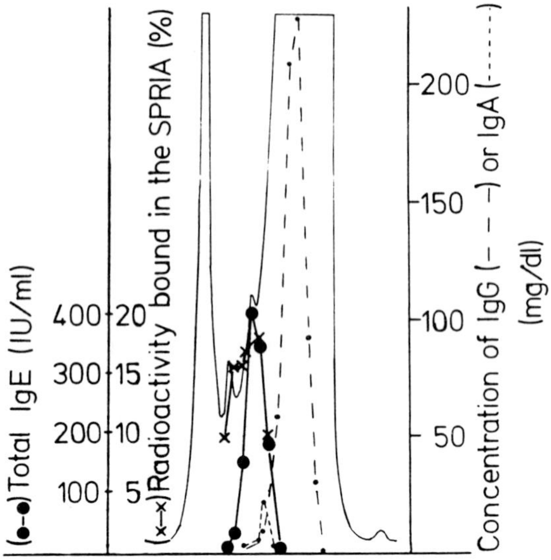

Fig. 2. Fractionation of serum F.H. by HPLC on a TSK 4000 column. For details, see materials and methods section.

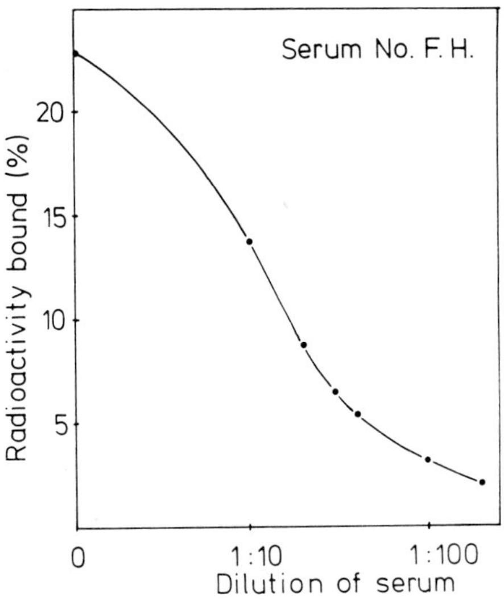

Fig. 3. Binding of serum F.H. in the SPRIA described by Schopfer et al. (1).

In the present report we demonstrated that reported binding of "specific IgE to S. aureus cells or cell walls" was not due to IgE antibodies. The nature of the binding caused in the SPRIA is yet unknown, but binding is dependent on serum dilution (Fig. 3). It is noteworthy that "specific IgE to S. aureus cells or cell walls was hitherto only described for patients with hyperimmunoglobulinemia E (1,2). Recently, nonspecific adsorption of IgM from certain pathological sera with raised IgM was reported (9) and parallelity may be discussed. Further detailed studies are, however, necessary to elucidate the nature of the reported binding (1,2).

Acknowledgement

This work was supported by DFG grant Se 427/1-1.

References

1. Schopfer, K., K. Baerlocher, P. Price, U. Krech, P.G. Quie and S.D. Douglas. 1979. Staphylococcal IgE antibodies, hyperimmunglobulinemia E and Staphylococcus aureus infections. N. Engl. J. Med. 300, 835-838.

2. Schopfer, K., S.D. Douglas and B.J. Wilkinson. 1980. Immunoglobulin E antibodies against Staphylococcus aureus cell walls in the sera with hyperimmunoglobulinemia E and recurrent staphylococcal infection. Infect. Immun. 27, 563-568.

3. Schleifer, K.H. and O. Kandler. 1972. Peptidoglycan types of bacterial cell walls and their taxonomic implications. Bact. Rev. 36, 407-477.

4. Seidl. P.H., P. Zwerenz, J.R. Golecki, N. Franken and K.H. Schleifer. 1986. Isolation of specific antibodies to the glycan moiety of peptidoglycan and their application in the indirect immunoferritin technique. FEMS Microbiol. Letters, in press.

5. Hartree, E.F. 1972. Determination of protein: a modification of the Lowry method that gives a linear photometric response. Analyt.Biochem. 48, 422-427.

6. Bennich, H. and K. Darrington. 1972. Structure and Conformation of the Immunglobulin E System. In: The Biological Role of the Immunglobulin E System (K. Ishizaka and D.H. Dayton, eds.). National Institute of Child Health and Human Development, Bethesda, MD. p. 19.

7. Johansson, S.G.O., T. Berg and T. Foucard. 1972. Circulating IgE antibodies measured by RAST and their significance in allergic diseases. In: The Biological Role of the Immunglobulin E System (K. Ishizaka and D.H. Dayton, eds.). National Institute of Child Health and Human Development, Bethesda, MD. pp. 211-219.

8. Killingworth, L.M. and Savory J. 1972. Manual nephelometric methods for immunochemical determination of immunoglobulins IgG, IgA and IgM in human serum. Clin. Chem. 18, 335-342.

9. Palusuo, T. and K. Aho. 1983. Technical falsely positive rheumatoid factor by ELISA in sera with elevated IgM levels. Med. Biol. 61, 203-207.

SPECIFIC IMMUNOGLOBULIN G ANTIBODIES IN MAN AGAINST THE GLYCAN STRAND OF PEPTIDOGLYCAN

E. Zauner, S. Reissenweber, G. Leitherer, K.H. Schleifer, and P.H. Seidl

Lehrstuhl für Mikrobiologie, Technische Universität München, Arcisstr. 21, D-8000 München 2, FRG

Introduction

The peptidoglycan molecule reveals at least five antigenic epitopes (for review see Seidl and Schleifer, this book). In man, specific antibodies to the non-crosslinked pentapeptide subunit with C-terminal R-D-Ala-D-Ala-OH and to the glycan strand have been hitherto detected, and several studies reported on a diagnostic role of antibodies against the pentapeptide subunit (1,2). The applicability of these assay systems is restricted, however, since many eubacteria including almost all gram-negative organisms lack such pentapeptides. On the other hand, the glycan strand of peptidoglycan is a rather uniform chemical component of all eubacteria. Therefore, an enzyme-linked immunosorbent assay for routine quantification of glycan specific antibodies should be useful.

Materials and Methods

Antigen albumin-(glycopeptide)$_n$. Glycopeptides were isolated from peptidoglycan of *Bacillus subtilis* NCIB 8060 by treatment with hen egg-white lysozyme (3), dialysis and fractionation of the dialysable material on a Sephadex G-25 column (compare Ref. 4, Fig. 2 and Results section). Glycopeptides were covalently coupled to albumin as a carrier (Sigma A-1887) by reacting 7.17 mg protein and 21 mg glycopeptides in 1.5 ml 0.1 M sodium carbonate buffer pH 9.6 with 100 µl 25% glutardialdehyde for 18 hours at 4°C. After addition of 0.2 moles lysine, the reaction mixture was dialysed and lyophilized.
ELISA. The ELISA was performed according to published procedure (5) with the following modifications: Polystyrene microtiter wells were coated with 50 µl of the antigen (2 µg/ml); 0.02 M phosphate-buffered saline pH 7.2 containing 0.2% Tween 20 was employed for

Biological Properties of Peptidoglycan

dilution of sera and washings.

Inhibition of ELISA. Glycan strands, unsubstituted by peptides, were obtained by action of thiophenol (6) on peptidoglycan of *S. aureus* 52A5 (glycan I). Aliquots were N-acetylated (glycan II) and, in addition, the carboxyl groups of muramic acid were reduced (glycan III) (6). Glycans I, II, III, diacetyl-chitobiose, triacetyl-chitotriose, MDP (Sigma), diacetyl-L-Lys-D-Ala_2-OH (Serva) and diacetyl-L-Lys-D-Ala-OH (UCB-Bioproducts, Brussels, Belgium) were employed as inhibitors according to published procedures (7), the methodology for preadsorption of sera to peptidoglycans was recently reported (8).

Results and Discussion

Peptidoglycan from *Bacillus subtilis* NCIB 8060 was used for isolation of glycopeptides because of its particular primary structure (Fig. 1). Special emphasis was put on complete removal of protein impurities and of teichoic acids from the peptidoglycan employed for isolation of glycopeptides. After extensive incubation of cell walls with trypsin, subtilisin, proteinase K and protease Type V

Peptidoglycan of *Bacillus subtilis* NCIB 8060:

- Lacks an Interpeptide Bridge
- Lacks Non-Crosslinked Pentapeptide Subunits with C-Terminal R-D-Ala-D-Ala-OH
- Lacks Non-Crosslinked Tetrapeptide Subunits
- Reveals 62% of Peptide Subunits as Tripeptides with C-Terminal Diaminopimelic Acid
- The High Percentage (55%) of Meso-Diaminopimelic Acid Residues with Free ω-Amino Groups Is Important for Covalent Linkage to the Carrier
- Reveals Carboxyl Groups of Muramic Acid Completely Substituted by Peptide Subunits
- Reveals a Minor Content of De-N-Acetylated Glucosamine Residues

Amino Acid/ Amino Sugar	Peptido-glycan	Glyco-peptides
Muramic Acid	1.06	0.97
Glucosamine	1.08	1.0
L-Alanine	1.0	1.0
N-Terminal L-Alanine	0	N.D.
D-Glutamic Acid	1.0	1.0
Meso-Diaminopimelic Acid	1.05	0.98
C-Terminal Meso-Diaminopimelic Acid	0.62	N.D.
N-Terminal Meso-Diamonopimelic Acid	0.55	N.D.
D-Alanine	0.38	0.22
C-Terminal Alanine	0	N.D.
Unsubstituted Amino Group of Glucosamine	0.09	N.D.

Fig. 1. Structural features of peptidoglycan from *Bacillus subtilis* NCIB 8060.

from <u>Streptomyces griseus</u>, the protein content was reduced to 0.7%. Teichoic acids were quantitatively removed by 70% hydrogen fluoride (9), as was demonstrated by the phosphate content of 0.05% and not detectable polyols by GLC (compare Fig. 2).

Detection of Glycan Specific IgG Antibodies

Employing albumin-(glycopeptide)$_n$ as an antigen in the ELISA described above, specific IgG antibodies to the glycan moiety could be detected in human sera. Fig. 3a shows the binding curves of

Fig. 2

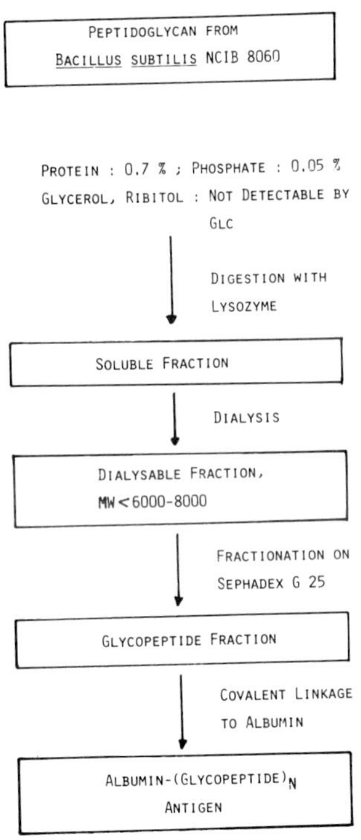

Fig. 3

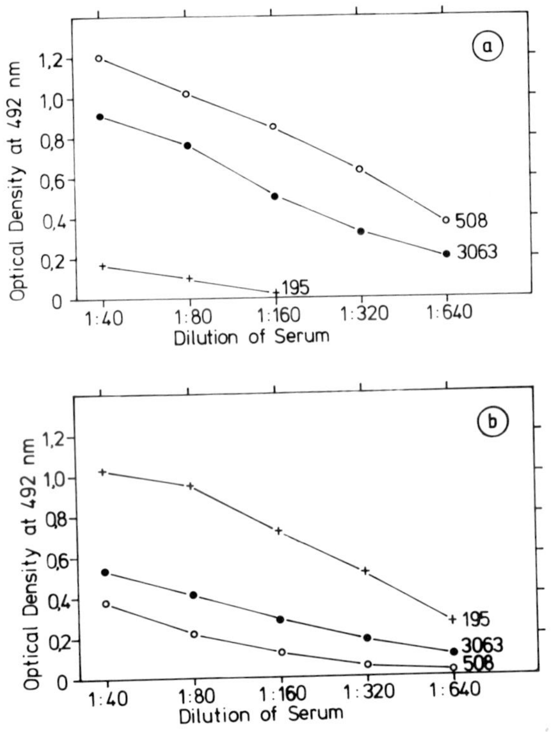

Fig. 2. Scheme of the preparation of albumin-(glycopeptide)$_n$ used as an antigen in the ELISA.

Fig. 3. Binding of specific IgG antibodies in an ELISA a) to the glycan strand of peptidoglycan. Albumin-(glycopeptide)$_n$ was adsorbed to polystyrene microtiter plates, and specific IgG antibodies to the glycan were detected with goat anti-human IgG conjugated to peroxidase; b) to the peptidoglycan sequence R-D-Ala-D-Ala-OH (7).

three human sera to albumin-(glycopeptide)$_n$, and in comparison (Fig. 3b), to albumin-(D-Ala$_3$)$_9$ (7). Serum 508 revealed specific antibodies to the glycan strand and lacked antibodies to the pentapeptide subunit. In contrast, serum 195 contained only specific antibodies against C-terminal R-D-Ala-D-Ala-OH, whereas serum 3063 revealed antibodies to both, the glycan moiety and the pentapeptide subunit. Thus, it was evident that the glycan moiety and the pentapeptide subunit represent independent antigenic epitopes of peptidoglycan in man.

Antigenic Specificity of the ELISA

Various glycan strands no more substituted by peptide subunits were prepared by treatment of peptidoglycan with thiophenol (6), compare Fig. 4; Glycan I, completely de-N-acetylated; Glycan II, completely N-acetylated; and completely N-acetylated Glycan III with free carboxyl groups of N-acetyl muramic acid reduced to -CH_2OH groups. It was important that binding of glycan strand specific human antibodies to the albumin-(glycopeptide)$_n$ antigen was completely inhibited by oligosaccharides composed of alternating ß-1,4-linked N-acetylglucosamine and N-acetyl muramic acid residues (Glycan II). The specificity of the ELISA for detecting antibodies directed against the glycan moiety of peptidoglycan was thus clearly proved. It was obvious from comparing the nearly identical inhibition curves obtained with Glycan II and Glycan III (carboxyl groups of N-acetyl muramic acid reduced to the -CH_2OH group) that

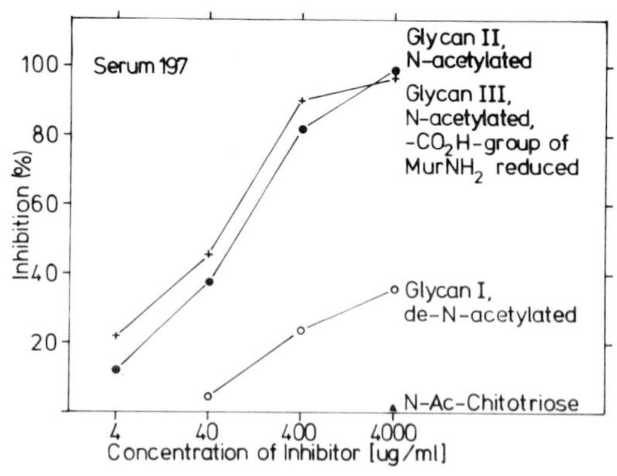

Fig. 4. Binding inhibition in the ELISA for measuring glycan strand specific antibodies, employing as inhibitors several glycan strand preparations no more substituted by peptides. For details, see text.

free carboxyl groups of N-acetyl muramic acid did not contribute to the antibody combining site. Complete removal of N-acetyl groups (Glycan I) affects binding of glycan strand specific antibodies, whereas in contrast, partial de-N-acetylation has no effect (Seidl and Schleifer, this book). Synthetic peptides with structural similarity to the C-terminal sequences of the stem peptide (Ac_2-L-Lys-D-Ala-OH, Ac_2-L-Lys-D-Ala_2-OH) or MDP did not inhibit, thus excluding binding specificity for these structures.

Further proof for the specificity of the ELISA was furnished by appropriate inhibition studies, employing as inhibitors culture filtrates from bacterial cells. It is well established that growing bacteria secrete peptidoglycan fragments into the medium (10-13; Rosenthal and Fleming, this book; Zwerenz et al., this book). According to this, the ELISA for detecting glycan specific antibodies was specifically inhibited by bacterial culture filtrates taken from the logarithmic phase of growth (e.g. from Staphylococcus aureus, Streptococcus pyogenes, Bacillus subtilis, Escherichia coli, Pseudomonas aeruginosa, Salmonella typhimurium, Edwardsiella spec., Enterobacter aerogenes, Shigella spec., Proteus mirabilis; data not depicted). Culture filtrates from yeasts as well as uninoculated nutrient broth did not inhibit.

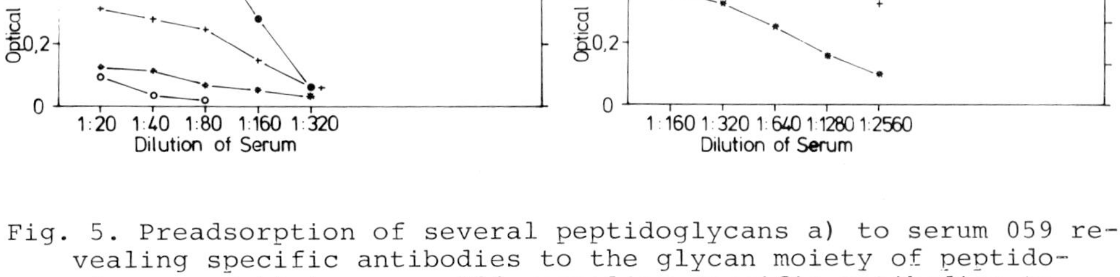

Fig. 5. Preadsorption of several peptidoglycans a) to serum 059 revealing specific antibodies to the glycan moiety of peptidoglycan and b) to serum 004 revealing specific antibodies to the pentapeptide subunit of peptidoglycan, prior to subsequent performance of the ELISA a) for detecting glycan specific antibodies b) for detecting specific antibodies against the pentapeptide subunit (7).

Finally, the specificity of the ELISA for detecting antibodies against the glycan strand of peptidoglycan was again corroborated by preadsorption studies (Fig. 5). Human serum 059 revealing glycan moiety specific antibodies was preadsorbed to bacterial peptidoglycan, prior to performance of the ELISA for measuring glycan specific antibodies (Fig. 5a). All peptidoglycans examined (several species of staphylococci and streptococci, E. coli, Enterobacter aerogenes, Bordetella pertussis, Klebsiella pneumoniae) markedly reduced subsequent binding in the ELISA for detecting glycan moiety specific antibodies. Cell walls from yeasts (e.g. Candida albicans, Saccharomyces spec.) or pseudomurein did not affect binding.

In comparison, specific antibodies to the pentapeptide subunit (serum 004) detected by an appropriate ELISA (7) were only adsorbed by peptidoglycans revealing such peptides (e.g. Staphylococcus aureus). Peptidoglycans from Bacillus subtilis or E. coli lacking pentapeptide subunits, or cell walls from yeasts did not affect binding (compare Fig. 5b).

In conclusion, due to its relative structural uniformity (14) and regarding the immunological data reported here, the glycan moiety of peptidoglycan may be considered an antigenic epitope shared by all eubacteria.

Acknowledgement

This research was supported by grant 01 ZR 112 from the Bundesminister für Forschung und Technologie (BMFT).

References

1. Heymer, B., K.H. Schleifer, S. Read, J.B. Zabriskie, R.M. Krause. 1976. J. Immunol. 117, 23

2. Zeiger, A.R., C.U. Tuazon, J.N. Sheagren. 1981. Infect. Immun. 33, 795.

3. Schleifer, K.H., L. Huss, O. Kandler. 1969. Arch. Microbiol. 68, 387.

4. Seidl, P.H., J.R. Golecki, N. Franken, P. Zwerenz, K.H. Schleifer. FEMS Microbiol. Lett. (in press).

5. Engvall, E., P. Perlmann. 1971. Immunochemistry 8, 871.

6. Kawagishi, S., Y. Araki, E. Ito. 1980. Eur. J. Biochem. 112, 273.

7. Franken, N., P.H. Seidl, E. Zauner, H.J. Kolb, K.H. Schleifer, L. Weiss. 1985. Mol. Immunol. 22, 573.

8. Franken, N., P.H. Seidl, T. Kuchenbauer, H.J. Kolb, K.H. Schleifer, L. Weiss.K.-D. Tympner. 1984. Infect. Immun. 44, 182.

9. Fiedler, F., M.J. Schäffler, E. Stackebrandt. 1981. Arch. Microbiol. 129, 85.

10. Glaser, L., B. Lindsay. 1977. J. Bacteriol. 130, 610.

11. Sinha, R.K., R.S. Rosenthal. 1980. Infect. Immun. 29, 914.

12. Mychajlonka, M., T.D. McDowell, G.D. Shockman. 1980. Infect. Immun. 28, 65.

13. Doyle, R.J., M.A. Motley, P.H.B. Carstens. 1982. Carbohydrate Res. 104, 147-152.

14. Schleifer, K.H., P.H. Seidl. 1985. Chemical Methods in Bacterial Systematics (M. Goodfellow, D.E. Minnikin, eds). Academic Press, London. p. 201.

RELEASE OF PENICILLIN-BINDING PROTEINS FROM ß-LACTAM TREATED BACTERIA: DETERMINATION BY ANTI-ß-LACTAM ANTIBODIES

R. Hakenbeck, H. Ellerbrok, Th. Briese
Max-Planck Institut für molekulare Genetik, Ihnestr. 63-73, D 1000 Berlin 33

N.F.Adkinson
The Good Samaritan Hospital, 5601 Loch Raven Bld., Baltimore, MD 21239

Introduction

Antibiotics that interfer with murein biosynthesis belong to the class of lytic antibiotics, i.e. they cause cellular lysis when added to a growing bacterial culture. However, in lysis-defective mutants or in bacteria with a suppressed autolytic system, it is possible to study biochemical events that are induced in the cells upon treatment with these drugs, since they respond only by growth inhibition without accompanied lysis. Tomasz and coworkers have documented in several gram-positive bacteria, that treatment with these antibiotics (under non-lytic conditions) leads to secretion of several cell wall components (murein and teichoiç acid precursors, as well as lipids and lipoteichoic acid) (1-4). Recently, we have analyzed also proteins that are released during bacitracin or penicillin treatment of autolysin-defective pneumococci. The proteins were mainly membrane proteins, and they were associated with membrane vesicles which could easily be recovered from the culture medium of the drug treated bacteria by high speed centrifugation. Among the membrane proteins, penicillin binding proteins (PBP) could be identified (5). These minor membrane components occur in all penicillin-sensitive bacteria. By definition, they are able to bind ß-lactams covalently, and it is this feature by which they can be visualized after incubation with radioactive penicillin, SDS-polyacryl-amide-gel-electrophoresis (PAGE) and fluorography (for review, see 6,7). At the same time, it is this very property which makes it difficult to quantitate them after the cells have been treated with ß-lactams. They do bind the (nonradioactive) antibiotics already in vivo by forming a PBP-penicilloyl-(BPO-)complex (Fig.1), are then partly released associated to the membrane vesicles. Postincubation of either cells or vesicles with radioactive penicillin, followed by PAGE and fluorography as described above, will then reveal only those PBP that had not reacted

Biological Properties of Peptidoglycan

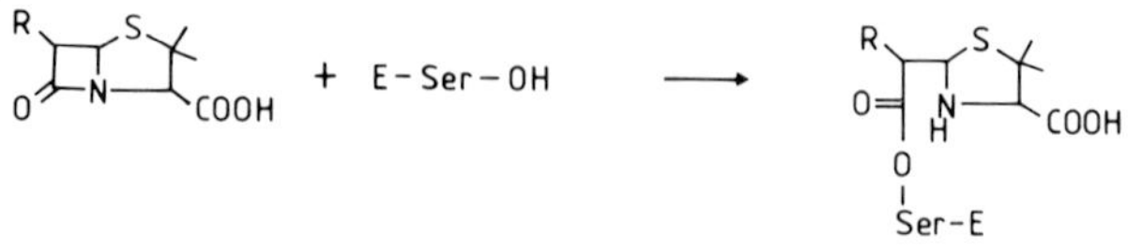

Penicillin + PBP ⟶ ⟶ (fast) penicilloyl-PBP ⟶ (slow) pen-derivative + PBP

Fig.1: Interaction of PBP with ß-lactam antibiotics

with the nonradioactive ß-lactam *in vivo*, or have lost it during the isolation or postincubation period. Thus, PBP could be studied only when cell wall inhibitors others than ß-lactams were used for induction of membrane secretion. Recently we discovered that ß-lactam/PBP-complexes can also be demonstrated after SDS-PAGE, blotting onto nitrocellulose membrane and immunostaining with anti-ß-lactam antibodies (submitted for publication; details of materials and methods are described in that manuscript). The method allows direct determination of penicilloyl-PBP without the need of radioactive penicillin. In the present communication, the release of PBP during penicillin- and aztreonam (AZ)-treatment of *Streptococcus pneumoniae* will be documented using the corresponding anti-ß-lactam antisera.

Results and discussion

1. Bacterial growth under non-lytic conditions. *S. pneumoniae*, either an autolysin defective mutant (cwl) or the wildtype (R6) grown under lysis-nonpermissive conditions, were used. It has been described, that the autolysin of pneumococci (an N-acetyl-L-alanine-amidase) (8) which is the only autolysin described in this species can be inhibited by growth in medium supplemented with 2% choline (9,10). This way, the amidase is obviously no longer capable of interaction with its proper substrate (11,12) and the cells do not lyse. Growth in 2% choline also interfers with cell separation (although proper cell division still occurs), and chain formation is the consequence. It has been argued, that this demonstrates a participation of the amidase in cell separation. However, the autolytic defective mutant cwl does not grow in chains under normal growth conditions, but it did so in 2% choline (Fig. 2B). This could mean, that either the mutant still contains residual amidase activity which is sufficient for cell separation, or that growth in 2% choline interfers also with other, still

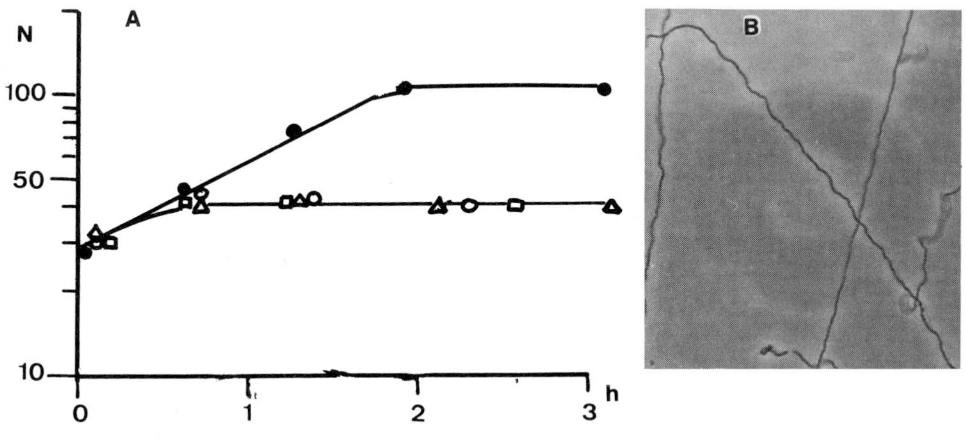

Fig.2: Growth of S. pneumoniae under non-lytic conditions. A. At time=0, benzylpenicillin (0.1 µg/ml) was added to exponentially growing cultures of cwl (○), cwl grown with 2% choline (□), and R6 grown with 2% choline (△). The control culture remained without antibiotic (●). Growth was monitored by nephelometry (N) B. Strain cwl grown with 2% choline for several generation.

undetected enzymes which participate in cell separation. In any case, wild type R6 as well as the mutant grown in 2% choline do not lyse upon treatment with ß-lactam antibiotics (Fig. 2A).

2. Determination of penicilloyl-PBP during penicillin treatment. S. pneumoniae contains six PBP ranging from a molecular weight of 94 kDa to 43 kDa (PBP 1a, 1b, 2x, 2a, 2b and 3). Strains cwl and R6 were grown in the presence of 2% choline, and cwl also without further choline-addition to the medium.Cells were treated with 0.1 µg/ml benzylpenicillin corresponding to approximately 10 x MIC, which is sufficient to saturate all PBP with the antibiotic in 5 min at 37°C. After 80 min, the cells were centrifuged and the supernatant (medium) recovered. PBP in the medium were measured either after acetone precipitation (which denatures) the BPO-PBP complex immediately), or alternatively, membrane vesicles were collected from the medium after high speed centrifugation. Although this step is carried out at 4°C, binding and release of penicillin could occur during this period. This method might therefore not be as accurate as the precipitation procedure, although larger quantities of the vesicles can be collected this way. Fig. 3A shows the outcome of this experiment. One can see, that under all conditions equivalent amounts of PBP are released from the various bacterial cultures. Judging from the staining intensities, approximately 25% of the total cellular PBP ends up outside the cell after penicillin treatment, which corresponds fairly well with values obtained by determination of penicillin-binding

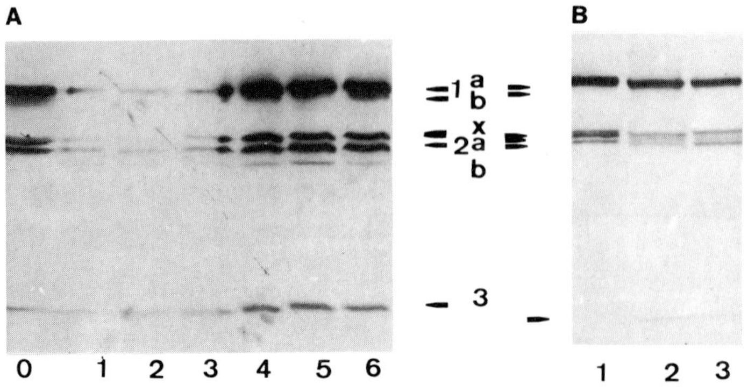

Fig. 3A. PBP in supernatants of penicillin- treated S. pneumoniae detected on immunoblots with anti-BPO-antibodies. Cultures of the mutant cwl (1,4), cwl grown with 2% choline (2,5), and wild type R6 grown with 2% choline (3,6) received 0.1 µg/ml benzylpenicillin for 80 min. Cells were then centrifuged. Either 0.5ml of the supernatant were mixed with 2ml acetone in order to precipitate proteins (1-3), or membrane vesicles from 4ml supernatant were recovered by centrifugation for 14h, 48,000 rpm, 50 Ti rotor (4-6). An R6 control culture received no drug during growth; after lysis of 1ml culture, PBP were labeled in vitro with penicillin (0). Proteins were separated on SDS-PAGE, blotted onto nitrocellulose and stained after incubation with anti-BPO-antibodies using phosphatase-conjugated anti-IgG. PBP are indicated.
B. PBP in cells and supernatant of penicillin treated R6. Strain R6 was grown in the presence of 2% choline, and 0.1 µg/ml penicillin was added for 90 min. PBP were separated and visualized as described above. 1: PBP in 5ml medium after 90 min Pen-treatment of the cells; 2: PBP in cells (1ml culture equivalent) after Pen-treatment for 10 min; 3: PBP in control cells (1ml culture equivalent) after labeling with Pen of cell-lysate.

activity (manuscript in preparation). Fig. 3B shows another aspect of the protein release, that is a quantitative difference of the PBP in the vesicles compared to those found in whole cell lysates. PBP 2a is present in larger amounts in the medium than PBP 2x and 2b, whereas the opposite is true in the lysed cell. Anti-BPO-antibodies do not react with PBP 3-complex for still unknown reasons, therefore the amount of PBP 3 can not be accurately determined with this method. However, it can be determined using anti-aztreonam-antibodies (13) as the AZ-PBP complex shown below.

3. Determination of aztreonam-PBP during aztreonam treatment. Fig. 4a shows PBP in the medium of AZ-treated cultures of cwl. The MIC for AZ is 50 µg/ml. PBP are saturated under these conditions with the exception of PBP 2b, which binds AZ only at 0.5 mg/ml and higher. Nevertheless, one can see that the higher the concentration of AZ the more PBP appear in the medium.

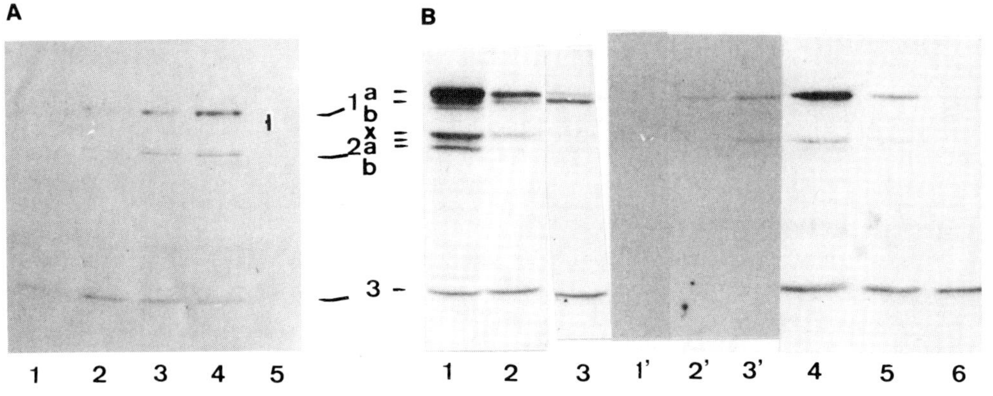

Fig. 4. PBP in cells and medium of aztreonam-treated pneumococci.
A. To aliquots of an exponentially growing culture of cwl, different amounts of AZ were added. After 25 min, cells from 1ml culture were removed by centrifugation and proteins of 0.5ml supernatant acetone precipitated. PBP in the samples were revealed after SDS-PAGE and immunoblotting using anti-AZ-antibodies. Concentrations of AZ used (mg/ml): 1-0.05; 2-0.2; 3-0.8; 4-2.0; 5-0.
B. Strain R6 was grown in the presence of 2% choline and various concentrations of AZ for 5h. Cells were recovered by centrifugation and lysed with detergent. One part was incubated with radioactive penicillin and radioactive BPO-PBP detected after fluorography (1´-3´). The other part was directly prepared for SDS-PAGE (0.2ml culture equivalents. AZ-PBP complexes in the cells (1-3) and in 0.5 ml medium (4-6) were visualized after blotting with anti-AT-antiserum. Concentrations of AZ used (µg/ml): 1,1´,4:10; 2,2´,5:1; 3,3´,6:0.1.

Fig. 4B shows PBP in cells and supernatants of R6 grown with 2% choline after treatment for several hours with subinhibitory concentrations of AZ. Binding of AZ to the cell-associated PBP was determined either directly with anti-AZ-antiserum (Fig.4B, 1-3), or indirectly by postlabeling PBP in cell-lysates with radioactive Pen (1´-3´). After SDS-PAGE and autoradiography, only those PBP show up that have not reacted with AZ during growth. Even at these low concentrations PBP appear to be labeled with AZ, and secretion of PBP occurs although at a much slower rate when compared to growth in the presence of high concentrations of AZ. Therefore, the samples were taken after several hours of AZ-treatment. Samples 4-6 show AZ-PBP secreted into the medium. It is again striking as has been pointed out already in Fig. 3B, that PBP 2x and 2b are preferentially associated with the cells, whereas almost all of PBP 2a appears in the medium. That finding confirms former results, in which the total protein composition of the secreted material was compared to membrane and mesosomal proteins (the term mesosomes refers to a membrane fraction which can be recovered from the medium after spheroplasting of the cells). In summary it appears that only a subfraction

of the membrane proteins, which might correspond to a subfraction of the membrane itself, participate in the secretion process which is induced by antibiotics that interfere with murein biosynthesis (5).
Another aspect concerns the situation in a patient whose infection is treated with ß-lactam antibiotics. Our results show, that massive amounts of proteins (BPO-PBP) that contain the antigenic site for anti-ß-lactam antibodies are released from the pathogenic bacteria, not only during cellular lysis but also from tolerant, non-lytic bacteria. Production of anti-ß-lactam antibodies is associated with the antibody-dependent penicillin-allergic reaction against ß-lactams (see 14 for review). In which way the BPO-PBP as penicilloyl-antigens influence this reaction remains to be clarified.

Acknowledgement

The technical assistance of Spassena Tornette is greatfully acknowledged.

References

1. Horne, D., A. Tomasz. 1977. Antimicrob. Agents Chemother. 11, 888.
2. Horne, D., R. Hakenbeck, A. Tomasz. 1977. J. Bacteriol. 132, 704.
3. Waks, S. A. Tomasz. 1978. Antimicrob. Agents Chemother. 39, 293.
4. Hakenbeck, R., S. Waks, A. Tomasz. 1978. Antimicrob. Agents Chemother. 13, 302.
5. Hakenbeck, R., C. Martin, G. Morelli. 1983. J. Bacteriol. 155, 1372.
6. Waxman, D.J., J.L. Strominger. 1983. Ann. Rev. Biochem. 52, 825.
7. Frère, J.-M., B. Joris. 1985. CRC Crit. Rev. in Microbiol. 11, 299.
8. Mosser, J.L., A. Tomasz. 1970. J. Biol. Chem. 245, 287.
9. Briese, Th., R. Hakenbeck. 1983. In: The target of penicillin (R. Hakenbeck, J.-V. Höltje, H. Labischinski, eds.). DeGruyter, Berlin, p.173.
10. Giudicelli, S., A. Tomasz. 1984. J. Bacteriol. 158, 1188.
11. Höltje, J.-V., A. Tomasz. 1975. J. Biol. Chem. 250, 6072.
12. Briese, Th., R. Hakenbeck. 1985. Eur. J. Biochem. 146, 417.
13. Adkinson, N.F., E.A.Swabb, A.A. Sugerman. 1984. Antimicrob. Agents Chemother. 25, 93.
14. Dewdney, J.M. 1977. In: The antigens (M. Sela, ed.). Academic Press New York, p. 82.

ANTI-PEPTIDOGLYCAN SEROLOGY IN PATIENT SERA AND EXPERIMENTAL PRODUCTION OF ANTI-PEPTIDOGLYCAN ANTIBODY BY IMMUNISATION WITH RHEUMATOID FACTOR

H.B. Evans, K.K. Phua, P.M. Johnson

Department of Immunology, University of Liverpool, P.O. Box 147, Liverpool, L69 3BX, U.K.

Introduction

There is some association between bacterial immunity and the genesis of rheumatoid factor (RF) autoantibody in man (1), e.g. in bacterial endocarditis; RF being characterised in the laboratory by its weak interaction with a determinant in the Fc region of IgG immunoglobulin (2). Serum antibodies reactive with Gram-positive bacterial cell wall structures, including peptidoglycan-polysaccharide (PG-PS), are elevated in rheumatoid arthritis (RA) and there is no direct cross-reaction of RF as anti-PG-PS, i.e. RF does not itself act as an anti-PG-PS antibody (3). We have further investigated the anti-PG-PS response in man with respect to the predominant IgG subclass of anti-PG-PS antibody and the nature of the PG-PS antigen with which such antibodies interact.

Experimental animals immunised with streptococcal cell walls produce substantial amounts of serum RF-like anti-IgG(Fc) antibodies (4). The possibility of some idiotypic complementarity between RF and antibody reactive with streptococcal PG-PS has therefore been investigated. Should RF represent an auto-anti-idiotypic response to a common idiotype of anti-PG-PS antibodies, as part of the normal immunoregulation of such antibody responses, then it ought to be possible experimentally to produce the reverse, i.e. some antibodies raised against a common RF idiotype could themselves bear idiotypic activity related to those of some anti-PG-PS antibodies. This concept is represented in Fig. 1.

Biological Properties of Peptidoglycan

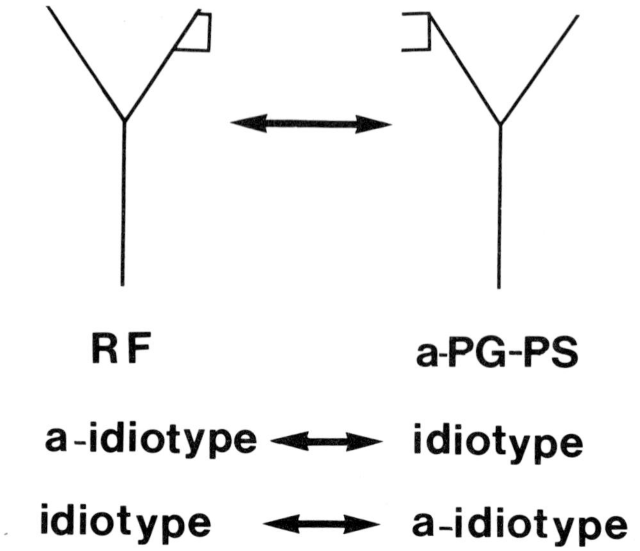

Fig. 1 Possible idiotypic interrelationship between RF and anti-PG-PS

Materials and Methods

Cell wall PG-PS polymer preparations were derived from Group A Streptococcus pyogenes as described previously (3). An aliquot of this preparation was further purified by extraction with 1% sodium dodecyl sulphate (SDS) at 100°C for 15 minutes. This antigen was used in a modification of the anti-PG-PS ELISA technique (3). Wells were coated with SDS-treated PG-PS at 2 g/ml. The IgG subclass distribution of anti-PG-PS antibodies was assessed using monoclonal anti-human IgG1, IgG2, IgG3 and IgG4 reagents (Unipath) at 1/1000 dilution, followed by peroxidase-conjugated rabbit anti- mouse Ig (Dakopatts) at 1/1000 dilution. Results were expressed as binding indices (5).

Details of repeated immunisation of Balb/c mice with purified RF and non-RF immunoglobulin preparations are given elsewhere (6). In an additional experiment, a purified RF preparation was separated into IgG-RF and IgM-RF components on Sephadex G-200 (Pharmacia) using 0.1M citric acid/sodium citrate buffer, pH 3.6. Groups of Balb/c mice were given repeated sterile intraperitoneal immunisations with either IgM-RF, IgG-RF or non-RF IgG preparations in Freund's incomplete adjuvant. Anti-PG-PS antibody activity in sera from these mice was measured by ELISA.

Results

Our previous studies have shown anti-PG-PS levels to be raised in juvenile chronic arthritis and both seronegative and seropositive rheumatoid arthritis (3). We have now shown that, of such sera with raised anti-PG-PS activity, 36% of these exhibit an IgG isotype restriction to the IgG2 subclass. In addition, of 10 tuberculosis patients, all with raised anti-PG-PS, 2 showed restriction to IgG2 antibody (Evans et al., in preparation). In contrast, in a study of serial samples from rheumatic fever patients (n=10), all of whom had high anti-PG-PS levels, 20% of patients exhibited restriction of antibody to the IgG3 subclass. However, preliminary experiments comparing PG-PS preparations isolated in the presence of SDS have revealed that antibody levels consistently appear higher using the SDS-treated PG-PS preparation and, with this antigen, anti-PG-PS antibodies of the IgG1, IgG2 and IgG3 subclasses could be detected in rheumatic fever patient sera.

Detailed results of anti PG-PS antibody levels in mice immunised with isolated polyclonal RF, non-RF IgG and non-RF IgM preparations have been given elsewhere (6). Anti-PG-PS levels for mice immunised with separated IgM-RF and IgG-RF preparations are shown in Table 1. Binding indices (B.I.) of >3.0 are considered positive.

TABLE I Anti-PG-PS Levels in Mice Immunised With IgM-RF, IgG-RF or Non-RF Ig

Mouse No.	Immunogen	B.I.
1	IgM-RF	3.02
2	IgM-RF	1.87
3	IgG-RF	3.74
4	IgG-RF	4.70
5	non-RF IgG	2.45
6	non-RF IgG	1.39
7	PG-PS	10.80
Pre-immune mouse sera (n=6)	-	0.74

Discussion

Anti-PG-PS levels have been shown to be raised in rheumatic fever, rheumatoid arthritis, juvenile chronic arthritis and tuberculosis, although it is not certain exactly with which antigenic components of PG-PS that these antibodies react. In our hands, higher antibody levels were detected using SDS-treated PG-PS, suggesting that this preparation may contain a higher density of relevant antigenic epitopes than an equivalent preparation not treated with SDS. IgG2 antibodies are known frequently to be associated with responses to carbohydrate antigens (7), and this may well be the case for those human anti-PG-PS antibodies showing IgG-isotype restriction to the IgG2 subclass. The possibility of an idiotypic complementarity between RF and an antibody reactive with streptococcal PG-PS has also been investigated. Repeated immunisation of mice with human RF, of either IgG or IgM class, produces an anti-PG-PS antibody response (6). As this anti-PG-PS activity can be absorbed by either PG-PS or RF, but not non-RF Ig, a cross-reactive antibody which reacts with both PG-PS and with a RF idiotype appears to have been produced. This interpretation is consistent with the view that RF from unrelated RA patients share common idiotypic determinants. Such RF may carry the internal image of a streptococcal antigenic epitope, since both RF and PG-PS appear to react with anti-PG-PS antibody. As this structure is not present on non-RF Ig, it is assumed to be an idiotypic determinant of RF; the fact that an anti-PG-PS response can be raised with either IgM-RF or IgG-RF would further support this supposition.

This data adds weight to the concept that RF is part of the normal immunoregulatory system for modulating certain anti-bacterial responses and would put forward an explanation, other than autoreactivity with IgG(Fc), for the genetic conservation of common RF idiotypes. Uncontrolled RF production in RA may represent an imbalance in this system, perhaps due to persistent infection or continued presence of peptidoglycan. Since Gram-positive peptidoglycans exhibit some common antigenicity, this could explain the lack of association of any single bacterial agent with RF production in RA (8).

Acknowledgements

This work was supported by the Arthritis and Rheumatism Council. We also thank Professor H.R. Perkins for advice in the production of streptococcal PG-PS preparations.

References

1. Williams, R.C. Jr. 1977. In: Autoimmunity: Genetic, Immunologic, Virologic and Clinical Aspects (Talal N. ed.), New York, Academic Press, pp. 457.

2. Johnson, P.M. 1981. Clin. Immunol. Allergy 1, 103.

3. Johnson, P.M., Phua, K.K., Perkins, H.R., Hart, C.A., Bucknall, R.C. 1984. Clin. exp. Immunol., 55, 115.

4. Bokisch, V.A., Chio, J.W., Bernstein, D., Krause, R.M. 1973. J. exp. Med., 138, 1184.

5. Barnes, R.M.R., Barton, P.G., Doig, J.E., Finn, R., Harvey, M.M. and Johnson, P.M. 1983. J. Clin. Lab. Immunol., 12, 175.

6. Johnson, P.M., Phua, K.K., Evans, H.B. 1985. Clin. exp. Immunol., 61, 373.

7. Hammarström, L., Granström, M., Oxelius, V., Person, M.A.A., Smith, C.I.E. 1984. Clin. exp. Immunol. 55, 593.

8. Bennett, J.C. 1978. Arthr. Rheum., 21, 531.

ANTIBODIES AGAINST A SYNTHETIC PEPTIDOGLYCAN-PRECURSOR PENTAPEPTIDE CONTAINING LYSINE CROSS-REACT WITH SOLUBLE PEPTIDOGLYCAN CONTAINING DIAMINOPIMELIC ACID

Allen R. Zeiger

Department of Biochemistry, Thomas Jefferson University, Philadelphia, PA 19107

Introduction

Antibodies have been found in man that are reactive towards D-Ala-D-Ala sequences (1-4). The latter sequences have been found in peptidoglycan-precursors and in the soluble peptidoglycans (SPG) that are secreted by some gram-positive bacteria grown in the presence of beta-lactam antibiotics (5). These antibodies may be of considerable medical importance since their titers are elevated in the sera of patients with rheumatic fever (1), Staphylococcus aureau-caused endocarditis and bacteremia (2), juvenile rheumatoid arthritis (1) and ankylosing spondylarthritis (3). There is mounting serological evidence implicating SPG as the major natural immunogen (5).

Recently, a unique enzyme-linked immunosorbent assay (ELISA) has been employed to show the presence of SPG in the urine of man following an oral dose of penicillin V (6). The ELISA utilized two reagents which were able to distinguish SPG in the presence of hydrolyzed or transpeptidated bacterial peptidoglycan-like products; e.g., vancomycin, a low molecular weight antibiotic (7) and an antibody population elicited by a completely synthetic immunogen, $(Glu^{60}\ Ala^{40})_n$–$(Ala\text{-}\gamma\text{-}D\text{-}Glu\text{-}Lys\text{-}D\text{-}Ala\text{-}D\text{-}Ala)^{5.1}$ (8).

SPG from different bacteria, like the cell wall-linked peptidoglycans (9), contain structural differences such as the nature of the diaminoacid present next to the penultimate D-alanine (5). Reagents which distinguish these structural differences in SPG could narrow the search for the bacteria responsible for their presence in man. The question to be answered in this manuscript

Biological Properties of Peptidoglycan

is whether the reagents in the ELISA are capable of recognizing SPG that contain diaminopimelic acid (A2pm) in place of lysine.

Materials and Methods

ELISA. The ELISA for the detection of SPG that has been described previously (5-6) has been modified slightly. The same affinity-purified rabbit antibody to the synthetic polypeptide, $(Glu^{60}\ Ala^{40})_n$ (Ala-γ-D-Glu-Lys-D-Ala-D-Ala)$^{5.1}$, was used. However, instead of using biotin-avidin complexes, the second antibody was an alkaline phosphatase labeled goat anti-rabbit immunoglobulin purchased from Sigma Chem. Co. (St. Louis, MO) and used at a 1:500 dilution. The p-nitrophenyl phosphate substrate (Sigma Chem. Co.) was added at 1 mg per ml of 10% diethanolamine buffer, pH 9.8, containing 0.01% $MgCl_2 \cdot 6H_2O$ and 0.02% NaN_3. Absorbances were read periodically at 405 nm in a Titertek Multiskan MC (Flow Labs).

SPG. The SPG from Staphylococcus aureus which contains lysine (10) and the SPG from Bacillus subtilis (11) which contains A2pm were prepared as described. Trace labels of ^{14}C-alanine were included in the minimal growth medium for the estimation of concentration.

Results and Discussion

Both SPG were tested for binding with the ELISA. Table 1 shows that the SPG from S. aureus binds well in this assay as expected. The SPG from B. subtilis also is recognized by the reagents in the ELISA and thus either SPG in human urine would be detected with this assay. However, the sensitivity of this ELISA for the DAP-containing SPG appears to be somewhat less than that for the Lys-containing SPG. In order to distinguish the two SPG in urine, another ELISA using a second antibody recognizing the DAP-containing SPG better than the Lys-containing SPG must be developed. The ratio of absorbances using the two antibodies should indicate which SPG is present.

Table 1. Detectability of Lys-containing SPG from S. aureus and A2pm-containing SPG from B. subtilis with Affinity-purified Rabbit Antibody to a Synthetic Immunogen.

	SPG from *S. aureus*		SPG from *B. subtilis*	
	Absorbance at 405 nm			
Conc (ng/ml)	Exp. I	II	Exp. I	II
5000	1.740	1.594	1.046	1.002
500	1.265	1.271	0.619	0.504
50	0.676	0.653	0.269	0.195
5	0.273	0.165	0.043	0.043
0.5	0.043	0.095	0.000	0.005

Acknowledgements

This research was supported by a grant from the Dean's Overage Research Fund. I would like to thank Mr. Brad Needleman for excellent technical assistance.

References

1. Heymer, B., K.H. Schleifer, S. Read, J.B. Zabriskie and R.M. Krause. 1976. Detection of antibodies to bacterial cell wall peptidoglycans in human sera. J. Immunol. 117, 23-26.

2. Zeiger, A.R., C.U. Tuazon and J.N. Sheagren. 1981. Antibody levels to bacterial peptidoglycan in human sera during the time course of endocarditis and bacteremic infections caused by Staphylococcus aureus. Infect. Immun. 37, 1112-1118.

3. Park, H., H.R. Schumacher, A.R. Zeiger and J.T. Rosenbaum. 1984. Antibodies to peptidoglycan in patients with spondylarthritis: A clue to disease aetiology? Ann. Rheum. Dis. 43, 725-728.

4. Franken, N., P.H. Seidl, E. Zauner, H.J. Kolb, K.H. Schleifer and L. Weiss. 1985. Quantitative determination of human IgG antibodies to the peptide subunit determinant of peptidoglycan by an enzyme-linked immunosorbent assay. Mol. Immunol. 22, 573-579.

5. Zeiger, A.R. 1985. Soluble peptidoglycans: Lymphocyte-activating bacterial products found in man. These proceedings.

6. Park, H., A.R. Zeiger and H.R. Schumacher. 1984. Detection of soluble peptidoglycan in urine after penicillin administration. Infect. Immun. 43, 139-142.

7. Perkins, H.R. 1969. Specificity of combination between mucopeptide precursors and vancomycin or ristocetin. Biochem. J. 111, 195-205.

8. Zeiger, A.R. and P.H. Maurer. 1973. Immunochemistry of a synthetic peptidoglycan-precursor pentapeptide. Biochemistry 12, 3387-3394.

9. Schleifer, K.H. and O. Kandler. 1972. Peptidoglycan types of bacterial walls and their taxonomic implications. Bacteriol. Rev. 36, 407-477.

10. Zeiger, A.R., W. Wong, A.N. Chatterjee, F.E. Young and C.U. Tuazon. 1982. Evidence for the secretion of soluble peptidoglycans by clinical isolates of Staphylococcus aureus. Infect. Immun. 37, 1112-1118.

11. Waxman, D.J., W. Yu and J.L. Strominger. 1980. Linear, uncross-linked peptidoglycan secreted by penicillin-treated Bacillus subtilis. J. Biol. Chem. 255, 11577-11587.

ANTIBODIES TO VARIOUS BACTERIAL CELL WALL PEPTIDOGLYCANS IN PATIENT SERA AND IN RABBIT SERA

Heidrun I. Wergeland, Curt Endresen

Department of Microbiology and Immunology, The Gade Institute, University of Bergen, Bergen, Norway

Introduction

The peptidoglycans (PG) present in almost all bacteria have considerable similarities in chemical composition. Bacterial PG is composed of a matrix of glycan polymers consisting of N-acetyl-glucosamine-β-1.4-acetyl-muramic acid units covalently bound to short peptide side chains (1). In gram-positive bacteria the peptide subunits are cross-linked by interpeptide bridges. Studies of the antigenic determinants have shown that the N-acetyl-glucosamine is the immunodominant sugar of the glycan polymer (2), and the C-terminal D-Ala-D-Ala is the immunodominant site of the pentapeptide (3). The L-Lys-D-Ala or the meso-diaminopimelic acid-D-Ala in the tetrapeptide (4,5) and the interpeptide bridge (4) are also immunogenic. Since all bacteria contain a rather uniform sugar moiety, and many bacteria also have similar tetra and penta peptides with common determinants, cross-reactive antibodies to these antigenic determinants are produced during infection or vaccination and can be detected in human and in animal sera. Most studies involving PG cross-reactive antibodies have been performed using streptococcal and staphylococcal PGs as antigens (6), but there is also a study of common PG peptide determinants in gram-negative bacteria (5). Analysis of sera from patients with various non-staphylococcal infections have shown that most of the sera contained high levels of antibodies reactive to S.aureus PG(7). This paper presents the analysis of cross-reactive antibodies present in patient sera and in rabbit immune sera.

Results

Antibodies to various bacterial cell wall peptidoglycans were ELISA analysed. Analysis of sera from patients with staphylococcal infection, patients with non-staphylococcal infections and blood donor sera showed that all sera contained antibodies to S.aureus PG (fig. 1).

Biological Properties of Peptidoglycan

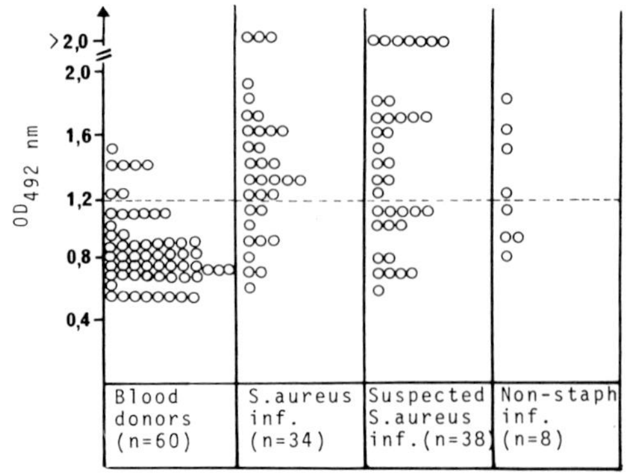

Fig. 1 Antibody values (OD) to S.aureus PG determined by ELISA in sera from blood donors, patients with verified and suspected S.aureus infection and patients with non-staphylococcal infections. The broken line represents the mean value plus 2 SD according to analysis of 60 blood donor sera.

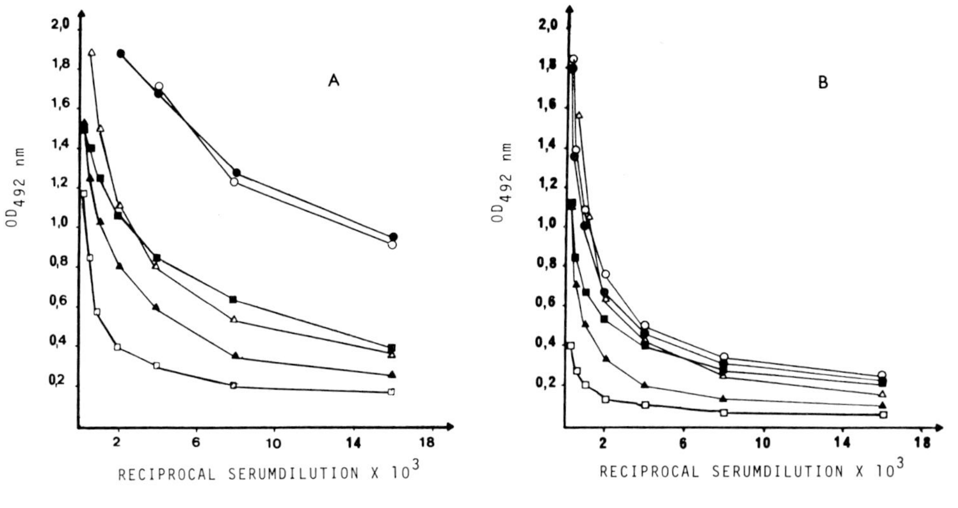

Fig. 2 Antibody levels to PG from S.aureus (○), S.epidermidis (●), S.Pyogenes (△), M.lysodeikticus (▲), B.subtilis (□) and E.coli (■) determined by ELISA in A) serum from a patient with S.aureus septicaemia

B) serum from a blood donor.

Four of eight sera from patients with non-staphylococcal infection had elevated antibody levels to S.aureus PG indicating the presence of cross-reacting antibodies. Sera from three patients with staphylococcal infection verified by positive blood cultures and blood donor sera were examined for antibodies to PG from S.aureus, S.epidermidis, S.pyogenes, M.lysodeikticus, B.subtilis and E.coli. All the patient sera contained antibodies to all PG tested (fig. 2A) at higher levels and/or affinity than in the blood donor sera (fig. 2B). This was most probably due to a S.aureus infection. Sensitive ELISA methods allow detection of S.aureus antibodies reactive against other PGs. Immunization of rabbits with S.aureus PG resulted in high amounts of specific antibodies, however cross-reactive antibodies in various amounts could also be detected. Rabbit pre-immune sera contained small amounts of antibodies reactive to PG from S.aureus, S.epidermidis, S.pyogenes, B.subtilis, M.lysodeikticus and E.coli. ELISA analysis of S.aureus reactive antibodies in the rabbit immune sera at a dilution of 1:640 gave OD values ranging from 1.4 - 1.8 compared to 0.6 - 0.8 in pre-immune sera. The antibody activity to heterologous PGs was also raised after immunization (fig. 3). The increase in antibody activity was highest to homologous antigen, however the elevated antibody levels to all PGs must be due to presence of common antigenic determinants in the PGs.

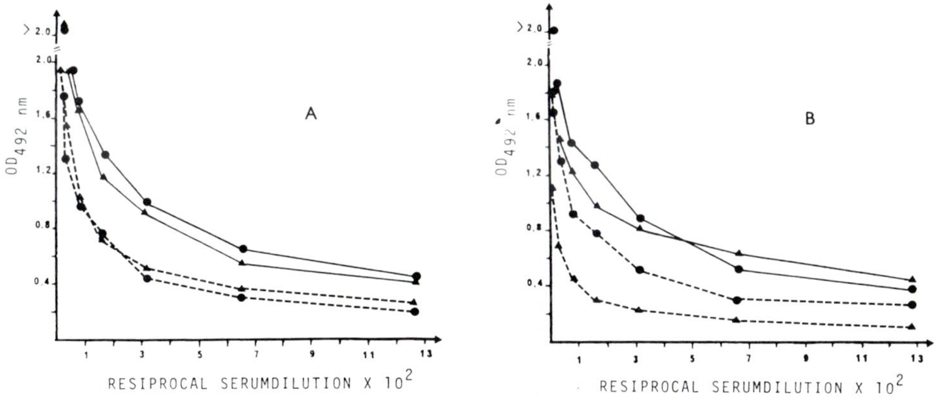

Fig. 3 Antibody levels (OD) in rabbit pre-immune serum (---) and rabbit anti S.aureus PG immune serum (——) determined by ELISA
Antigens: A) PG from M.lysodeikticus (●) and B.subtilis (▲)
B) PG from S.pyogenes (▲) and E.coli (●).

The specificities of the PG ELISA's were assessed by using pure PG preparations. The amino acid and carbohydrate contents corresponded to earlier analysis (1). The PGs were not contaminated by teichoic acids and lipoteichoic acids since sugar alcohols were not detected, and the phosphate content was less than 1% (w/w). The specificity of the PG ELISA's were further proved by inhibition studies. It could be shown that the antibody activity was inhibited by PG homologous to the PG antigen used in ELISA. A serum dilution which gave a OD value near 1.0 was used. Increasing consentrations of sonicated PG were added to aliquots of serum dilutions. ELISA values were reduced and complete inhibition of the anti PG activity was obtained (fig. 4).

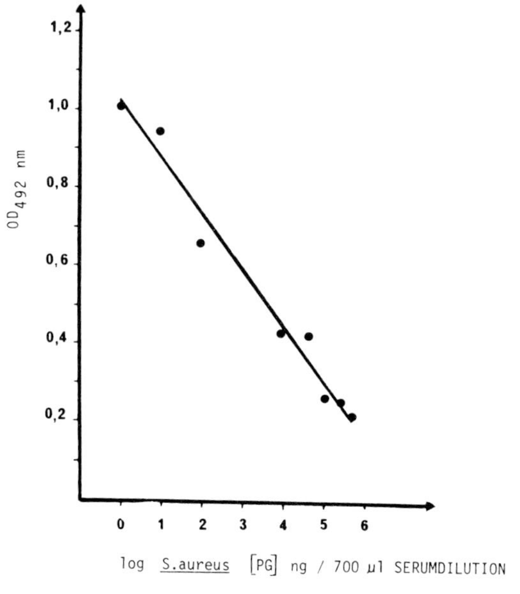

Fig. 4 Inhibition of S.aureus reactive antibodies measured by ELISA in serum from a patient suffering from S.aureus septic arthritis.

Elevated amounts of cross-reactive antibodies were found in the patient sera. The results from a study of cross-reactive antibodies in a patient serum are presented in table 1. Inhibition experiments were performed with serum dilutions corresponding to the linear component of the serum titration curves for the target antigens. The analysis showed that homologous PG antigen and PG inhibitor always resulted in the highest inhibition. All PGs always reduced the antibody activity in ELISA to any of the PG antigens. The PG from S.aureus and S.epidermidis seem to share common antigenic determinants. The chemical analysis of the two PGs also showed very similar composition. The inhibition experiments showed that a fraction of the antibodies present in the patient sera were reactive against all PGs tested, indicating at least one common epitope.

Table 1 Antibody levels to various bacterial PGs determined by ELISA in serum from a patient with S.aureus septic arthritis prior to and after inhibition with homologous and heterologous PGs. Aliquots of diluted serum were incubated with homologoues and heterologous PGs. The non-inhibited antibody activities were analysed by ELISA.

ELISA VALUES ($OD_{492\ nm}$) x 100

TARGET ANTIGEN IN ELISA	NO INHIBITOR	PG INHIBITOR 50 µg / 700 µl SERUMDILUTION					
		S.aureus	S.epidermidis	S.pyogenes	M.lysodeikticus	B.subtilis	E.coli
S.aureus	115 1)	27	38	79	76	73	73
S.epidermidis	107 1)	48	32	80	67	74	68
S.pyogenes	119 1)	99	106	17	106	101	107
M.lysodeikticus	145 2)	110	107	118	39	110	89
B.subtilis	82 3)	79	75	87	69	28	67
E.coli	143 2)	99	92	117	94	90	37

Serum dilutions: 1) 1:3000, 2) 1:500, 3) 1:250

Concluding remarks

The presence of cross-reactive antibodies to various bacterial PGs in human sera as well as in rabbit immune sera have been demonstrated. During a bacterial infection, or by immunization with PG, antibodies were raised to homologous as well as heterologous PG. The presence of cross-reactive antibodies in patient sera may therefore cause problems using whole PG as antigen in a diagnostic test for identification of the causative bacterial agent.

Further analysis is needed in order to elucidate the epitop specificity of the cross-reactive antibody. The results indicate, however, that they are reactive against the sugar moiety, since PG from a gram-negative bacteria such as E.coli can inhibit the antibody activity against PG from gram-positive bacteria. In PG from E.coli no pentapeptide has been found (5) and the tetrapeptide, with meso-diaminopimelic acid and D-Ala terminal as the immunodominant part of the PG peptide (5) is different from the tetrapeptide found in most gram-positive bacteria. Antibodies to staphylococcal and streptococcal PGs are also reactive against the tetrapeptide L-Lys-D-Ala and the pentapeptide D-Ala-D-Ala determinants. These also are found in PG from other gram-positive bacteria. The results obtained with S.pyogenes PG showed that the serum contained rather high amounts of reactive antibodies which are not inhibited by S.aureus or S.epidermidis PGs. The result is difficult to explain, and was also observed when testing other patient sera. The cross-reactive antibodies as well as specific antibodies may have opsonic activity. It is well known that PG is exposed on the surface of the bacteria and both specific antibodies and cross-reactive antibodies may have a function in the defence against bacterial infections.

References

1. Schleifer, K.H., O. Kandler. 1972. Bact.Rev. 36, 407.
2. Karakawa, W.W., H. Lackland, R.M. Krause. 1976. J.Immunol. 99, 1178.
3. Schleifer, K.H., R.M. Krause. 1971. J.biol.Chem. 246, 986.
4. Helgeland, S.M., A. Grov, K.H. Schleifer. 1973. Acta path.microbiol. immunol.scand.Sect.B 81, 413.
5. Nguyen-Huy, H., C. Nauciel, C.-G. Wermuth. 1976. Europ.J.Biochem. 66, 79.
6. Karakawa, W.W., D.G. Braun, H. Lackland, R.M. Krause. 1968. J.exp.Med. 128, 325.
7. Wergeland, H., C. Endresen, O.B. Natås, P. Aasjord, P. Oeding. 1984. Acta path.microbiol.immunol.scand.Sect.B 92, 265.

ANTIBODIES TO THE PENTAPEPTIDE SUBUNIT OF PEPTIDOGLYCAN IN HUMAN SERA

T. Kuchenbauer, K.-D. Tympner

Kinderklinik, Städtisches Krankenhaus München-Harlaching, D-8000 München 90, FRG

P.H. Seidl, K.H. Schleifer

Lehrstuhl für Mikrobiologie, Technische Universität München, D-8000 München 2, FRG

Introduction

The pentapeptide subunit H-L-Ala-D-Glu(L-Lys-D-Ala-D-Ala-OH)-OH represents one of the antigenic epitopes of peptidoglycan hitherto described in man (1-5; Seidl and Schleifer, this book). Due to lacking D,D-carboxypeptidases splitting off the terminal D-alanine (6), pentapeptides with C-terminal D-alanyl-D-alanine are in particular found among streptococci and staphylococci (5,7; Seidl and Schleifer, this book). This paper reports on specific antibodies against the pentapeptide subunit in human sera, determined by a modified radioactive hapten binding assay originally described by Heymer et al. (1).

Materials and Methods

Sera and immunoglobulin preparations for intravenous application. Sera were collected at the City Hospital at Munich-Harlaching. Immunoglobulin preparations for intravenous application were obtained from Behringwerke, Marburg, FRG (Beriglobin®); from Biotest Pharma, Frankfurt, FRG (Intraglobin®); from Deutsche Kabi-Vitrum, Munich, FRG (Gammonativ®); from Sandoz AG, Nürnberg, FRG (Sandoglobulin®); from Travenol, Munich, FRG (Gammagard®) and from Troponwerke, Cologne, FRG (Polyglobin®).
Radioactive hapten binding assay. The assay is based on measuring the binding of synthtic pentapeptide subunit (8) labelled with ^{125}iodine (1) to specific antibodies. The methodology of Heymer et al. (1) was modified as follows: ^{22}Na as an internal standard was omitted; after incubating sera with the radiolabelled pentapeptide (total volume 20 µl) globulins were precipitated at 4°C

Biological Properties of Peptidoglycan

for 30 min by addition of 20 µl saturated ammonium sulfate. Quantification was achieved employing a reference serum with known specific antibody content (1,5).
Assays for antistreptolysin O, IgG and IgM. Antistreptolysin O was measured by a commercial test kit ('Blue-ASO-Test', Mast-Diagnostics, Hamburg, FRG). The total amounts of IgG and IgM were quantified by kinetic nephelometry (9).

Results and Discussion

Specific antibody levels were determined of 160 healthy children in age groups up to fifteen years (Fig. 1). Each group consisted of 20 probands. The criteria for selecting a normal proband were carefully stated according to clinical investigations and to the data of laboratory diagnosis. In particular, the following probands were excluded: healthy persons with known hyperimmunglobulinemia E or hypersensitivity reactions (due to an interrelationship reported between skin reactions to bacterial peptidoglycan and serum peptidoglycan antibody titer (10)), healthy persons with recent bacterial, viral and mycotic infections. The figure shows

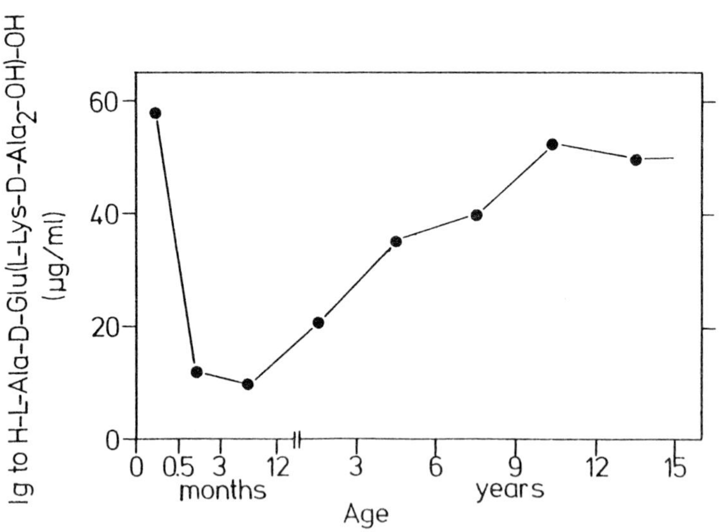

Fig. 1. Age dependency of specific antibodies to the pentapeptide subunit of peptidoglycan in normal sera from children (Ig, immunoglobulins).

a clear age dependency of the mean specific antibody content. No correlation between antibodies to the pentapeptide subunit and antistreptolysin O titres, total IgG and IgM was shown within the respective age groups. The surprisingly high levels of specific antibodies were corroborated investigating commercial immunoglobulin preparations for intravenous applications (Fig. 2). Calculating a mean content of 1200 mg/100 ml for total IgG, specific antibodies against the pentapeptide H-L-Ala-D-Glu(L-Lys-D-Ala-D-Ala-OH)-OH amounted from 48 µg/ml (Intraglobin®) to 70 µg/ml (Gammonativ ®). It should be pointed out, however, that specific antibodies of all immunoglobulin classes are measured by the radioactive hapten binding assay employed.

Time course studies with healthy adults demonstrated very constant amounts of specific antibodies to the pentapeptide subunit of peptidoglycan over a three years' period. This was surprising regarding the half-life time for IgG antibodies being 7 days (IgG3) or 21 days for IgG1, IgG2 and IgG4 (11). We therefore performed time course studies in sera from 5 patients with agammaglobulinemia type Bruton, during therapy with an immunoglobulin preparation for intravenous application. Fig. 3 shows for 2 patients the typical results obtained, immunoglobulin substitutions are indicated by arrows. Regarding the constant time course in healthy persons at unexpected high levels and comparing it with the data from immu-

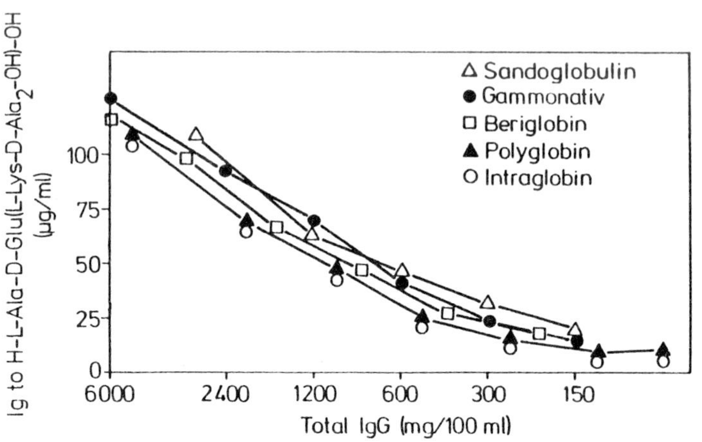

Fig. 2. Specific antibodies to the pentapeptide subunit of peptidoglycan in immunoglobulin preparations for intravenous application.

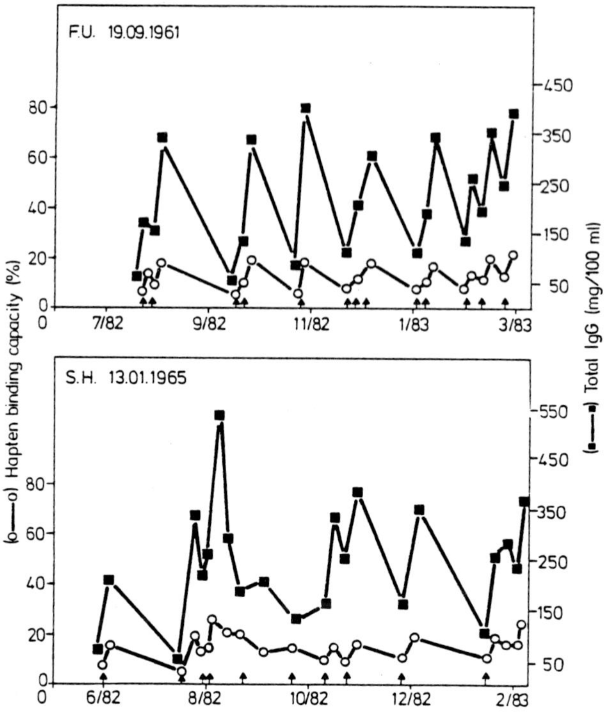

Fig. 3. Time course studies of specific antibodies to the pentapeptide subunit in sera from patients with agammaglobulinemia type Bruton during therapy with an immunoglobulin preparation for intravenous application. Arrows: substitution by immunoglobulins.

nodeficient persons, a continuous antigen stimulation by circulating antigen (peptidoglycan) resulting from colonization may be postulated for healthy probands.

In contrast to the constant time course levels typical for healthy probands, antibody levels to the pentapeptide subunit increased significantly in sera from patients with streptococcal or staphylococcal infections (compare Figs. 4 and 5, p. 109).

It is well known that secondary infections of cutaneous lesions by bacteria are very common in patients with atopic dermatitis. Most often, staphylococci or group A streptococci are recovered from infected lesions. According to this, 8 from 12 sera from patients with atopic dermatitis revealed positive titres with respect to the age related levels (Fig. 6, p. 109). High titres in

Fig. 4

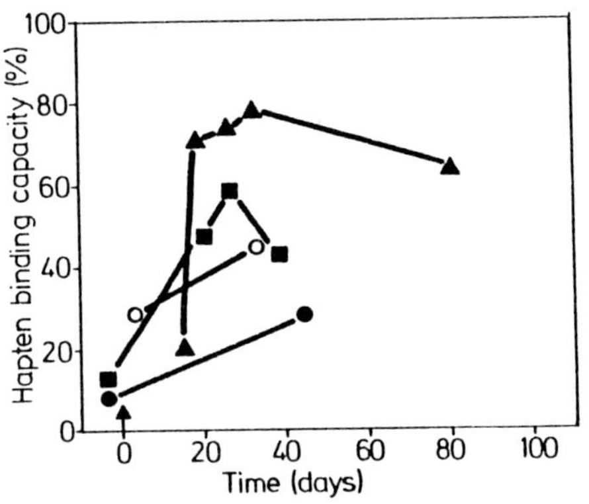

Fig. 5

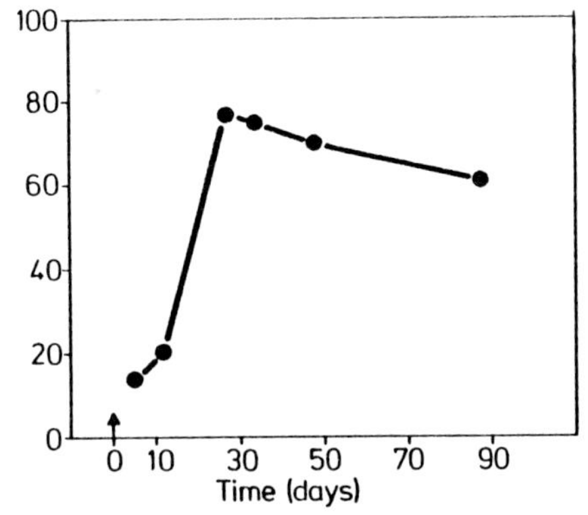

Fig. 4. Time course of specific antibody levels to the pentapeptide subunit of peptidoglycan in sera from 4 patients with scarlet fever. The arrow indicates the beginning of the disease. Sera (●) and (■) were already surveyed before the onset of scarlet fever.

Fig. 5. Time course of specific antibody level to the pentapeptide subunit of peptidoglycan in a serum from a patient with staphylococcal scaled skin syndrome.

Fig. 6

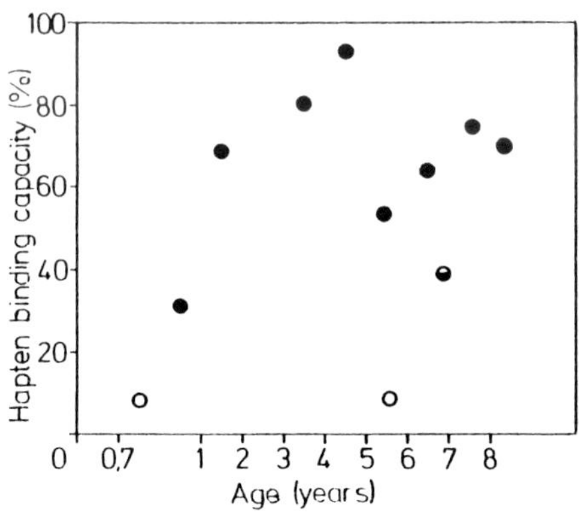

Fig. 7

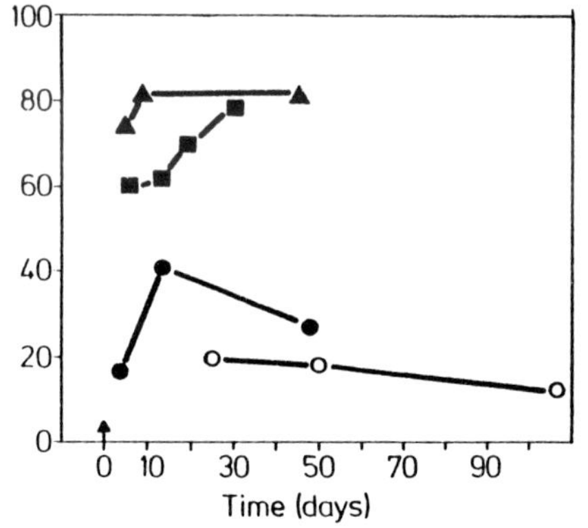

Fig. 6. Specific antibodies to the pentapeptide subunit in sera from patients with atopic dermatitis. 'Positive' sera with respect to the age related levels are symbolized by full circles and, 'negative' sera by open circles.

Fig. 7. Time course of specific antibody levels to the pentapeptide subunit in sera from 4 patients with 'Kawasaki-syndrome' diagnosis. The arrow indicates the beginning of the disease.

specific antibodies seemed to correlate with the severity of the disease.

Surprising data were obtained with sera from the 4 children with 'Kawasaki-syndrome', a disease of still unknown etiology (12). Altering antibody levels to the pentapeptide subunit either indicate wrong diagnosis or bacteria as the causative agents. This topic is under current research.

The pentapeptide subunit of peptidoglycan is an antigenic epitope shared by many streptococci and staphylococci. Measuring specific antibodies to this determinant may therefore be of value in diagnosing streptococcal or staphylococcal diseases.

References

1. Heymer, B., D. Bernstein, K.H. Schleifer and R.M. Krause. 1975. A radioactive hapten-binding assay for measuring antibodies to the pentapeptide determinant of peptidoglycan. J. Immunol. 114, 1191-1196.

2. Heymer, B., K.H. Schleifer, S. Read, J.D. Zabriskie and R.M. Krause. 1976. Detection of antibodies to bacterial cell wall peptidoglycan in human sera. J. Immunol. 117, 23-26.

3. Zeiger, A.R., C.U. Tuazon and J.N. Sheagren. 1981. Antibody levels to bacterial peptidoglycan in human sera during the time course of endocarditis and bacteremic infections caused by Staphylococcus aureus. Infect. Immun. 33, 795-800.

4. Franken, N., P.H. Seidl, T. Kuchenbauer, H.J. Kolb, K.H. Schleifer, L. Weiss and K.-D. Tympner. 1984. Specific immunoglobulin A antibodies to a peptide subunit sequence of bacterial cell wall peptidoglycan. Infect. Immun. 44, 182-187.

5. Franken, N., P.H. Seidl, E. Zauner, H.J. Kolb, K.H. Schleifer, L. Weiss. 1985. Quantitative determination of human IgG antibodies to the peptide subunit determinant of peptidoglycan by an enzyme-linked immunosorbent assay. Mol. Immunol. 22, 573-579.

6. Rogers, H.J., H.R. Perkins and J.B. Ward. 1980. Microbial cell walls and membranes. Chapman and Hall, London, U.K.

7. Seidl, P.H. and K.H. Schleifer. 1985. Secretion of fragments from bacterial cell wall peptidoglycan. In: Environmental regulation of microbial metabolism (Kualev IS, Severin AT, Dawes EA, eds). Academic Press, London, 443-450.

8. Schleifer, K.H. and R.M. Krause. 1971. The immunochemistry of peptidoglycan. The immunodominant site of the peptide subunit and the contribution of each of the amino acids to the binding properties of the peptides. J. Biol. Chem. 246, 986-993.

9. Killingworth, L.M. and Savory J. 1972. Manual nephelometric methods for immunochemical determination of immunoglobulins IgG, IgA and IgM in human serum. Clin. Chem. 18, 335-342.

10. von Mayenburg, J., B. Heymer, K.H. Schleifer, P.H. Seidl, A. Neiss and S. Borelli. 1982. Studies on bacterial hypersensitivity in man. Allergy 37, 249-258.

11. Riesen, W. 1980. Struktur und biologische Eigenschaften von Immunglobulinen und γ-Globulin-Präparaten. I. Struktur und Funktion von Immunglobulinen. Schweiz. med. Wschr. 110, 74-79.

12. Kato, H., O. Inone, Y. Koga and M. Shingu. 1983. Variant strain of Propionibacterium acnes: a clue to the etiology of Kawasaki disease. Lancet II, 1383-1387.

PROBLEMS WITH 'TEICHOIC ACID' ANTIGEN USED IN STAPHYLOCOCCUS AUREUS SEROLOGY

Ch. Herzog

Department of Clinical Research, F. Hoffmann-La Roche & Co. Ltd, CH-4002 Basel, Switzerland

V. Just, R. Berger, M. Just

Laboratory for Microbiology and Immunology, University Childrens Hospital, CH-4005 Basel, Switzerland

Introduction

Serological follow-up of 'teichoic acid' antibodies can be of great help in the management of patients with Staphylococcus aureus bacteremia, as has been shown by a number of investigators [1-6]. Reviewing the literature on antistaphylococcal serology, starting with the first clinical use of "teichoic acid antigen" by Crowder in a immunodiffusion assay [1] one notices a marked variation in the specificity and sensitivity reported for the various teichoic acid antibody tests published over the last ten years. There are two main reasons: the unknown specificity and nature of the antibodies measured and the impossibility to compare the published laboratory and clinical data. This is due to

1) the use of a diversity of poorly standardized and often crude cell wall preparations - mostly labelled 'teichoic acid' [1-5,8,12,14].
2) the use of different test systems such as immunodiffusion (ID), [1,4,14], counterimmunoelectrophoresis (CIE) [2,5,14], indirect hemagglutination (IHA) [7], radioimmunoassay (RIA) [2] and enzyme-linked immunosorbent assay (ELISA) [2,5,6]
3) the use of different, arbitrarily selected antibody titre cut-offs and
4) the use of different clinical case definitions for the various S. aureus infections investigated.

Biological Properties of Peptidoglycan

In the following we are presenting clinical data on 'teichoic acid' antibodies measured by ELISA, CIE and ID using two different antigens and preliminary data on the analysis of different 'teichoic acid' antigens prepared from S. aureus according to published methods. The problems caused by the use of different test systems, cut-offs and case definitions cannot be discussed in this short presentation.

Methods

Patient and controls: Sera from 47 patients with serious S. aureus infections, 63 patients with septicemia caused by other organisms and 147 healthy controls [5] were examined for antistaphylococcal antibodies. More details on the groups of patients are indicated in table 1.

Antigen preparation from S. aureus H and S. aureus Wood 46: The organisms were grown aerobically at 37 °C in a tryptone-yeast extract medium containing 0.5% tryptone, 0.25% yeast extract, 0.4% NaCl, 0.25% K_2HPO_4, pH 7.2 [11].

a) A crude staphylococcal antigen was prepared from mechanically disrupted cells according to the method described by Crowder and White (C-TA) by pelleting untreated cell wall fragments and using the supernatant [1].
b) A different crude antigen was prepared from mechanically disrupted cells according to the method described by Sheagren et al. (S-TA) by pelleting trypsin treated, boiled cell wall fragments and using the supernatant after dialyzing it against distilled water [8].
c) Teichoic acid (TA) was prepared by a trichloro acetic acid extraction from trypsin treated cell walls [9].

Antibody measurement in the clinical samples: Antistaphylococcal antibodies were measured by CIE and ELISA as described previously [5], using for both tests a crude staphylococcal antigen (S-TA/1981) prepared from S. aureus H, and by a commercial ID test (Endo-Staph®, Meridian Diagnostics Inc., Cincinnati, U.S.) [4] using partially purified teichoic acid [10].

Table 1: Diagnostic Groups of Patients Studied for 'Teichoic Acid' Antibodies [5]

Group	No. of cases	Bacteriological diagnosis and clinical features
A	6	S. aureus: endocarditis
B	26	S. aureus: deep-seated infections[1)]
C	15	S. aureus: uncomplicated bacteremia
D	40	Septicemia or meningitis with gram-positive organisms
E	23	Septicemia or meningitis with gram-negative organisms
F	147	Adult healthy controls

[1)]Blood culture positive (8) and negative (18) infections

Table 2: 'Teichoic Acid' Antibodies in S. aureus Infections, Infected and Healthy Controls Measured by ELISA, CIE and ID

Diagnostic groups[1)]	A (n=6)	B (n=26)	C (n=15)	D (n=40)	E (n=23)	F (n=147)
ELISA[2)]						
mean OD	1.055	0.341	0.221	0.151	0.076	0.140
(SD)	(0.484)	(0.311)	(0.107)	(0.152)	(0.081)	(0.075)
pos.result	5	6	0	2	0	0
CIE[3)]						
pos.result	5	6	1	1	0	0
ID[4)]						
pos.result	5	5	2	5	1	–[4)]
Overall agreement:						
ELISA-CIE	100%	100%	93%	95%	100%	100%
ELISA-ID	100%	81%	87%	93%	96%	–[4)]

[1)]Bacteriological diagnosis and clinical features in table 1
[2)]Cut-off for ELISA = 2.5 SD above mean of healthy controls
[3)]Cut-off for CIE ≥ 1:2 = above 100% of healthy controls
[4)]Cut-off for ID ≥ 1:4 (test specification) = above 88% of healthy blood donors (n=128)

Results

Antistaphylococcal serology: The results of a previous study with ELISA and CIE [5] and the comparison with a commercially available ID-test kit are summarized in table 2. The group of patients with S. aureus endocarditis had significantly higher single ELISA readings than the patients with other serious S. aureus infections ($p<0.001$, Student t test). The mean OD value for deep-seated S. aureus infections (group B) was significantly higher ($p<0.001$) than the values for the infected or healthy control groups (D-F), but only the eight blood culture positive cases among group B had higher OD values ($p<0.05$) than the uncomplicated S. aureus bacteremias (group C). Using the cut-offs defined in table 2 and the results of 29 patients with bacteremic S. aureus infections the sensitivity, specificity and predictive value for metastatic-septic complications of a positive result were 50/100/100% for ELISA, 50/93/88% for CIE and 57/87/80% for ID; the predictive value of a negative result was 68% for all three test methods.
Staphylococcal antigen: A preliminary analysis of the two different crude 'teichoic acid' antigens was carried out using CIE, ID and measurement of phosphat content. The antigens prepared from S. aureus Wood 46 were tested in a CIE assay against a high titre control serum in order to demonstrate the different optimal antigen contents : 250 µg/ml for C-TA, 125 µg/ml for S-TA and 30 µg/ml for TA. Table 3 shows a marked variation in sensitivity and specificity of the same antigens tested in an ID assay against a high titre control serum and against a serum from a patient with coagulase negative staphylococcal septicemia: whereas C-TA is two dilutions less sensitive, if used against the positive control compared to S-TA and TA, the latter two antigens show somewhat more crossreaction if used against the infected control. The content in teichoic acid varied considerably between the two crude 'teichoic acid' preparations, between the two strains used as antigen source, and between two batches of the same type of antigen prepared previously [5] and now (table 4).

Table 3: Antistaphylococcal Antibodies in S. aureus Endocarditis: Results with Immunodiffusion Using Various 'Teichoic Acid' Antigen Preparations from S. aureus Wood 46

Antigen Preparation[1] (1000 µg/ml)	Precipitation reaction[2] at increasing serum dilutions: neat	1:2	1:4	1:8	1:16
S. aureus Endocarditis:					
C-TA	++	++	+	(+)	-
S-TA	++	++	++	++	+
TA	++	++	++	++	+
Infected Control[3]:					
C-TA	+	-	-	-	-
S-TA	++	+	+	-	-
TA	++	+	-	-	-
Healthy Control:					
C-TA	+	-	-	-	-
S-TA	+	-	-	-	-
TA	+	-	-	-	-

1) for specification of antigens see method section
2) ++ (strong), + (moderate), (+) (weak), - (negative)
3) Coagulase negative staphylococcal septicemia

Table 4: Teichoic Acid Content (% dry weight) of Two Different Antigen Preparations from S. aureus Wood 46 and S. aureus H Estimated from Phosphat Content[1]

Strains	Antigen-Preparations[2] C-TA	S-TA	S-TA/1981
S. aureus H	0.6 %	2.6 %	0.0 %
S aureus Wood 46	0.2 %	1.6 %	-

1) Analysis and quantitative determination carried out by H.P. Seidl, Munich [11]
2) for specification of antigens see method section

Discussion

The discovery of teichoic acid in the early 1960ies as an important and antigenically distinct cell wall component of staphylococci has helped to coin 'teichoic acid' as a term for any type of crude cell wall extract of S. aureus. In consequence studies have been published in respected journals, which used a variety of antigen preparations such as live cell suspensions [12], lysostaphin extracts of whole cells [8,14], crude sonicates [2,4], supernatants (after centrifugations with 10-15'000 rpm) of mechanically disrupted cells [1], and supernatants of disrupted cells after an additional tryptic digestions of the cell wall fragments and dialysis against distilled water [5,8] ... all termed 'teichoic acid'. Knowing the multitude of up to 55 antigens detectable with crossed immunoelectrophoresis in a sonicate of a protein A-free strain of S. aureus [13] the lack of standardization of the above mentioned cell wall antigen preparations [3] doesn't need any further comment. Although some authors found an unacceptably high prevalence of antibodies in up to 44% of healthy controls [7,14] the majority of these crude 'teichoic acid' preparations have done amazingly well in clinical practice if used in a suitable test system.This was also the experience of one of us (Ch.H.) using a crude antigen in an ELISA and in CIE (table 1 & 2); both tests have been proven useful in monitoring consecutive samples from patients with acute, deep seated and bacteremic S. aureus infections [5]. In comparison a commercial ID test [4] using partially purified teichoic acid surprisingly showed a lower specificity and had a high rate of 12% positive results among healthy blood donors (table 2).

The attempt to reproduce our published results [5] with a newly prepared batch of the same type of crude S. aureus cell wall antigen (S-TA) or a "slightly" different antigen (C-TA) failed for obvious reasons. The low teichoic acid content of our 'teichoic acid' preparations can easily be explained by the fact that in S. aureus ribitol-teichoic acid is covalently bound to peptidoglycan and goes without chemical extraction in small amounts only into a cell wall suspension. The failure to detect any teichoic acid in our previous cell wall extract (S-TA/1981) which gave excellent results in the ELISA and CIE was rather surprising, but can be explained by the unavoidable variations in the preparation process and by the presence of an other, undefined but obviously important antigen. Therefore we don't think that standardization of crude antigen preparations against purified ribitol-teichoic acid using radial immunodiffusion [3,4] or in relation to the protein content [14] is sufficient to overcome the variation in antigenically important 'impurities'. It is outside the scope of this short contribution to discuss the value of alternative antigens or combinations of antigens [6] in S. aureus serology. However, it has to be made clear that only standardized and reproducible antigen preparations should be used in future.

Acknowledgement

We are indebted to H.P. Seidl and collaborators, Lehrstuhl für Mikrobiologie, Technische Universität München, FRG for the measurement of phosphate in the antigen preparations. The work was supported by a grant from the Schweizerischer Nationalfonds (No. 3.900-0.93).

References

1. Crowder, J.G. and A. White. 1972. Teichoic Acid Antibodies in Staphylococcal Endocarditis. Ann.Int.Med. 77, 87-90.

2. Christensson, B., F. Espersen, S.A. Hedström and G. Kronvall. 1982. Staphylococcus aureus Antibodies in Patients with Staphylococcal Septicemia. Scand.J.Infect.Dis. 14, 155-156.

3. White, A., L.J. Wheat, R.B. Kohler. 1983. Diagnostic and Therapeutic Significance of Staphylococcal Teichoic Acid Antibodies. Scand.J.Infect. Dis.Suppl. 41, 105-116.

4. Wheat, J., R.B. Kohler, M. Garten, A. White. 1984. Commercially Available (ENDO-STAPH) Assay for Teichoic Acid Antibodies. Arch.Int.Med. 144, 261-264.

5. Herzog, Ch., H.C. Wood, I. Noel, J.C. Booth. 1984. Comparison of a New Enzyme-Linked Immunosorbent Assay Method with Counterimmunoelectrophoresis for Detection of Teichoic Acid Antibodies in Sera from Patients with Staphylococcus aureus Infections. J.Clin.Microbiol. 19, 511-515.

6. Christensson, B., F. Espersen, S.A. Hedström, G. Kronvall. 1985. Serological Assays against Staphylococcus aureus Peptidoglycan, Crude Staphylococcal Antigen and Staphylolysin in the Diagnosis of Serious S. aureus Infections. Scand.J.Infect.Dis. 17, 47-53.

7. Oeding, P., O.B. Natas and J. Fleurette. 1981. Detection of Staphylococcal Antibodies in human sera. In: The Staphylococci. Proc. Alexander Ogston Centennial Conference, (A. Macdonalds and G. Smith eds). Aberdeen Univerisity Press. 94-108

8. Sheagren, J.N., B.L. Menes, D.P. Han, J.L. Sanders, M.A. Sheagren. 1981. Technical Aspects of the Staphylococcus aureus Teichoic Acid Antibody Assay: Gel Diffusion and Counterimmunoelectrophoretic Assays, Antigen Preparation, Antigen Selection, Concentration Effects, and Cross-Reactions with Other Orgnisms. J.Clin.Microbiology 13, 293-300.

9. Archibald, A.R., J. Baddiley, G.A. Shaukat. 1968. The Glycerol Teichoic Acid from Walls of Staphylococcus Epidermidis I2. Biochem.J. 110, 583-588.

10. Grov, A. and R. Sidsel. 1967. Immunochemical Examination of Phenylhydrazine-treated Staphylococcus aureus Cell Walls. Acta Path. Microbiol Scand. 71, 417-421.

11. Endl, J., H.P. Seidl, F. Fiedler, K.H. Schleifer. 1983. Chemical Composition and Structure of Cell Wall Teichoic Acids of Staphylococci. Arch.Microbiol. 135, 215-223.

12. Nagel, J.G., J.N. Sheagren, C.U. Tuazon, T.A. Cardella. 1977. Teichoic Acids in Pathogenic Staphylococcus aureus. J.Clin.Microbiol. 6, 233-237.

13. Schiøtz, P.O., N. Høiby, J.B. Hertz. 1979. Cross-Reactions beetween Staphylococcus aureus and fifteen other Bacterial Species. Acta Path. Microbiol.Scand.Sect.B, 87, 329-336.

14. Leffell, M.S., J.D. Folds, B. Wasilauskaus. 1978. Counterimmunoelectrophoretic Detection of a High Incidence of Precipitin Reactions in Normal Human Sera Against Staphylococcal Teichoic Acids and Protein A.J.Clin. Microbiol. 8, 591-597.

LIPID A ANTIBODIES IN MAN

W. Marget, P. Mar, H. Haslberger, M. Huber

Universitäts-Kinderklinik, Division of Infectious Diseases, Lindwurmstr.4, 8000 München 2, FRG

Fig. 1

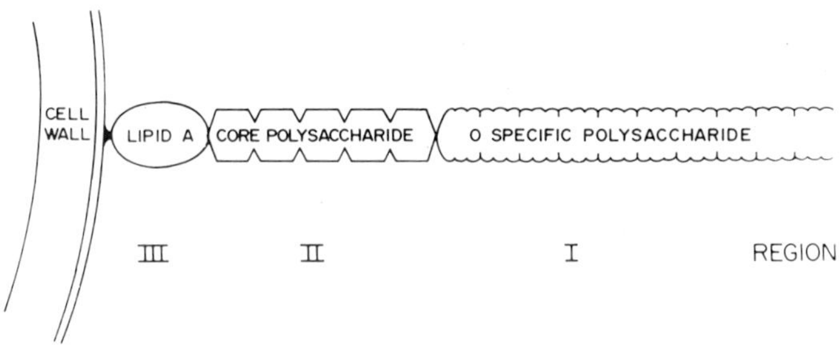

Lipid A is the toxic substance of the lipopolysaccharide of gram-negative bacteria in man. For many species it is almost identical, eg. E.coli, salmonellae and Pseudomonas spp.. We became aware of this substance a long time ago due to the fact that it causes a large number (about 30) of clinical/toxic symptoms, the most important being toxic shock. It goes without saying that a substance like this, which we carry around in large quantities, not only interests but also worries physicians. For simplicity's sake, the lipopolysaccharide of gram-negative bacteria was formerly described as the endotoxin complex.

Biological Properties of Peptidoglycan

After lipid A had been clearly shown by Westphal, Galanos et al. at the Max-Planck Institute in Freiburg and after Galanos had demonstrated its hemolytic antibody, albeit labile, our own group started to investigate it from the clinical point of view in conjunction with Westenfelder, Simon, Schüßler, Madsen and others (1-4). For many years and for various reasons, we were the only ones to investigate lipid A antibodies in man although we made only slow progress due to the lack of precision in the antibody determinations and the very low antigenicity of lipid A. However, we are now relatively well informed about the antibody reaction of this substance in man, particularly since our most important results have been confirmed in the past few years.

Table 1

Group	Lipid A µg/kg	Incubation with human erythr.= 1µg lipid A +1ml erythr. concentrate	Total no.of animals	no. that died	no. that surv.
Test Group A	100	100 ml erythr.+ 100 µg lipid A	6	0	6
Test Group B	150	100 ml erythr.+ 100 µg lipid A	5	0	5
Control Group C	100	-	7	6	1
Control Group D	70	-	4	0	4

Lipid A is an aggressive proteinophilic substance which can only be determined in serum under certain conditions. Moreover, it binds very rapidly to HDL and cells, including erythrocytes, as can be seen in the above table. In this way 100 ml of erythrocyte suspension were sufficent to counteract the toxicity of 100 µg lipid A in rabbits.

Table 2 : Preliminary study for high-titer serum applications in septicemia without lipid A antibody titers. Antibody does not seem to be completely bound up by endotoxin.

	Lipid A Titer
a) Serum from septicaemia case (1 day after onset)	Ø
b) Serum from recurr. UTI case (1 day after onset)	1:512
<u>Mixture</u> 50% serum(a) + 50% serum(b)	1:64
c) Serum from septicaemia case (12 days after onset)	Ø
d) Serum from recurr. UTI case (12 days after onset)	1:256
<u>Mixture</u> 50% serum(c) + 50% serum(d)	1:32

We were also able to show that detoxication in man is probably possible using serum with high lipid A antibody titers.

In 1974 we used the passive hemolysis test to look for lipid A antibodies in patients and healthy individuals (2). We found that healthy adults are largely free of lipid A antibodies. On the other hand, patients with chronic or long-lasting infections with gram-negative bacteria showed a relatively high number of theses antibodies, eg. patients with UTI and renal pelivc catheter.

Fig. 2

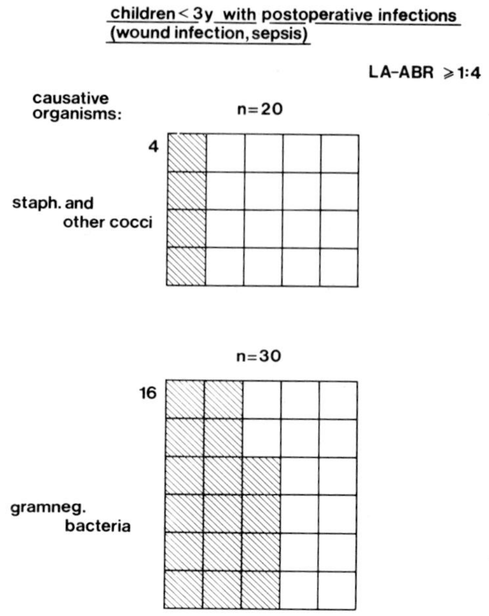

Figure 2 is taken from a publication of ours from 1974 on postoperative wound infections of varying severity in infants (2). In comparison to gram-positive bacteria, gram-negative bacteria led to an increase of over 50% of lipid A antibodies.

Table 3

THE EFFECT OF INCREASING LIPID-A-DSA(DOG SERUM ALBUMIN) ON THE DEVELOPMENT OF L-A-ANTIBODY TITERS. BESIDE THE HEMOLYTIC ACTIVITY OF THE ANTI-LIPID-A-ANTIBODY, THE PRESENCE (A) OR ABSENCE (a) OF AUTORADIOGRAPHICALLY DETECTABLE L-A IS SHOWN.

DOG No.	TITER BEFORE INJECTION	L-A-DSA µg/kg	WEEK 1	2	3	4	5	6
1	0	200	16 A	32 A	64 a	32 a	32 a	- A
2	16	200	128 A	256 A	16 a	0 a	16 A	16 A
3	0	300	2048 A	512 a	256 A	128 a	64 a	32 A
4	0	300	2048 A	2048 A	512 a	512 A	64 a	-
5	0	400	0 a	0 a	0 a	-	16 A	32 A
6	0	400	8 A	16 A	64 a	64 A	0 A	- A
7	16	500	32 a	0 A	0 A	32 a	32 a	-
8	8	500	0 a	8 A	0 A	0 A	0 A	0 a
9	0	600	256	128	32	-	16	-
10	0	600	64	32	8	0	-	-

M.Westenfelder, Freiburg/Br., 1977,W-Germ.

This effect was also observed in UTI patients, who differed from the control group in that the kidney tissue was obviously severely affected. Thereafter we carried out a joint investigation in puppies with Madsen from Wisconsin (4). By doing this we were trying to understand the pattern of lipid A antibody formation. We injected radiolabelled lipid A into the kidney tissue and observed that certain concentrations, ie. up to 300 µg/kg, led to a rise in lipid A antibodies. This was again determined by the passive hemolysis test. Lipid A could still be traced in the kidney tissue 6 weeks later.

Table 4 *Comparison of lipid A titers in Crohn's disease, ulcerative colitis, acute enteritis and control group.*

RECIPR. TITER	CONTROL GROUP	ENTERITIS GROUP	ULCERATIVE COLITIS	CROHN'S DISEASE
512	--	--	--	1
256	--	--	2	--
128	--	--	--	3
64	--	2	--	5
32	5	2	--	5
16	7	6	2	3
8	11	4	10	--
4	14	3	6	1
< 4*	31	7	8	--
Total	68	24	28	18

* not detectable

We then investigated other infections related to gram-negative bacteria, ie. Salmonella and Shigella enteritis as well as ulcerous colitis and Crohn's disease (5). To our surprise, we only observed a clear titer rise in the case of Crohn's disease, although this did in fact attack only a small section of the digestive tract, but for a long time.

This rise in titer was closely correlated to the rise in O-titers of various species, expecially Pseudomonas (6). In the latter case we used the most common mixed antigens.

Table 5

	LIPID A TITER	COLI TITER	KLEBS. TITER	PROTEUS TITER	PS.AERUG. TITER
RECIPROCAL MEAN TITER IN THE CONTROL GROUP	4 (n=68)	640 (n=26)	160 (n=26)	80 (n=26,	80 (n=26)
VARIANCE (IN TITER STAGES)	1,69	1,03	1,54	2,08	2,1
PATIENT NO.					
1	+5	+3	+3	+4	+8
2	+4	+2	0	0	+2
3	+5	+3	+3	+3	+4
4	+2	0	+2	+2	+2
5	+3	0	+1	+3	+1
6	+4	+2	0	+1	+3
7	+3	0	+1	0	+2
8	+4	+3	+2	+3	+4
9	+5	+2	+3	+4	+3
10	+7	n.d.	n.d.	n.d.	n.d.
11	+2	n.d.	n.d.	n.d.	n.d.
12	+4	+5	+4	+1	+2
13	+4	-2	+1	0	0
14	+2	+4	+3	+2	+3
15	+3	+1	+2	+1	+1
16	+2	0	+1	+3	+2
17	+2	+1	+3	+3	+4
18	0	+3	+2	+3	+4

Deviation in titer stages measured in 18 patients with Crohn's disease from the mean titers in the control group. The underligned figures represent the highest rise of O antibodies in regard to mean titers and variance.

Thanks to the ELISA method available today, we were in a much better position than previously and were able to trace specific IgG and IgM titer rises. In our first investigation, in which we only differentiated into three categories, the results were basically the same as before (7).

Table 6: IgG

Group	(n)	+	+-?	-	Only patients over three months of age (n)	+	+-?	-
I. UTI	(17)	7	3	7	(16)	7	3	6
II. Pyelonephritis	(22)	10	8	4	(21)	9	8	4
III. Sepsis	(19)	4	6	9	(10)	1	3	6
IV. Tracheal secretion	(17)	4	4	9	(7)	2	1	4
V. Wound infection	(6)	4	1	1	(4)	3	0	1
VI. Control group	(12)	11	1	0	(12)	11	1	0
	(93)	40	23	30	(70)	33	16	21

At this point it should be noted that tracheal colonization is not regarded as an infection. In sepsis patients examined 1-5 days after the outbreak of infection, there was only one IgG-positive child in the group over three months of age (in infants below this age IgG antibodies can be transmitted from the mother). However, the child in question was only just over three months of age, meaning that IgG shouldn't really have been determined in view of its half-life.

Table 7: IgM

Group	(n)	+	+-?	-	Only patients over three months of age (n)	+	+-?	-
I. UTI	(16)	7	6	3	(15)	7	6	2
II. Pyelonephritis	(18)	17	0	1	(17)	16	0	1
III. Sepsis	(16)	6	5	5	(8)	3	4	1
IV. Tracheal secretion	(16)	2	2	12	(7)	2	2	3
V. Wound infection	(6)	2	1	3	(4)	2	1	1
VI. Control group	(12)	6	2	4	(12)	6	2	4
	(84)	40	16	28	(63)	36	15	12

As expected, IgM titer could not be determined as exactly as IgG titers. However, in the case of chronic tissue infection in pyelonephritis, they were naturally somewhat higher.

Table 8: DIFFERENCES IN LIPID A ANTIBODY TITER WITH AGE.

Group	IgG Titer $\geq$ 1		IgM Titer $\geq$3	
	0-5 Years	6-17 Years	0-5 years	6-17 Years
I. LUTI	1/6	6/18	1/6	9/18
II.PN	3/6	4/7	3/6	6/7
IV.Control	0/3	2/22	0/3	7/22

A more precise differentiation of UTI and PN in infants led to the same results as before, although the differences were no longer so pronounced in the case of the lower limit. The titer-limits seem to be a little bit too low and the titer level seems also to depend on the age.

Table 9: ELISA OF IgG AND IgM ANTIBODY TO LIPID A IN HEALTHY ADULTS AND PATIENTS WITH GRAM-NEGATIVE SEPSIS

	IgG titer 0	1	IgM titer 0	1	2	3	4
Patients(adults) (n=6)	3	3	1	2	0	1	2
Healthy adults (n=113)	100	13	95	18$\geq$ 1			

Table 9 shows the lipid A antibody titers in healthy individuals. The determinations were made only a short time ago.

In order to investigate the antitoxic effect of the anti-lipid A titers, we are now performing a double-blind study im Klinikum Großhadern. This is being carried out with high-titered anti-lipid A sera. We were able to show that this effect is pharmacokinetically possible using lipid A antibody sera in sepsis patients and controls - one case with placebo and the other a healthy individual. By means of this study we hope to be able to show the same improved antitoxic effect which is being shown in the USA and Switzerland using the J5 mutant of E.coli.

References

1. Galanos, C., Lüderitz, O., Westphal O. 1971. Preparation and properties of antisera against the lipid A component of bacterial lipopolysaccharides. Eur.J.Biochem. 24, 116-122

2. Simon, G., Reindke, B., Marget, W.: Lipoid-A-Antikörpertiter bei Pyelonephritis und anderen Infektionen mit gram-negativen Bakterien. Infection 2 (174) 178-184

3. Marget, W., Schüßler, P., Kruis, W., Weinzierl, M., Rindfleisch, G.: Is the pathogenesis of Crohn's disease similar to that of juvenile recurrent pyelonephritis? Infection 4 (1976) 2-4.

4. Westenfelder, M., Galanos, C., Madsen, P.O., Marget, W.: Pathological activities of lipid A: Experimental studies in relation to chronic pyelongephritis. In: Schlesinger D. (ed.): Microbiology. American Society for Microbiology, Washington, D.C.1977, pp.277-279

5. Schüßler, P., Kruis,W. Marget, W.: Lipoid-A-Antikörper bei Morbus Crohn. Klin.Wschr.54 (1976) 1055

6. Schüßler, P., Kruis, W., Marget, W.: Lipoid-A-Antikörpertiter und O-Antikörpertiter bei Enterocolitis Croh, Colitis ulcerosa und akuter Enteritis. Med.Klin.71(1976, 1898-1902

7. Marget, W., Weiß, M., Ruhland, B.: Lipid A antibody determinations using ELISA on patients at a children's hospital: a preliminary report.Infectic 11 (1983) 82-84

ELECTRON MICROSCOPIC LOCALIZATION OF PEPTIDOGLYCAN IN THE CELL WALL OF STREPTOCOCCUS PYOGENES BY MEANS OF LABELLED ANTIBODIES AND LYSOZYME

M. Wagner

Central Institute of Microbiology and Experimental Therapy, Jena
German Democratic Republic

M. Rýc

Institute of Hygiene and Epidemiology, Prague, Czechoslovakia

B. Wagner

Central Institute of Microbiology and Experimental Therapy, Jena
German Democratic Republic

Introduction

Based on the morphological appearance of the cell wall in ultrathin sections and some other evidences it has been believed for a long time that the cell wall of streptococci, as other Gram-positive bacteria too, is constructed from several layers, each of them consisting of one defined chemical component. The peptidoglycan was supposed to be the innermost layer forming the rigid basal structure of the wall (1 - 3). However, phage absorption experiments have shown, that at least parts of the peptidoglycan macromolecule should be accessible on the cell surface (4). The aim of our investigations was to localize the peptidoglycan in the cell wall of group A streptococci by electron microscopic labelling methods. The present review summarizes the results of our previous papers (5 - 8).

Materials and Methods

Bacteria. Group A Streptococcus strain NY5 (M type 12), lacking the IgG•Fc-binding receptor (9), from the collection of the Central Institute of Microbiology and Experimental Therapy, Jena was used for all experiments.

Cell Walls and Peptidoglycan. Cell walls were prepared by mechanical disintegration and differential centrifugation (6). In some experiments the walls were also treated with trypsin. Peptidoglycan was isolated from cell walls by extraction with 10% trichloroacetic acid (4h at 60°C) or by hot formamide extraction.

Peptidoglycan antibodies and ferritin conjugates. Peptidoglycan antibodies from antisera raised against Group A variant streptococci (strains T 12 and T 50) were isolated by absorption to group A peptidoglycan followed by extraction with 0.1M glycine-

Biological Properties of Peptidoglycan

HCl buffer, pH 2.5, and neutralization (6). Ferritin conjugation of the antibodies was performed with glutaraldehyde (6).

Lysozyme-peroxidase conjugates. Lysozyme was conjugated with horse-radish peroxidase by glutaraldehyde. Residues of uncoupled lysozyme were separated by chromatography on Con A-Sepharose (Pharmacia, Uppsala). The conjugate showed both the activity of lysozyme (lysis of Micrococcus flavus) and of the peroxidase.

Inhibition Experiments. The following peptides were used for antibody inhibition experiments: Dialanine(D-Ala-D-Ala), trialanine (D-Ala-D-Ala-D-Ala), tetrapeptide (L-Ala-D-iGln-L-Lys-D-Ala), pentapeptide (L-Ala-D-iGln-L-Lys-D-Ala-D-Ala) and the M 24 protein peptide L-Ala-L-Glu-L-Lys-L-Ala-L-Ala (10). Ferritin-conjugated antibodies against peptidoglycan were incubated with the peptides for 30 min at 37°C followed by 18 h at 4°C, and slight precipitates were removed by centrifugation. Inhibitory concentrations of the synthetic peptides were calculated as double those causing 50% inhibition of the precipitation with peptidoglycan (11).

Incubation of Cells and Cell Fragments with Conjugates. To about 0.1 ml of sedimented cells, cell walls or peptidoglycan inhibited or noninhibited ferritin-conjugates (0.2 - 0.3 ml) were added and incubated at 37°C with shaking for 1 h.
Incubation with lysozyme-peroxidase conjugates was done at 0°C in an ice bath overnight. After washing with ice-cooled veronal acetate buffer the samples were fixed with glutaraldehyde, incubated with 3,3'-diaminobenzidine and H_2O_2, and processed for transmission electron microscopy (6). As controls cells and cell fragments were incubated with peroxidase and further processed in the same way.

Electron Microscopy. Ultrathin sections were examined without post-staining. For the evaluation of the inhibition experiments on whole cells, walls and peptidoglycan fragments one hundred segments of 0.2 µm length were selected for counting the number of ferritin particles using parts of antitangentially sectioned objects. Inhibition was calculated in percent of the counts on noninhibited controls.

Results

Localization of peptidoglycan by ferritin-conjugated antibodies

All investigations were carried out on whole cells, isolated walls (treated with trypsin and RNase) and isolated peptidoglycan. Whole cells displayed ferritin particles directly on the cell surface as well as on filamentous protrusions (Fig. 1). The specificity of the labelling was confirmed by negative results obtained with ferritin-conjugated normal IgG. Isolated walls and peptidoglycan bound labelled antibodies on both surfaces (Figs. 2, 3). The ferritin particles labelled short filamentous structures protruding from both sides of the fragments.

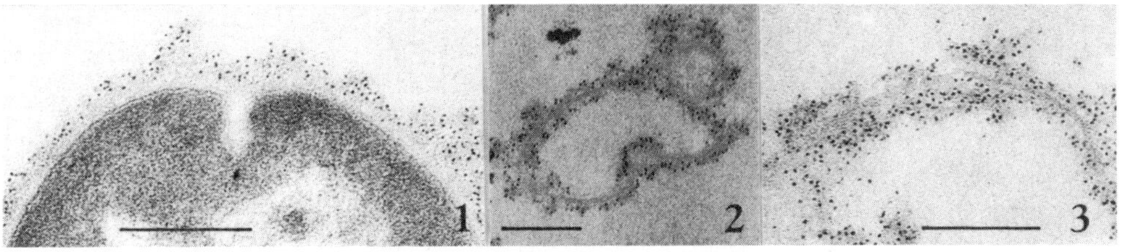

Figs. 1-3. Strept.pyogenes, strain NY5, labelled with ferritin-conjugated peptidoglycan antibodies. All bar markers represent 0.2 µm.
(1) Whole cell. (2) Cell wall. (3) Peptidoglycan

Characterization of the peptidoglycan immunodeterminants

The nature of the peptidoglycan immunodeterminants demonstrated by ferritin-conjugated antibodies was investigated in inhibition experiments (Table 1).

Table 1: Inhibition of Ferritin-Conjugated Peptidoglycan Antibodies Demonstrated on Group A Streptococcus Cells, Cell Walls and Peptidoglycan

Inhibition by	Reduction of Particle Count (%) in Comparison to Noninhibited Controls		
	Whole Cells	Cell Walls	Peptidoglycan
Dialanine	76.6	83.8	87.7
Trialanine	49.0	77.7	77.0
Tetrapeptide	41.2	64.5	44.7
Pentapeptide	92.6	98.2	97.1
M Protein Sequence	7.3	11.2	7.6

The results of these studies clearly show that the pentapeptide exhibited the highest inhibition activity. In the experiments with cell walls and peptidoglycan it inhibited practically all antibody activity. The tetrapeptide had only the half of this activity. The M protein sequence used in the same concentration as the pentapeptide was practically ineffective.
The figs. 4-6 show that after inhibition by synthetic peptides labelling was abolished or reduced on both sides of isolated cell walls.

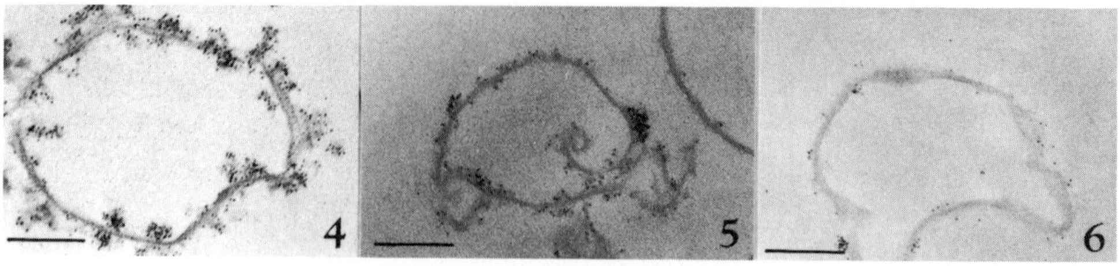

Figs. 4-6. Strept.pyogenes, strain NY5. Labelling of cell walls with antibody-ferritin conjugate was inhibited by synthetic peptides. Bar markers represent 0.2 µm. (4) No inhibition. (5) Inhibition by tetrapeptide. (6) Inhibiton by dialanine

Localization of peptidoglycan by peroxidase-conjugated lysozyme

Incubation of isolated peptidoglycan with lysozyme-peroxidase followed by the reaction with 3,3'-diaminobenzidine and H_2O_2 led to a heavy labelling on both sides of the fragments (Fig. 7). Similarly, isolated trypsinized cell walls were also labelled to a lesser extent on both sides (Fig. 8). Incubation of peptidoglycan or walls with peroxidase alone had no such effect (Fig.9). On the other hand, whole cells exhibited only a weak labelling after incubation with lysozyme-peroxidase (Fig. 10).

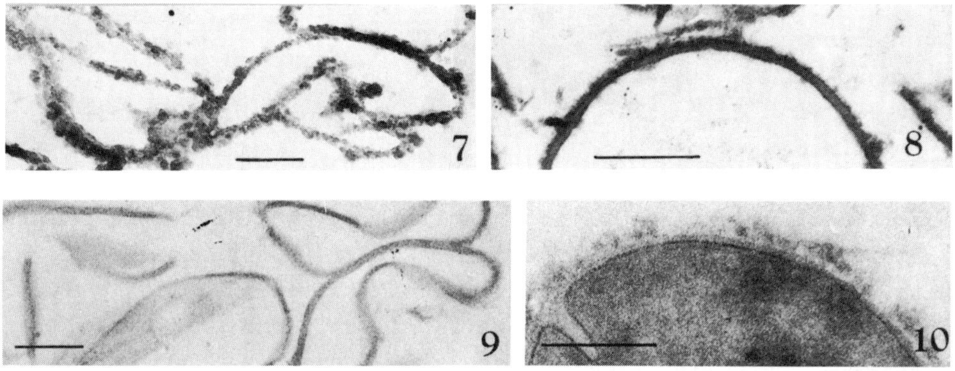

Figs. 7-10. Strept.pyogenes, strain NY5. Labelling with lysozyme-peroxidase. Bar markers represent 0.2 µm.
(7) Peptidoglycan. (8) Cell wall.
(9) Cell wall, control with peroxidase alone.
(10) Whole cell.

Discussion

The use of immunoelectron microscopic methods allows a direct proof of the location of antigenic cell components. In our investigations two approaches have been applied to study the location of peptidoglycan in the cell wall of group A streptococci: The accessible immunodeterminant groups were detected by antibodies raised in rabbits against group A variant streptococci and purified by absorption to group A streptococcal peptidoglycan. The location of the glycan portion of the molecule was demonstrated by means of lysozyme conjugated to peroxidase. Ferritin-conjugated peptidoglycan antibodies labelled isolated cell walls and peptidoglycan fragments on both sides. The ferritin particles were mainly bound to filamentous structures. Whole cells were also labelled on their surface. These results contradict models in which peptidoglycan forms the innermost layer of the cell wall (1-3) and support concepts of a more complicated mosaic structure.
Inhibition experiments with synthetic peptides demonstrated that the immunodeterminant groups demonstrated in this investigation are predominantly of pentapeptide nature. Interestingly, the distribution of these determinants was found to be similar on both sides of walls and peptidoglycan fragments. Recently, Seidl et al. (12) using antibodies against the D-alanyl-D-alanine moiety of the non-crosslinked pentapeptide also localized pentapeptide subunits of peptidoglycan on the surface of whole cells of Streptococcus pyogenes and some other Gram-positive bacteria. These results confirm our observation that the peptidoglycan of group A streptococci is accessible from the cell surface.
The use of lysozyme-peroxidase conjugate based on the finding that conjugates of lysozyme with peroxidase or fluorescent dyes (5, 13) remain the properties of lysozyme to bind to the glycan strand and to lyse the peptidoglycan. Near $0^{o}C$, however, only binding to the peptidoglycan takes place. Using this probe both peptidoglycan and trypsinized cell walls were heavily labelled on both sides whereas the surface of whole cells seems to be lesser accessible for the lysozyme.
Summarizing, our results show that the peptidoglycan of Streptococcus pyogenes does not form a separate inner layer of the cell

wall but must be present throughout the whole wall as a network. This conception is in agreement with contemporary models of the architecture of peptidoglycan of Gram-positive bacteria.

Acknowledgement

We thank Dr. R. Straka, Prague, for the synthesis of peptides and Dr.E.H.Beachey, Memphis, for providing us the M protein peptide.

References

1. Krause, R.M. 1972. In: Streptococci and Streptococcal Diseases. Recognition, Understanding and Management (L.W.Wannamaker and J.M.Matsen, eds.) Academic Press, New York, p.3.

2. Davis, B.D., R.Dulbecco, H.N.Eisen, H.S.Ginsburg, W.B.Wood, M.McCarty. 1973. Microbiology, 2nd edn., Harper and Row, Hagerstown, Maryland, p. 708.

3. Kasper, D.L., C.J.Baker, H.J.Jennings, 1985. In: Neonatal Group B Streptococcal Infections (K.K.Christensen, P.Christensen and P.Ferrieri, eds.) Antibiotics and Chemotherapy, 35, Karger, Basel, p. 90.

4. Cleary, P.P., L.W.Wannamaker, M.Fischer, N.Laible. 1977. J.Exp.Med. 145, 578.

5. Wagner, M., B.Wagner. 1978. Zbl.Bakt.Hyg., I.Abt.Orig.A 240, 302.

6. Wagner, M., B.Wagner, M.Rýc.1978. J.Gen.Microbiol. 108, 283.

7. Rýc, M., M.Wagner, B.Wagner. 1979. In: Pathogenic Streptococci (M.T.Parker, edit.) Reedbooks Ltd., Chertsey, England, p. 46.

8. Rýc, M., B.Wagner, M.Wagner, R.Straka. 1982. Current Microbiology 7, 187.

9. Rýc, M., M.Wagner, B.Wagner, J.Havlíček. 1982.Microbios 34, 7.

10.Beachey, E.H., J.M.Seyer, A.H.Kang. 1978. Proc.Natl.Acad.Sci. U.S. 75, 3163.

11.Schleifer, K.H., R.M.Krause. 1971. J.Biol.Chem. 246, 986.

12.Seidl, P.H., J.R.Golecki, N.Franken, K.H.Schleifer. 1985. Arch.Microbiol. 142, 121.

13.Gould, G.W., D.L.Georgala, A.D.Hitchins. 1963. Nature (Lond.) 200, 385.

IMMUNOELECTRON MICROSCOPIC STUDIES ON PEPTIDOGLYCAN FROM GRAM POSITIVE BACTERIA: SPECIFIC REACTIONS WITH THE GLYCAN MOIETY, THE PENTAPEPTIDE SUBUNIT AND THE INTERPEPTIDE BRIDGE

Norbert Franken

Boehringer Mannheim GmbH, Forschungszentrum Tutzing, 8132 Tutzing, FRG

Jochen R. Golecki

Institut für Biologie II, Mikrobiologie, Albert-Ludwigs-Universität, 7800 Freiburg, FRG

Peter H. Seidl, Peter Zwerenz, Karl H. Schleifer

Lehrstuhl für Mikrobiologie, Technische Universität München, Arcisstraße 21, D-8000 München 2, FRG

Introduction

The peptidoglycan molecule reveals at least five independent antigenic epitopes, i.e. a) the glycan moiety, b) the pentapeptide subunit, c) N-terminal and d) C-terminal sequences of the interpeptide bridge and e) the tetrapeptide subunit (1; Seidl and Schleifer, this book). This report deals with immunoelectron microscopic studies of the glycan moiety, the pentapeptide subunit and N-terminal interpeptide bridge epitopes in the cell wall of several gram positive bacteria.

Materials and Methods

Specific antibodies to peptidoglycan epitopes. The isolation of antibodies to the glycan moiety by affinity chromatography techniques was recently reported (2). Antisera to the C-terminal D-alanyl-D-alanine moiety of H-L-Ala-D-Glu(L-Lys-D-Ala-D-Ala-OH)-NH_2 were elicited in rabbits by immunization with synthetic immunogen albumin-(CH_2CO-Gly-L-Ala_2-D-Ala_2-OH)$_{39}$ (3). Antibodies to N-terminal sequences of the pentaglycine interpeptide bridge or of the tri-L-alanine interpeptide bridge were produced as previously described by immunization with (Gly_5-ε-Ahx)$_{20}$-albumin or (L-Ala_3-ε-Ahx)$_{22}$-albumin (4).

Radioactive hapten binding assays and enzyme immunoassays. Antibodies to the tetrapeptide or to the pentapeptide subunit were determined by radioactive hapten binding assays (5,6) and by an en-

Biological Properties of Peptidoglycan

zyme-linked immunosorbent assay (ELISA) (7). Glycan strand specific antibodies were measured by a radioactive hapten binding assay (2) or by an appropriate ELISA (Zauner et al., this book). Binding inhibition studies for proving antibodies' specificity were carried out as previously described for the radioactive hapten binding assays (5-7) or for the ELISAs (7; Zauner et al., this book).

Bacterial strains used. The origin of Streptococcus pyogenes A-variant A486, Staphylococcus aureus 52A5 and of Bacillus subtilis W23 is given in a recent paper (8).

Indirect immunoferritin technique and electron microscopy. Labelling of cells in the indirect immunoferritin technique and electron microscopy were carried out as recently described in detail (8).

Analytical methods. Gas chromatography, quantitative determination of phosphate and protein, amino acid analyses, determination of C-terminal amino acids and of the L-alanine/D-alanine ratio in peptidoglycans were carried out as described (8).

Results and Discussion

Specificity of Antibodies to the Glycan Moiety, Pentapeptide Subunit and Interpeptide Bridge Epitopes

Antibodies with specificity to the glycan moiety of peptidoglycan were isolated from antisera to whole cells of Micrococcus luteus CCM 169, known to contain high levels of glycan strand specific antibodies (9). Removal of minor antibody levels in those sera with specificity to the pentapeptide subunit (1,9) was achieved by affinity chromatography on Sepharose-$(Gly_2$-L-Ala-D-Ala-D-Ala-OH$)_n$ (10). Glycan strand specific antibodies in the eluent were then prepared by affinity chromatography (2). Peptidoglycan of Bacillus subtilis NCIB 8060 was used as an immunoadsorbent (2) for the isolation of glycan strand specific antibodies since it lacks as antigenic epitopes interpeptide bridges, pentapeptide subunits or non-crosslinked tetrapeptide subunits (1, 6-9,11). Specificity of the isolated antibody fraction for the glycan moiety was verified by immunoassays for measuring glycan strand specific antibodies (1,2, 9; Zauner et al., this book), and by excluding binding specificity to the tetrapeptide or to the pentapeptide subunit (1,2,5-7).

Specific antibodies to the C-terminal D-alanyl-D-alanine moiety of non-crosslinked pentapeptide subunit H-L-Ala-D-Glu(L-Lys-D-Ala-D-Ala-OH)-NH_2 (10) were elicited in rabbits using as an immunogen albumin-(CH_2CO-Gly-L-Ala_2-D-Ala-D-Ala-OH)$_{39}$ (3). Exclusive specificity of these antibodies for non-crosslinked peptide subunit pentapeptides was demonstrated in detail by appropriate precipitin and precipitin inhibition studies (3) and by Farr type hapten binding or binding inhibition studies (5,6,8).

Specific antibodies to the N-terminal sequences of the pentaglycine interpeptide bridge or to the tri-L-alanine interpeptide bridge were obtained by immunization with synthetic immunogens (Gly_5-ε-Ahx)$_{20}$-albumin or with (L-Ala_3-ε-Ahx)$_{22}$-albumin, respectively (4). Selective specificity of the particular antiserum for the pentaglycine interpeptide bridge of staphylococci or against the oligo-L-alanine interpeptide bridge typical for many micrococci or pyogenic streptococci was demonstrated in detail by appropriate precipitin and precipitin inhibition studies (4,12) and by latex agglutination (13).

Immunoelectron Microscopic Detection of the Glycan Moiety, the Pentapeptide Subunit and of the Interpeptide Bridge Epitope in Gram Positive Bacteria

The results obtained from the immunoelectron microscopic studies employing specific antibodies to the particular peptidoglycan epitopes are schematically summarized in Table 1, p. 138. Whole trypsinized cells of *Streptococcus pyogenes*, incubated with specific antibodies to the glycan moiety followed by incubation with ferritin-labelled anti-rabbit IgG were specifically labelled with ferritin particles as was evident from these sections comparing untreated controls (not depicted) with antibody coated cells (Fig. 1, p. 138). Heavy specific labelling with ferritin particles (data not depicted) was also obtained with trypsinized cells of *Staphylococcus aureus* and of *Bacillus subtilis*, compare Table 1 and (2).

Employing in the indirect ferritin technique specific antisera to the pentapeptide subunit (3), *Bacillus subtilis* remained completely unlabelled however (Fig. 2a, p. 139 and Table 1), due to missing pentapeptide subunits in its peptidoglycan (3,8). On the contrary, heavy specific labelling was obtained with cells of *Streptococcus pyogenes* (Fig. 2b, p. 139) or of *Staphylococcus aureus* (Fig. 2c, p. 139), compare Table 1. These results fully

Table 1. Labelling of trypsinized cells of Streptococcus pyogenes, Staphylococcus aureus and of Bacillus subtilis with specific antibodies to the glycan moiety, to the pentapeptide subunit and to the interpeptide bridge epitope (indirect immunoferritin technique). +++, heavy labelling; -, no labelling.

	Str. pyogenes A-variant A486	S. aureus 52A5	B. subtilis W23
Specific antibodies to the glycan moiety	+++	+++	+++
Specific antibodies to the R-D-Ala$_2$ moiety of the pentapeptide subunit	+++	+++	-
Specific antibodies to the Gly$_5$-interpeptide bridge	-	+++	-
Specific antibodies to the L-Ala$_3$-interpeptide bridge	+++	-	-

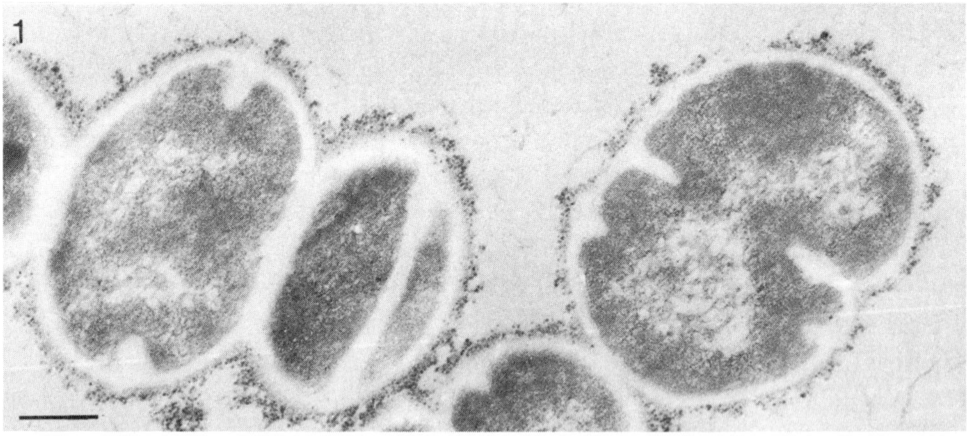

Fig. 1. Labelling of trypsinized cells of Streptococcus pyogenes with specific antibodies to the glycan moiety (2) of peptidoglycan (indirect ferritin technique). Bar represents 200 nm.

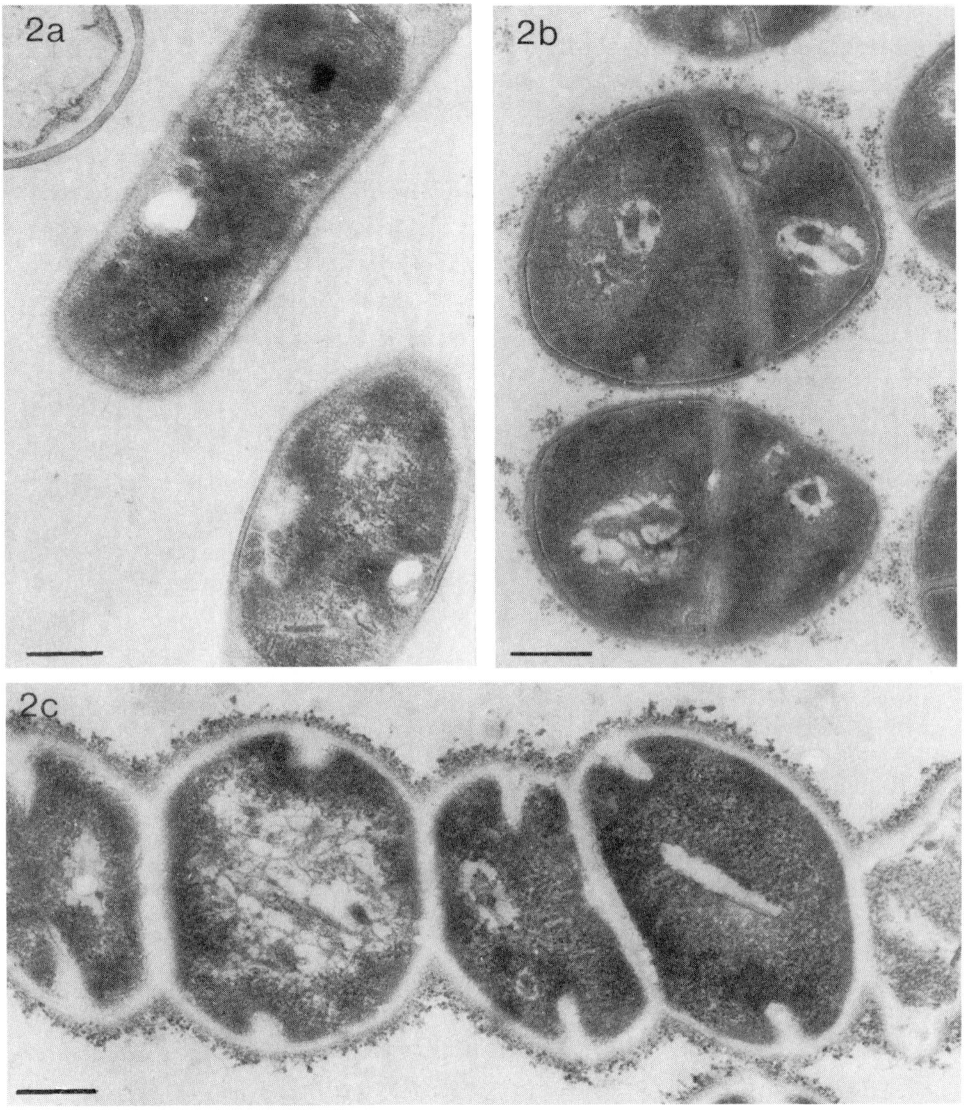

Fig. 2. Labelling of trypsinized cells of a) Bacillus subtilis W23, b) Staphylococcus aureus 52A5, c) Streptococcus pyogenes A-variant A486 with specific antibodies to the pentapeptide subunit (3) of peptidoglycan (indirect ferritin technique). Bar represents in all Figs. 200 nm.

correspond with our previous precipitin studies (3), and with the peptidoglycan structures of bacteria investigated. Due to missing D,D-carboxy-peptidases (14), pentapeptide subunits with C-terminal D-alanyl-D-alanine occur frequently among staphylococci or streptococci (6,7) and were chemically demonstrated in the peptidoglycans of Str. pyogenes A486 or S. aureus 52A5, used in this study (8).

Employing in the indirect ferritin technique specific antisera to the interpeptide bridge (4), trypsinized cells of Staphylococcus aureus were specifically labelled with antisera to the pentaglycine interpeptide bridge of staphylococci (Fig. 3a) but not with antisera to the tri-L-alanine interpeptide bridge (Fig. 3b). On the other hand, cells of Streptococcus pyogenes were heavily labelled with antiserum to the tri-L-alanine interpeptide bridge but not when using antisera to the pentaglycine interpeptide bridge (not depicted, compare Table 1). Neither interpeptide bridge specific antiserum resulted in labelling of Bacillus sub-

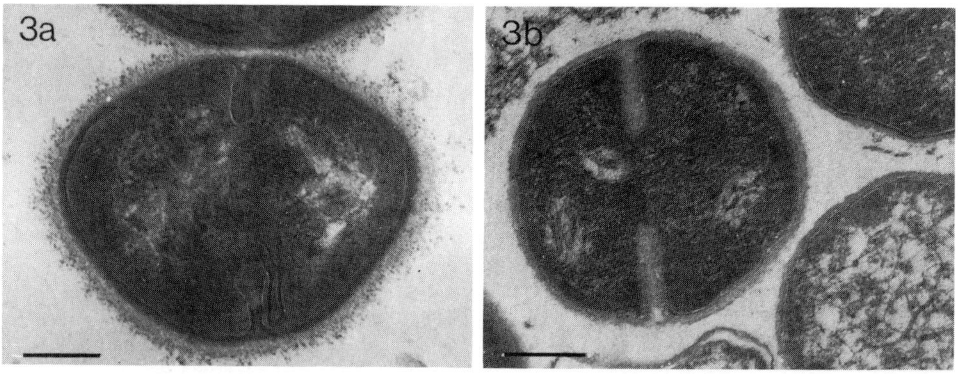

Fig. 3. Labelling of trypsinized cells of Staphylococcus aureus 52A5 with specific antibodies to the interpeptide bridge (4) of peptidoglycan (indirect immunoferritin technique). Bar represents 200 nm. a) ultrathin sections of S. aureus incubated with antiserum to $(Gly_5\text{-}\varepsilon\text{-}Ahx)_{20}$-albumin; b) ultrathin sections of S. aureus incubated with antiserum to $(L\text{-}Ala_3\text{-}\varepsilon\text{-}Ahx)_{22}$-albumin.

tilis. This selective binding was in excellent agreement with our previous data (4,13) and corresponded to the peptidoglycan structures of strains investigated (11).

Immunoelectron Microscopic Controls and Exclusion of Non-Specific Binding

Cells not incubated with specific antibodies missed any labelling (8), and incubation of cells with ferritin conjugate alone did not result in any labelling either (8). Non-specific binding of rabbit IgG to possible Fc-binding factor of Str. pyogenes A-variant A486 (15) or traces of protein A of S. aureus 52A5 was ruled out studying direct binding of ^{125}iodine labelled rabbit IgG to cells of Str. pyogenes A486 or S. aureus 52A5 (8), and could in addition be excluded from selective binding when using interpeptide bridge specific antibodies (compare Table 1).

Application of Specifically Detecting Particular Peptidoglycan Epitopes in the Electron Microscope

We have now a tool for specifically detecting particular peptidoglycan epitopes in the electron microscope. For example, pentapeptide subunits represent a precursor sequence from peptidoglycan biosynthesis and accumulate in the wall under sublethal concentrations of penicillin (16) or decrease under cycloserine (17). Recent immunoelectron microscopic studies to localize the points of attack of penicillin revealed for Streptococcus pyogenes the accumulation of pentapeptide subunits close to the septum (18). On the contrary, the rod-shaped Lactobacillus gasseri DSM 20243 (strain AM63) accumulates non-crosslinked pentapeptide subunits at the poles under sublethal concentrations of penicillin (unpublished).

Studies on growing cells of Streptococcus pyogenes revealed secretion or incorporation (?) of pentapeptide subunits near the wall band (unpublished) and studies with reverting spheroblasts are under current research.

References

1. Heymer, B., P.H. Seidl and K.H. Schleifer. 1985. Immunochemistry and biological activity of peptidoglycan. In: Immunology of the bacterial cell envelope (Stewart-Tull DES, ed). John Wiley, Chichester, U.K.

2. Seidl, P.H., P. Zwerenz, J.R. Golecki, N. Franken and K.H. Schleifer. 1986. Isolation of specific antibodies to the glycan moiety of peptidoglycan and their application in the indirect immunoferritin technique. FEMS Microbiol. Letters, in press.

3. Schleifer, K.H. and P.H. Seidl. 1974. The immunochemistry of peptidoglycan. Antibodies against a synthetic immunogen cross-reacting with peptidoglycan. Eur. J. Biochem. 43, 509-519.

4. Seidl, P.H. and K.H. Schleifer. 1978. Specific antibodies to the N-termini of the interpeptide bridges of peptidoglycan. Arch. Microbiol. 118, 185-192.

5. Heymer, B., D. Bernstein, K.H. Schleifer and R.M. Krause. 1975. A radioactive hapten-binding assay for measuring antibodies to the pentapeptide determinant of peptidoglycan. J. Immunol. 114, 1191-1196.

6. Seidl, P.H. and K.H. Schleifer. 1985. Secretion of fragments from bacterial cell wall peptidoglycan. In: Environmental regulation of microbial metabolism (Kualev IS, Severin AT, Dawes EA, eds). Academic Press, London, 443-450.

7. Franken, N., P.H. Seidl, E. Zauner, H.J. Kolb, K.H. Schleifer, L. Weiss. 1985. Quantitative determination of human IgG antibodies to the peptide subunit determinant of peptidoglycan by an enzyme-linked immunosorbent assay. Mol. Immunol. 22, 573-579.

8. Seidl, P.H., J.R. Golecki, N. Franken and K.H. Schleifer. 1985. Immunoelectron microscopic studies on the localization of peptidoglycan peptide subunit pentapeptides in bacterial cell walls. Arch. Microbiol. 142, 121-127.

9. Schleifer, K.H. and P.H. Seidl. 1977. Structure and immunological aspects of peptidoglycans. In: Microbiology (D. Schlesinger, ed). American Society for Microbiology, Washington, U.S.A.

10. Schleifer, K.H. and R.M. Krause. 1971. The immunochemistry of peptidoglycan. The immunodominant site of the peptide subunit and the contribution of each of the amino acids to the binding properties of the peptides. J. Biol. Chem. 246, 986-993.

11. Schleifer, K.H. and O. Kandler. 1972. Peptidoglycan types of bacterial cell walls and their taxonomic implications. Bacteriol. Rev. 36, 407-477.

12. Seidl, P.H. and K.H. Schleifer. 1979. The immunochemistry of peptidoglycan. Serological detection of a difference in a single N-terminal amino acid. Mol. Immunol. 16, 385-388.

13. Seidl, P.H. and K.H. Schleifer. 1978. Rapid test for the serological separation of staphylococci from micrococci. Appl. Environ. Microbiol. 35, 479-482.

14. Rogers, H.J., H.R. Perkins and J.B. Ward. 1980. Microbial cell walls and membranes. Chapman and Hall, London, U.K.

15. Wagner, B., M. Wagner, M. Ryc. 1983. Morphological evidence for different types of IgG-Fc receptors in group A streptococci. Zbl. Bakt. Hyg. I. Abt. Orig. A 256, 61-71.

16. Munoz, E., J.M. Ghuysen, M. Leyh-Bouille, J.F. Petit, H. Heymann, E. Bricas, P. Lefrancier. 1966. The peptide subunit N^{α}-(L-alanyl-D-isoglutaminyl)-L-lysyl-D-alanine in cell wall peptidoglycans of Staphylococcus aureus, strain Copenhagen, Micrococcus roseus R27, and Streptococcus pyogenes group A type 14. Biochemistry 5, 3748-3763.

17. Spahn, K.F. 1977. Die Wirkung von D-Cycloserin auf die Biosynthese des Mureins. Thesis. University of Munich.

18. Seidl, P.H., P. Zwerenz, J.R. Golecki and K.H. Schleifer. 1985. Streptococcus pyogenes grown under sublethal concentrations of penicillin G accumulates close to the septum pentapeptide subunits of peptidoglycan. FEMS Microbiol. Lett. 30, 325-329.

SOLUBLE PEPTIDOGLYCANS: LYMPHOCYTE-ACTIVATING BACTERIAL PRODUCTS FOUND IN MAN

Allen R. Zeiger

Department of Biochemistry, Thomas Jefferson University, Philadelphia, Pennsylvania 19107

Introduction and definition of SPG

We are all thankful to Dr. Schleifer and Dr. Seidl for their effort in organizing this Workshop. Our knowledge of bacterial cell wall polymers has progressed markedly since the First Workshop over ten years ago. One measure of the progress since then is that the topic of this review, the soluble peptidoglycans (SPG), had not yet been reported in the literature. It is important from the outset to differentiate between cell wall peptidoglycan (PG) degradation products and what I am calling SPG. SPG is defined as the high molecular weight (non-dialyzable), water-soluble PG-like products that are secreted into the medium by some bacteria when incubated in the presence of penicillin. As will be shown, SPG can be distinguished from PG-turnover products by its physiological, structural and immunological properties.

Detection of SPG

In vitro, SPG is generally produced by incubation of the bacteria in a minimal cell wall growth medium containing a beta-lactam antibiotic, the amino acids found in the PG, glucose, salts and cofactors (1), including a labeled PG-synthetic precursor. SPG is usually identified by its chromatographic immobility, its exclusion from a Sephadex G-100 gel, its amino acid and amino sugar analysis and its lysozyme sensitivity. Figure 1 shows the elution profile through Sephadex G-100 of the spent minimal cell wall growth medium from *Staphylococcus aureus* cells that had been incubated one hour at 37°C with 50 μg/ml of penicillin G (2). The label in this experiment was ^{14}C-alanine.

Biological Properties of Peptidoglycan

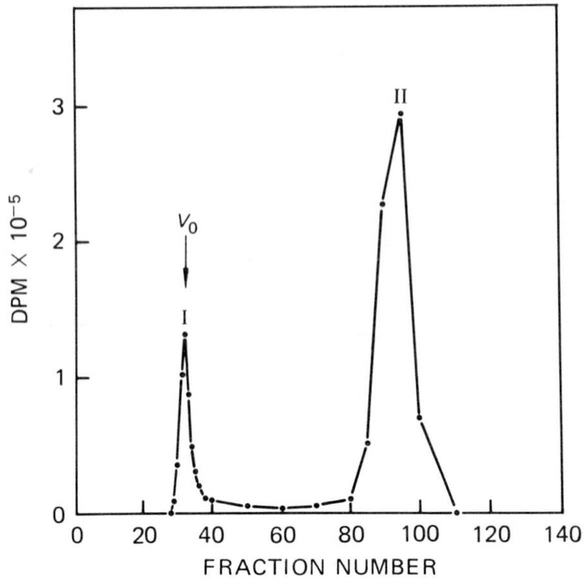

Figure 1. Fractionation through a Sephadex G-100 column of spent minimal cell wall medium (4 µCi of L-(14C)alanine) from S. aureus (2×10^8 CFU/ml) incubated with penicillin G (50 µg/ml) at 37°C for 60 min. The first peak coincides with the void volume, V_0. Elution was with distilled water.

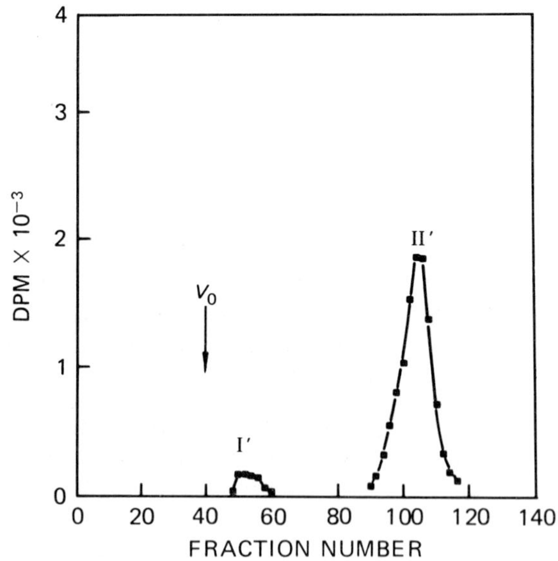

Figure 2. Fractionation through a Sephadex G-100 column of 14C-alanine-labeled SPG (peak 1 in Fig. 1) which had been digested overnight with lysozyme. Elution was with 0.15 M phosphate buffered saline (pH 7.2).

Figure 2 is an elution profile of the same SPG through the same gel following lysozyme digestion. At least 85% of the label now migrated as low molecular weight components. By virtue of its intact D-Ala-D-Ala residues, SPG has been shown to bind completely to immobilized vancomycin or to D-Ala-D-Ala specific antibodies. Figure 3 shows an elution profile of the 14C-alanine labeled SPG of S. aureus through a vancomycin-Sepharose 4B column.

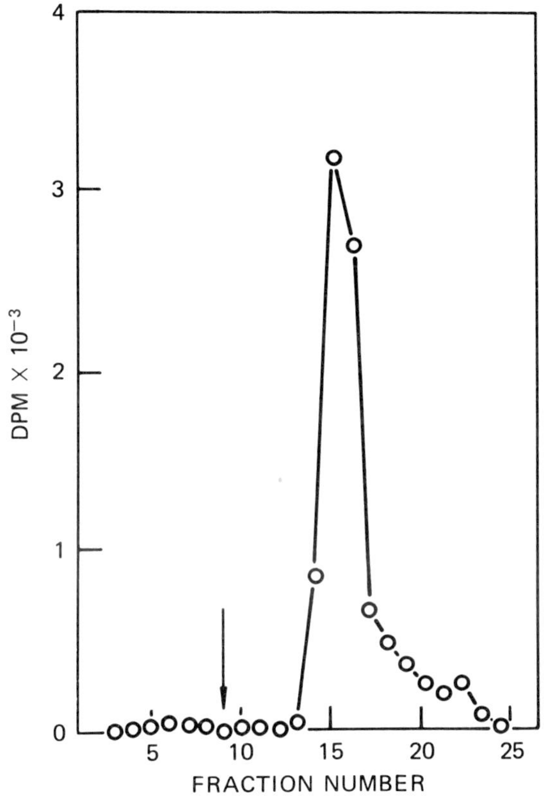

Figure. 3. Fractionation through a vancomycin-Sepharose 4B column of 14C-alanine-labeled SPG from S. aureus. The arrow represents a change from 0.15 M phosphate-buffered saline (pH 7.2) to NH_4OH (pH 10.2).

These specificities have allowed the detection of SPG *in vivo* by an enzyme-linked immunosorbent assay (ELISA) (3) whose steps are illustrated in Figure 4. Plastic wells are coated successively with vancomycin, the fluid such as urine containing an unknown amount of SPG, antibody to D-Ala-D-Ala (in this case, affinity-purified rabbit antibody to $(Glu^{60}\ Ala^{40})_n$—(Ala-γ-D-Glu-Lys-D-Ala-D-Ala)$^{5.1}$), a second antibody against immunoglobulin of the first antibody species to which biotin is covalently bound, avidin plus biotinylated horseradish peroxidase and substrate.

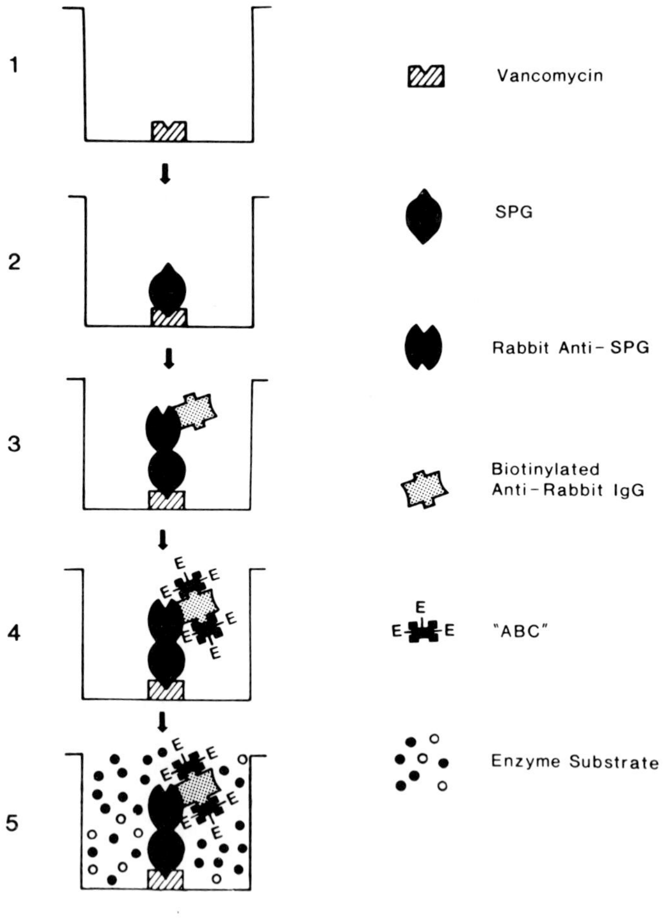

Figure. 4. Sequential steps in the ELISA to detect SPG.

Physiological properties of SPG

SPG is secreted by a variety of gram-positive bacteria when exposed to penicillin G. These bacteria include Micrococcus luteus (4), Brevibacterium divaricatum (5), Streptococcus faecium (6-7), S. aureus (2), Bacillus licheniformis (8), Bacillus subtilis (9), Streptococcus pyogenes (10), Streptococcus pneumoniae (11) and Peptococcus anaerobis (10). As yet, no detectable SPG has been reported to be secreted by any of the gram-negative bacteria. The bacterial source is one way in which PG-turnover products are clearly differentiatable from SPG. Str. faecium, for instance, has been reported to have no detectable PG turnover (12). Yet, in the presence of penicillin G, this organism has been reported capable of secreting SPG (6-7). Escherichia coli, on the other hand, has been reported to have PG turnover (13). Yet, little or no SPG is secreted into the incubation medium by these bacteria when grown in the presence of penicillin (10).

Besides penicillin G, a variety of beta-lactam antibiotics can trigger SPG secretion from S. aureus cells in approximately the same amounts. These include semisynthetic penicillin-class antibiotics like nafcillin, oxacillin, methicillin and ampicillin and cephalosporins like cephalothin and cefoxitin. One beta-lactam antibiotic that was unable to cause SPG secretion was mecillinam, which is reported to act mainly on gram-negative bacteria by a mechanism different from that of most beta-lactam antibiotics (14). Vancomycin did not trigger the secretion of SPG by M. luteus (4) or S. aureus (2) cells.

Secretion of SPG by S. aureus cells is dependent on time as shown in Figure 5 (2). Maximal synthesis of SPG in 50 µg/ml of penicillin occurred in approximately 1 hr. Maximal SPG secretion has also been shown to be dependent on the density of bacteria although not very dependent on the penicillin G concentration (10-200 µg/ml) (2).

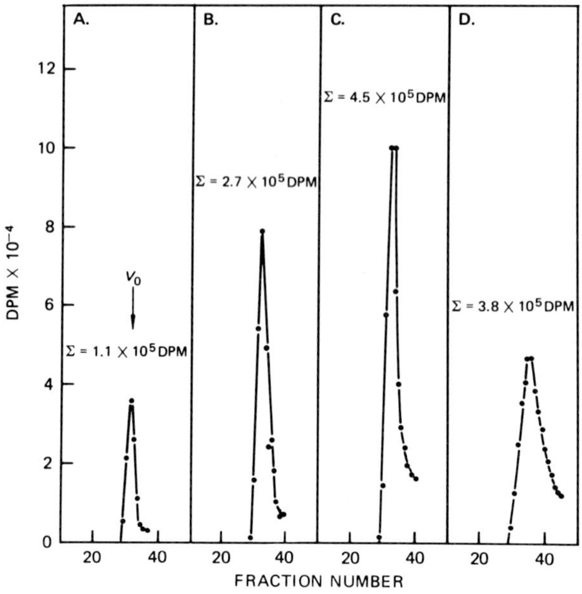

Figure. 5. Fractionation through a Sephadex G-100 column of equal portions of spent minimal cell wall medium (4 µCi of L-(14C)-alanine) from S. aureus ($2x10^8$ CFU/ml) incubated with penicillin G (50 µg/ml) at 37°C for different times. Only the initial peak (peak 1) from each incubation is shown. Panel A, 15 min; Panel B, 30 min; panel C, 60 min; panel D, 120 min. The V_0 is indicated in panel A. The total radioactivity recovered in each peak is shown in each panel. Elution was with distilled water.

Studies with S. aureus (2), B. subtilis (9) and Str. pneumoniae (11) have shown that SPG is synthesized de novo following penicillin G treatment. In one experiment, S. aureus was grown for at least four generations in the presence of L-14C-alanine, filtered and incubated in minimal cell wall growth medium containing 3H-labeled alanine and penicillin G. Only the 3H-label was incorporated into SPG over a 120 minute period with kinetics of synthesis similar to that in Figure 5. This provides further evidence that SPG is not a degradation product of cell wall-derived PG.

Structure of SPG

Figure 6 depicts a typical SPG. The size of SPG has been reported as 40,000 d for the product from *M. luteus* (4) and 50,000 d for that from *Str. faecium* (7). The SPG from *Br. divaricatum* (5), *M. luteus* (4), *S. aureus* (2), *B. subtilis* (9), *Str. faecium* (6) and *Str. pneumoniae* (11) have been shown to be lysozyme-sensitive, which indicates little modification of the glycan chain. At relatively high penicillin concentrations (≥10µg/ml), the peptide portion of the SPG appears to be highly, if not completely, uncross-linked (4,7,9). Shockman has evidence, however, that lowering the penicillin concentration from 50 µg/ml to 5 µg/ml results in some hydrolysis of the D-Ala-D-Ala sequence in the SPG secreted by *Str. faecium* (7).

```
- NAG - NAM - NAG - NAM - NAG - NAM - NAG - NAM - (d)
         |          (a)          |            |
        Ala                     Ala          Ala
         |                       |            |
        D-Glu - γ - DA(b,c)     etc.         etc.
         |           |
        (e)         D-Ala
                     |
                    D-Ala

            NAG = N-acetyl-D-glucosamine
            NAM = N-acetyl-D-muramic acid
            DA  = Diamino acid
```

Figure. 6. The general structure of SPG.

There are some structural differences among the SPG (a-e in Fig. 6), many of which reflect differences in the PG in the bacterial cell wall. Most SPG contain lysine at position 3 of the peptide chain, although the SPG from *B. subtilis* (9) and *Br. divaricatum* (5) contain diaminopimelic acid. Most SPG have either a free alpha-carboxyl or amide at the D-glutamic acid residue at position 2, although the SPG from *M. luteus* is peptidated with a glycine (4). The SPG from *M. luteus* (4) and *B. subtilis* (9) do not appear to have a "peptide bridge" attached to the diamino acid. However, the SPG from *S. aureus* (2) and *Str. faecium* (7) have been reported to have pentaglycine and D-isoasparagine respectively in their "peptide bridge".

The muramic acid moieties of SPG from Str. faecium appear to be fully substituted with peptide chains (7). This is in contrast to the SPG from M. luteus and B. subtilis, whose muramic acids are only about half substituted with peptide (4,9). To my knowledge, L-alanine amidases have not been described in Str. faecium, whereas these enzymes have been reported for M. luteus (15-16) and B. subtilis (17). Since the PG for M. luteus has been shown to be poorly substituted with peptide (18), it is possible that the SPG mimic their respective PG in terms of their overall peptide substitution.

The SPG from M. luteus does not appear to be either randomly substituted nor half-fully substituted, half-unsubstituted. T4 lysozyme, which requires peptide substitution for activity (19), was able to digest about 95% of the peptide into small molecular weight products, leaving about 35% of the glycan as a chromatographically immobile product (19). This would suggest that there was an almost fully unsubstituted glycan among the SPG strands. However, as seen in Figure 7, affinity chromatographic studies showed that all of the SPG from M. luteus was bound by an antibody specific for D-Ala-D-Ala sequences (20). The first change in eluant was to a relatively low concentration of tripeptide containing Lys-D-Ala-D-Ala. The second change in eluant was to a hundred-fold higher concentration of the peptide. This indicates that all of the glycan strands were at least monosubstituted with peptide and that most were multi-substituted. The arrangement of the peptide strands in the SPG and PG from M. luteus must await further study.

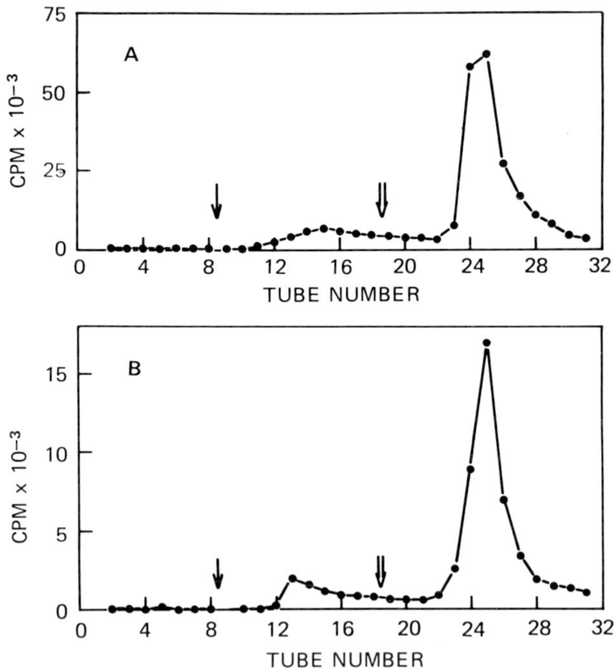

Figure 7. Affinity chromatography on purified rabbit antibodies to synthetic PG-precursor pentapeptide bound to Sepharose 4B. The top panel is an elution profile of 580,000 cpm of 14C-alanine-labeled SPG from M. luteus and the bottom panel is an elution profile of 62,000 cpm 14C-glycan-labeled SPG from M. luteus. The first arrow represents a change to 0.02 mM t-butyl-oxycarbonyl-Lys-D-Ala-D-Ala. The two pronged arrow represents a change to 2 mM of this tripeptide.

Digestion of the SPG from M. luteus with hen egg white lysozyme left 10-15% undigested material (4). The SPG from S. aureus, when produced in a minimal medium containing 3H-labeled glycerol and 14C-labeled alanine, had a teichoic acid-like molecule covalently attached (21). Even 1.5% sodium dodecyl sulfate could not separate the two portions of the molecule from an affinity column containing vancomycin. Although Barrett and Shockman originally reported that no non-SPG was attached to the SPG from Str. faecium (7), their laboratory has recently obtained SPG preparations from Str. faecium with covalently-attached rhamnose-containing saccharide (22). In investigating the biological and immunological properties of SPG, precautions must be taken to ensure that the properties belong to the SPG portion of the molecule. The review of the inflammatory properties of PG-polysaccharide complexes by Dr. Schwab should reinforce the need for this precaution (23). It remains an open field for investigation as to how the structure of SPG will influence their properties.

SPG in man

The ELISA mentioned above was used to detect SPG in the urine of man (24). Of 13 healthy volunteers that ingested 250 mg doses of penicillin V, seven had positive titers for SPG in their urines within 6 hours (test column in Table 1). The binding was completely abolished with a synthetic tripeptide, t-butyloxycarbonyl-Lys-D-Ala-D-Ala (control column). Before penicillin ingestion, none of the patients had detectable levels of SPG in their urine.

Table 1. Study of SPG in Urine Samples Six Hours after Penicillin V Treatment.

Subject Sex and Age (yr)	Absorbance at 490 nm			
	Pre-treatment		Post-treatment	
	Test	Control	Test	Control
M 41	.010	.009	.043[a]	.011
F 36	.014	.013	.167[a]	.015
F 13	.010	.010	.110[a]	.012
M 40	.009	.010	.013	.012
F 37	.010	.007	.157[a]	.013
F 50	.015	.017	.042[a]	.011
M 36	.012	.012	.009	.011
M 27	.010	.013	.012	.016
F 47	.011	.014	.233[a]	.021
M 34	.013	.012	.013	.012
M 36	.015	.014	.013	.015
F 58	.010	.011	.028[a]	.010
M 26	.007	.009	.008	.009

[a] Considered positive.

The diversity observed among individuals in SPG detection reflects what has been found in several reports of anti-PG antibodies in humans (25-27). It is interesting to speculate that the two observations are related. Indeed, there is evidence that SPG has many of the criteria expected for a natural immunogen. Dr. Seidl has already discussed the immunodominance of the D-Ala-D-Ala portion so prevalent in SPG. In addition, as seen in Figure 8, titers of antibodies to D-Ala-D-Ala are elevated in the sera from patients with S. aureus-caused endocarditis and bacteremia who were treated with beta-lactam antibiotics but not in the sera from patients who were treated with vancomycin (27). Furthermore, there was an increase in the titers of these antibodies that correlated with antibiotic treatment, rather than with the onset of infection. Figure 9 shows the time course of antibody titers to D-Ala-D-Ala among patients with S. aureus-caused bacteremia.

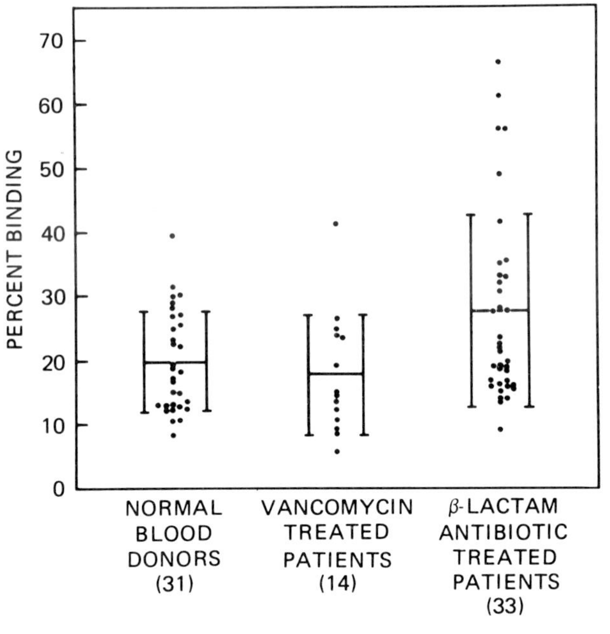

Figure 8. Percentage of binding of PG-precursor antigen by human sera. The horizontal bar represents the average percentage of binding. The vertical line shows the standard deviation.

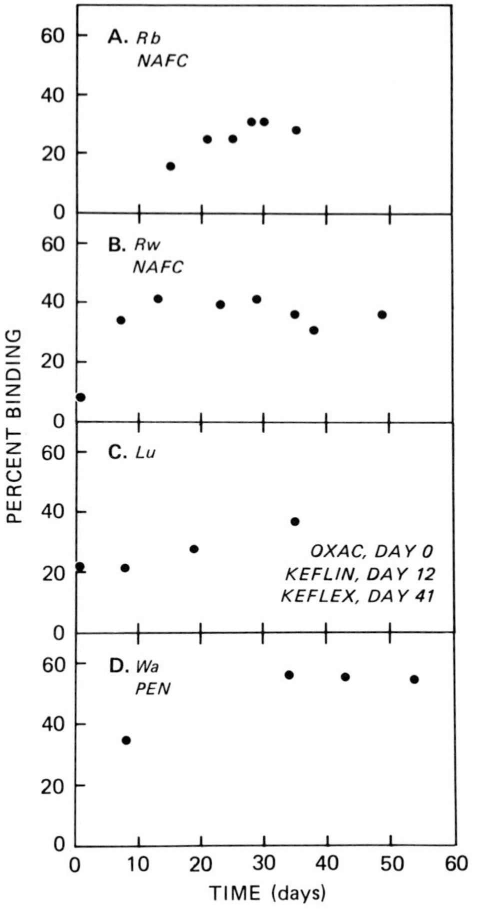

Figure 9. Percentage of binding of PG-precursor antigen by sera from four staphylococcal bacteremia patients at various times after the initiation of antibiotic therapy. NAFC is nafcillin; OXAC is oxacillin; Keflin is cephalothin and Keflex is cephalexin.

At least three possibilities can account for the diversity of the SPG found in the urine. There may be a diversity of the indigenous bacteria that can secrete SPG. Not all humans have appreciable colonization by S. aureus in their intestinal tract (28). On the other hand, virtually the entire population does have group D streptococci which have been reported able to secrete SPG (28). A variation of this possibility is that not all bacterial strains of a given organism can secrete the same amount of SPG. Such diversity in the ability to secrete SPG *in vitro* has been observed with clinical isolates of *S. aureus* (2).

The location of the SPG-secreting bacteria may be important. Six of the seven individuals with positive levels of SPG in their urine were female, suggesting that vaginal bacteria may have played a role. Alternatively, the location of the SPG-secreting bacteria is in the intestinal tract, but the passage of SPG across the mucosal barrier is dependent on its structure.

One also cannot exclude the interplay of the host. The concentration of potential SPG-degrading enzymes such as lysozyme may vary from host to host or perhaps the macrophage may be more or less active in opsonization or enzyme secretion. These possibilities need to be investigated further.

Antibodies cross-reactive with SPG in man

The possibilities for the diversity of the antibody response to PG include the diversity of SPG for reasons mentioned above, genetic control of the immune response in the host to PG, and tolerance resulting from early exposure to SPG-like immunogens. At the first Workshop on Peptidoglycans, Dr. Krause noted the importance of investigating the possible genetic control of the immune response to PG-precursor peptides (29). I have begun to examine this question with synthetic PG-like immunogens containing the PG-precursor pentapeptide covalently bound to the random sequence carriers, poly (Glu^{60} Ala^{40}) and poly (Glu^{90} Tyr^{10}). The former random polymer is immunogenic in mice of the H2-b and d haplotypes, but not in mice of the H2-q haplotype (30). The latter random polymer is not immunogenic in any inbred mice (30). Table 2 shows the humoral response in H2-b mice to the PG-precursor on the poly (Glu^{60} Ala^{40}) polymer. Low titers of antibody to the PG-precursor peptide were obtained. The higher titers obtained against the random polymer correlated with that found for the humoral response to the immunogenic random polymer. No increase in titers was observed in the secondary bleeds.

Table 2. Immunization of C57BL/6 Mice (H-2b) with $(Glu^{60}\ Ala^{40})_n$ —(Ala-γ-D-Glu-Lys-D-Ala-D-Ala)$^{5.1}$.

Dose (μg)	Bleed	Anti-pentapeptide (%)	Anti-carrier (%)
100	primary	22.6 + 3.6	56.9 + 4.7
1	primary	14.6 + 8.0	24.5 + 14.0
100	secondary	19.2 + 5.5	60.8 + 8.2
1	secondary	7.5 + 6.3	33.8

The results with the same immunogen in H2-d inbred mice is shown in Table 3. Higher titers against the PG-precursor peptide were observed in these H2-d mice at 100 μg and 1 μg doses than in H2-b mice. The humoral response to the random polymer portion of the immunogen was again equivalent to that of the random polymer immunogen alone. Secondary responses to the PG-precursor portion especially at the 1 μg dose appeared to be higher than those of the primary bleedings. The H2-q mice did not mount an immune response to the PG-precursor portion of the immunogen. This was similar to what has been reported for the immunogenicity of the random polypeptide alone.

Table 3. Immunization of BALB/c Mice (H2-d) with $(Glu^{60}\ Ala^{40})_n$ —(Ala-γ-D-Glu-Lys-D-Ala-D-Ala)$^{5.1}$.

Dose (μg)	Bleed	Anti-pentapeptide (%)	Anti-carrier (%)
100	primary	36.5 + 10.9	69.5 + 4.5
1	primary	26.4 + 12.3	63.6 + 3.0
100	secondary	47.8 + 14.3	74.9 + 3.4
1	secondary	53.2 + 11.6	69.9 + 4.4

With a 100 μg dose of the PG-precursor peptide linked to poly (Glu^{90} Tyr^{10}), no detectable antibody titers were observed in the sera of H2-b and H2-d mice. This was also similar to what had been observed for the random polymer in these mice.

It appears that the PG-precursor peptide is acting as a classic hapten to the random polypeptide carrier. By analogy with the SPG molecule, one would expect the immunogenicity to be governed by the glycan and/or any non-PG material associated with SPG.

As previously mentioned, there are reports of anti-PG titers in human sera. These reports have indicated a relationship of the titers of anti-SPG in human sera to various diseases, such as *S. aureus* infections, rheumatic fever, juvenile and adult rheumatoid arthritis, and ankylosing spondylarthritis. Thus, the serology of PG would appear to be of some medical importance. Some of the above diseases have been linked to bacterial infection, either by gram-positive or gram-negative organisms. Often, these infections are treated with beta-lactam antibiotics. Since indigenous bacteria have been shown to be quite capable of secreting SPG when exposed to beta-lactam antibiotics, it is important to know and specify the therapeutic history of the patient when correlating anti-PG titers with a particular disease. This has not been done with most previous studies that associated elevated anti-PG-titers with pathogenic bacteria. Perhaps, this review will inspire a renewal of our efforts in this field.

Biological properties of SPG

The observation that SPG can be found in humans gives an added dimension to the study of its biological properties. The spectrum of biological properties of cell wall-derived PG and MDP should stand as testimony to the potential effects of SPG. Many properties of PG are related to interactions with lymphocytes. SPG from *S. aureus* has been shown capable of interacting with murine lymphocytes *in vitro* and *in vivo* (31). Table 4 shows the ability of SPG from *S. aureus* to act as a mitogen in outbred, inbred and congenic mice.

Table 4. Mitogenicity of SPG from S. aureus in murine splenocyte cultures.

Addition	Concn (μg/ml)	BALB/c cpm	BALB/c SI[a]	Swiss Webster cpm	Swiss Webster SI	B10.M cpm	B10.M SI
None		4,569		6,142		4,888	
Con A	1	183,152	40.1	Not done		Not done	
LPS	30	111,301	24.4	134,037	21.8	19,103	3.9
SPG	100	127,518	27.9	131,978	21.5	45,419	9.3
SPG	33	177,823	38.9	119,439	19.5	25,010	5.2
SPG	11	32,488	7.1	19,254	3.1	24,003	4.9
SPG	3.8	24,447	5.4	Not done		Not done	

[a] SI, Stimulation index.

Table 5 shows that the mitogenic effect is greater at day 3 than days 1 or 2. Treatment with anti-theta serum and complement to remove T lymphocytes did not result in any loss of mitogenic activity (Table 6). The combination of SPG from S. aureus with LPS (another murine B cell mitogen) gave synergistic responses, indicating that the B cell subpopulations stimulated by the two B cell mitogens are different (Table 7).

Table 5. Kinetics of the Mitogenic Stimulation of BALB/c Splenocytes by SPG from S. aureus.

Day in culture		1		2		3	
Addition	Concn (μg/ml)	cpm	SI	cpm	SI	cpm	SI
None		3,169					
Con A	1	5,851	1.9	58,109	5.5	150,371	9.6
LPS	30	8,549	2.7	36,239	3.4	126,886	8.1
SPG	100	3,532	1.1	22,049	2.1	85,782	5.5
SPG	50	Not done		Not done		101,125	6.5
SPG	25	2,319	0.7	24,097	2.3	33,635	2.2

Table 6. Mitogenicity of SPG from *S. aureus* in T-cell Depleted Swiss Webster Cultures.

Addi-tion	Concn (µg/ml)	Complement only[a] cpm	SI	Complement + anti-θ cpm	SI
None		1,478		2,069	
Con A	1	98,000	66.3	5,522	2.7
LPS	34	12,500	8.5	75,963	36.7
SPG	25	14,674	9.9	24,108	11.7

[a] Complement only resulted in a 10% loss of cellular viability.
[b] Complement + anti-θ (anti-theta serum) resulted in a 45% loss of cellular viability.

Table 7. Effects of B Lymphocyte Mitogens and Combinations Thereof on the 3H-TdR Incorporation by Swiss Webster Splenocytes.

Addition Mitogen	None cpm	SPG (10 µg/ml) cpm	SPG (10 µg/ml) additive	LPS (10 µg/ml) cpm	LPS (10 µg/ml) additive
None	5,333				
SPG	11,030			46,096	32,821
LPS	27,124	46,096[a]	32,821		
(T-G-A-G)	20,049	26,452	25,746	57,304	23,800
Dextran sulfate	17,212	21,077	22,909	61,696	39,003
Poly I. Poly C	36,218	23,598	41,915	72,510	58,009

[a] Underlined values signify a greater than additive response.

SPG from S. aureus is also a polyclonal B cell activator (PBA). Table 8 shows the effect of SPG on the number of immunoglobulin-secreting cells. The assay utilizes protein A-coupled sheep red blood cells (SRBC) and complement. Those lymphocytes secreting immunoglobulin will cause a lysis of the SRBC nearby giving a clear plaque on the SRBC field. The data in the left hand column shows that SPG and LPS are both equivalent as PBA. SPG from S. aureus also has a stimulatory effect on the number of immunoglobulin secreting cells among immunized splenocytes as well. In the right hand column are results with cells immune to SRBC. Antibody will bind directly to the SRBC and no protein A coupling is necessary. Once again, SPG and LPS produce rather significant increases in the number of immunoglobulin secreting cells.

Table 8. Effect of SPG from S. aureus on the PFC Response of BALB/c Splenocyte Cultures.

Addition	Concn (μg/ml)	PFC/10^6 cells					
		Expt 1	Expt 2	IgM (expt 1)	IgG (expt 1)	IgM (expt 2)	IgG (expt 2)
None		0	0	1,603	52	1,081	17
LPS	30	5,250	4,000	ND[a]	ND	ND	ND
SPG	33	7,750	5,250	49,725	4,492	38,076[b]	6,263[b]
PG	400	ND	ND	50,808	7,814	73,747	17,956

[a] ND, Not determined.
[b] The concentration of SPG for this experiment was 75 μg/ml.

The proliferation of murine splenocytes and their differentiation to plasma cells by SPG from S. aureus also occurs in vivo. Table 9 shows the results of iv inoculations of Swiss Webster mice with 100 μg of SPG or LPS. After three or four days, the splenocytes were examined. SPG and LPS were equivalent in their abilities as mitogens and B cell activators, giving about fivefold stimulation compared to the control.

Table 9. *In Vivo* Proliferation and Activation of Murine B Lymphocytes with SPG from *S. aureus* and LPS.

Treat-ment	Animal	Cells/spleen ($x 10^{-6}$)	3H-TdR-incorp. (cpm/2 x 10^5 cells)	PFC/2 x 10^5 cells
	1	84	1,989	105
Control	2	77	1,147	150
	3	81	< 200	35
	1	226	3,696	750
SPG	2	120	6,956	825
	3	154	5,745	715
	1	190	4,652	660
LPS	2	210	4,672	600
	3	200	7,322	775

Recently, Gold *et al*. (31) showed that the SPG from *Str. faecium* stimulates monocyte-macrophage inflammatory functions, including elevated interleukin 1 production, elevated colony stimulating factor production and reduced plasminogen activator production. Thus, SPG appears capable of mediating inflammatory reactions as well as of specific and non-specific lymphocyte stimulation. Further studies on the physiology of SPG are necessary to evaluate the pathological and immunological consequences of short term or prolonged treatment of patients with beta-lactam antibiotics.

References

1. Chatterjee, A.N. and J.T. Park. 1980. Proc. U.S. Acad. Sci. 51, 9-16.

2. Zeiger, A.R., W. Wong, A.N. Chatterjee, F.E. Young and C.U. Tuazon. 1982. Infect. Immun. 37, 1112-1118.

3. Park, H., A.R. Zeiger and H.R. Schumacher. 1984. Infect. Immun. 43, 139-142.

4. Mirelman, D., R. Bracha and N. Sharon. 1974. Biochemistry 13, 5045-5053.

5. Keglevic, D., B. Ladesic, O. Hadzija, J. Tomasic, A. Valinger and M. Pokorny. 1974. Eur. J. Biochem. 42, 389-400.

6. Rosenthal, R.S., D. Junkind, L. Daneo-Moore and G.D. Shockman. 1975. J. Bacteriol. 125, 398.

7. Barrett, J.F. and G.D. Shockman. 1984. J. Bacteriol. 159, 511-519.

8. Tynecka, A. and J.B. Ward. 1975. Biochem. J. 146, 253-267.

9. Waxman, D.J., W. Yu and J.L. Strominger. 1980. J. Biol. Chem. 255, 11577-11587.

10. Seidl, P.H. and K.H. Schleifer. 1985. In: Environmental Regulation of Micribial Metabolism, (I.S. Kulaev, E.A. Dawes and D.W. Tempest, eds.). Academic Press, London, pp. 443-513.

11. Fischer, H. and A. Tomasz. 1984. J. Bacteriol. 157, 507-513.

12. Boothby, D., L. Daneo-Moore, M.L. Higgins, J. Coyette and G.D. Shockman. 1973. J. Biol. Chem. 248, 2161-2169.

13. Goodell, E.W. and U. Schwarz. 1985. J. Bacteriol. 162, 391-397.

14. Spratt, B.G. 1977. Antimicrob. Agents Chemother. 11, 161-166.

15. Jensen, S.E. and J.N. Campbell. 1976. J. Bacteriol. 127, 319-326.

16. Monodane, T., Y. Matsushima, Y. Hirachi and S. Kotani. 1978. Microbiol. Immunol. 22, 57-66.

17. Fan, D.P. and M.M. Beckman. 1972. J. Bacteriol. 109, 1258-1265.

18. Ghuysen, J.M. 1968. Bacteriol. Rev. 32, 425-464.

19. Mirelman, D., G. Kleppe and H.B. Jensen. 1975. J. Biochem. 55, 369-373.

20. Zeiger, A.R., S.M. Eaton and D. Mirelman. 1978. Eur. J. Biochem. 86, 235-240.

21. Zeiger, A.R. and C.U. Tuazon. 1982. VIII Pan-Amer. Congr. Rheum. Abstr. A86, S83.

22. Kariyama, R., D.O. Dollinger and G.D. Shockman. 1985. Amer. Soc. Microbiol. Abstr. K85, 185.

23. Schwab, J.H. 1985. These Proceedings

24. Park, H., A.R. Zeiger and H.R. Schumacher. 1984. Infect. Immun. 43, 139-142.

25. Zeiger, A.R. 1985. These Proceedings.

26. Seidl, P.H. 1985. These Proceedings.

27. Zeiger, A.R., C.U. Tuazon and J.N. Sheagren. 1981. Infect. Immun. 33, 785-800.

28. Sommers, H.M. 1980. In: The Biologic and Clinical Basis of Infectious Diseases, (G.P. Youmans, P.Y. Patterson and H.M. Sommers, eds.). W.B. Saunders Co., Philadelphia, pp. 83-94.

29. Krause, R.M. 1975. Z Immunitatsforsch 149, 136-150.

30. Merryman, C.F. and P.H. Maurer. 1976. J. Immunol. 116, 739-742.

31. Babu, U.M. and A.R. Zeiger. 1983. Immun. 42, 1013-1016.

32. Gold, M.R., C.L. Miller and R.I. Mishell. 1985. Infect. Immun. 49, 731-741.

HOW ARE CELL-WALL COMPONENTS OF PATHOGENIC MICROORGANISMS DEGRADED IN INFECTIOUS AND INFLAMMATORY SITES?: FACTS AND MYTHS

Isaac Ginsburg

Department of Oral Biology,
Hebrew University-Hadassah Faculty of Dental Medicine,
Jerusalem, Israel.

Introduction

Although a voluminous literature exists today which describes, in great detail, the role played by "professional" phagocytes and by serum components in the killing of pathogenic bacteria *in vitro* and in vivo(1-7) very surprisingly, however, little is actually known about the fate and mode of disposal of microorganisms once they had been rendered non-viable by the defence systems of the host. It is expected that the rich arsenal of lysosomal hydrolases, including the key muralytic enzyme lysozyme (LYZ), present in leukocytes and in body fluids might be adequate to biodegrade the complex structures of the microbial cells. Paradoxically, however, the majority of bacteria are highly refractory to LYZ action. There is also some confusion in the literature concerning the distinction between bactericidal and bacteriolytic processes. It is conceivable that while a major degradation of microbial cell walls may be followed by a bactericidal reaction, the mere killing of bacteria either by oxygen radicals (2) or by complement-dependent cytlytic antibodies (7) may not necessarily be accompanied by a significant cell wall degradation. Many experimental models, with laboratory animals, have distinctly shown the persistence, for very long periods, of non-viable bacteria and of undegraded microbial cell wall components, within macrophages, in chronic inflammatory sites (8-18). Thus, one should categorically differentiate between bactericidal and bacteriolytic phenomena. It is apparent, therefore, that mammalian tissues fail, for still not fully known reasons, to biodegrade and eliminate microbial cell wall components. Peptidoglycan (PPG)-polysaccharide (PS) complexes derived from microbial cell walls possess distinct pathobiological and

Biological Properties of Peptidoglycan

and pathophysiological properties (19-21). These include the capacity to activate the complement cascade and to generate chemotactic agents, to induce fever, to activate the respiratory burst in leukocytes and to modulate the immune responses (19-24), to mention only a few of the plethora of functions ascribed to PPG. These properties may also explain the very complex interrelationships which exist between the parasite and the host during microbial infections and the possible reasons for the development of certain post-infectious sequelae, which involve the prolonged persistence of bacterial cell wall components in tissues (10-15).

Why are microbial cells resistant to biodegradation <u>in vivo</u>?

The biodegradation of microbial cells probably takes place after these had been rendered non-viable either by serum complement (7), by oxygen radicals generated by leukocytes (2) or by cationic proteins (25,26). This may be followed by an attack by proteases and nucleases which may induce plasmolysis and subsequently by PPG-degrading enzymes, which form a breach in the cell wall. This then may lead to the rupture of the protoplast due to its high osmotic pressure. Many microbial species, however, possess a variety of outer coats which shield the PPG against the action of LYZ (7,10,12,15,26,27). Furthermore, both humoral and cellular factors may also markedly affect the biodegradation of bacteria either by serum components or by leukocyte hydrolases (11-15,28,29).(Table 1)

A) Bacteria may be coated by highly negatively charged outer layers (hyaluronate, poly-D-glutamate, acid polysacchaides, teichoic acid, lipopolysaccharide-LPS, waxes and complex lipids) which deter the interaction of LYZ (a distinct cationic agent) with its PPG target (see comprehensive reviews 10,12,15,26,27).

B) Bacteria may be resistant to LYZ action because their cell walls may contain 0-acetyl groups (30,9C), free amino acids in the peptide portion of the wall (31), a high degree of cross linking (32) and a high content of N-unacetylated glucosamine in the PPG (33).

C) Inflammatory exudates which develop in infectious sites may contain large amounts of anionic polyelectrolytes (e.g. sulfated mucopolysaccharides, DNA, hyaluronate) which may bind LYZ prior to its interaction with the bacterial walls. Furthermore, anionic polyelectrolytes may strongly bind to bacterial surfaces to block the interaction of LYZ with the cells (28). Anionic polyelectrolytes may directly interact with and inhibit certain lysosomal enzymes of leukocytes (34, Also see Reviews 12,15).

D) Gram positive bacteria may be resistant to PPG-PS degradation because their autolytic wall enzymes may be tightly regulated by endogenous agents like lipoteichoic acid (35,36). Furthermore, certain sulfated polysaccharides may interfere, in a still unknown way, with the activation or activity of autolysins which can be triggered by cationic proteins (28,37,38) (See below). We ascribe a great importance to the balance which may exist in inflamed tissues between cationic polyelectrolytes (activators of autolytic wall enzymes) and anionic polyelectrolytes which may function as modulators of autolysis (24,28,29,37,38) as this may affect microbial degradation in vivo.

E) Proteases of leukocyte origin (both acid and neutral) and proteases of tissue and serum origin (plasmin) may destroy autolytic wall enzymes of certain microbial species, and thus retard their biodegradation (39).

F) An overwhelming phagocytosis of bacteria may "exhaust" the lysosomal arsenal and retard the intracellular degradation of bacteria (12,15). This may also lead to a "storage" phenomenon, and to excessive exocytosis of lysosomal enzymes which may cause tissue damage (12,15).

G) Immunoglobulins which markedly enhanced phagocytosis of pathogenic bacteria may paradoxically delay their biodegradation within phagocytes (11,12,15,40). Similarly cationic polyelectrolytes may aggregate bacteria (41) and interfere with the accessibility of LYZ to its targets.

H) Microbial toxins (hemolysins, coagulase) may retard microbial killing and degradation by inducing degranulation of lysosomes or by coating the bacterial surfaces (5,13).

I) Certain intracellular bacteria and parasites may inhibit the fusion of lysosomes with phagosomes (42,43) and thus facilitate the survival of these agents within the otherwise hostile environment of the phagolysosome (see Review 44).

J) Certain lysomortopic agents and drugs may interfere with the normal cellular function of macrophages to retard microbial degradation (15).

K) The formation of granulomas by certain microorganisms may retard the penetration of enzymes into such tight structures (10).

L) Certain antibiotics (of the beta lactam series) which are known to interfere with biosynthesis of the cell wall, may induce the formation of large aggregates or long filaments due to a faulty separation of the daughter cells (36,45). Such aggregated bacteria may become more resistant to LYZ action either because of the formation of a larger bulk of PPG or because LYZ fails to infiltrate in between the clumped cells. (Ginsburg et al, 1985 In:Symposium on the Influence of Antibiotics on the Host-Parasite Relationship II. Adam, Hahn, Opferkuch, ed Springer Verlag. Heidelberg. P. 225.) Alternatively penicillin-treated cells have been shown paradoxically to have suppressed autolytic wall enzymes which may retard autodegradation (45A). Bacteria, which had been cultivated in the presence of chloramphenicol have been shown to synthesize excess of peptidoglycan and to undergo a much slower degradation by lysozyme (46).

M) A variety of drugs (steroids, non-steroidal anti-inflammatory drugs, anti-tumor agents) may interfere with microbial degradation by affecting chemotaxis, phagocytosis, antibody synthesis, intracellular killing, the activation of the respiratory burst (47).

N) Defects in leukocyte functions affecting chemotaxis, phagocytosis and bacteria killing may also indirectly affect the elimination of microbial cell wall components from tissues (5,48,49).

The reader is referred to a series of comprehensive reviews on these subjects, which had appeared in the literature in the last few years (1,5-7,10-18,22,24,26,29,43,44,47-50).

Role of autolytic wall enzymes in microbial degradation

Since most microbial species are extremely resistant to degradation either by lysosomal hydrolases or by serum factors, one may postulate that endogenous microbial autolytic wall enzyme may also participate in bacterial elimination from tissues. The autolytic wall enzymes of bacteria are thought to regulate normal microbial cell division (51-53). In staphylococci, at least, amidases, peptidases, muramidases and N-acetylglucosaminidases may be involved (54). Lipoteichoic acid and Forsmann-like substances (phospholipids) have been implicated as the naturally occurring regulators of autolytic wall enzymes in Gram positive bacteria (35). Less is known about the nature of regulators of autolysins in Gram negative bacteria (36). Any agent which will modulate such regulators may lead to uncontrolled digestion, by autolysins, of the bacterial own wall components, leading to the generation of peptidoglycan fragments, some of which have been shown to possess distinct biological properties and to affect host and parasite interrelationships (19-21).

Spontaneous autolysis of Pneumococci, Bacilli and Streptococcus faecalis may occur under laboratory conditions, due to changes in growth rates, and to lack of certain essential nutrients (36,47, 52). Induced autolysis can, however, be achieved by actively removing lipoteichoic acid (35,55,56). This can occur either during growth in the presence of penicillin (57) or following treatment by certain cationic polyelectrolytes (58). Lysis of very resistant microorganisms (e.g. streptococci, myobacteria) can also be achieved, under laboratory conditions, either by altering the ionic milieu, by supplementing lysozyme, proteases, lysolecithin and detergents (12,15,59-65), by treatment with muramidases derived from bacteriophage-infected group C streptococci (96) or by muramidases from Streptomyces species (97).

Employing Staphylococcus aureus which had been internally labeled with ^{14}C-N-acetyl glucosamine (NAGA) and electron microscopic techniques to study ultrastructural alteration in the bacterial cells, we found that a series of cationic agents (histone, poly-L-arginine, spermine, spermidine, polymyxin B, phospholipase A_2, pancreatic ribonuclease, cytochrome C, leukocyte cationic proteins) induced autolysis in the bacterial cells at pH 5.0 (28,29,37,38,58, 63,64,66,67). A breach appeared near to the cross wall, which lead to the pealing off of the whole rigid structure of the staphylococci but left apparently intact but coagulated protoplasts (37). Incubation at pH 7.4 also leads to the complete rupture of the protoplast.

Both the spontaneous and the induced cell wall lysis can be totally inhibited by a variety of anionic polyelectrolytes (28,64,66). The most potent inhibitors of autolysis found by us were polyanethole sulfonate (liquoid), dextran sulfate and heparin. Bacteria which had been cultivated in the presence of such anionic agents failed to undergo either spontaneous or cationic poly-electrolyte-induced lysis (28). Staphylococci, which had been grown in the presence of liquoid showed normal rates of incorporation of ^{14}C-NAGA into the walls. Paradoxically, however, no radiolabeled turnover products were released from such cells following cultivation in non-radioactive medium (To be published). The findings suggested that liquoid interfered, by a still unknown manner, with metabolic pathways connected with the degradation of the cell walls.

Electron microscopical analysis of liquoid-grown staphylococci revealed a defect in cell separation, similar to that induced by penicillin (36). Since penicillin at certain critical concentrations induced bacteriolysis (56), and since lipoteichoic acid is suggested to inhibit or control lysis (35,56,66,67) it was postulated that penicillin might have acted by promoting the activation of autolysins, which were responsible for the degradation of the microbial cells (38). We, therefore, postulated that liquoid might also inhibit bacteriolysis, which had been induced by penicillin. Indeed, a marked inhibition of cell wall lysis occurred when liquoid was added to the growth medium together with penicillin (38). It, thus, appears that autolytic wall enzymes may, under certain conditions, also aid in the biodegradation of microbial cell wall components.

Role of microbial degradation products in tissue damage. Many investigators have analyzed the role played by a variety of microbial cell wall components in the pathogenesis of inflammation. Live or heat-killed bacteria, as well as fragments derived from viable bacteria by mechanical breakage (sonication, disruption with glass beads), have been injected via different routes to laboratory animals (8-21,29,66,69,70). Some of the microbial cell wall components have been shown to be translocated, via macrophages, to remote tissue sites, which might perhaps explain the mechanisms of "dissemination" of chronic inflammatory lesions (11,15,71-73). Very prolonged inflammatory responses developed apparently as a result of the inability of the tissues of the host to biodegrade and eliminate the highly phlogistic cell-wall components of a variety of microorganisms.

In some cases a very long lag period occurred between the injection of certain viable bacteria and the development of post-infectious sequelae, pointing to the very slow biodegradation of cell wall components in vivo (10,16-19,69,70,74,75). Several investigations have also pointed out that the size of bacterial cell wall fragments, which had been generated either by mechanical breakage (76,77) or following spontaneous autolysis (29) determined their phlogistic effects in vivo. Employing group A streptococci, it was found that fragments of PPG-PS complexes with a molecular weight greater than 5.3×10^6 daltons induced both acute and chronic arthritis. Similarly, fragments derived from staphylococci with a molecular weight $>10^6$ daltons, induced very severe chronic arthritis in the rat knee joint, but smaller fragments (10^4-10^5) were ineffective (29) (see below).

Activation of both macrophages and neutrophils, in culture, by microbial cell wall components, was monitored by the capacity of the cells to secrete tissue-damaging lysosomal enzymes (plasminogen activators, cathepsins, elastases, collagenases, etc.) (16,78) and oxygen radicals (79) also thought to participate in tissue destruction (1,2,3).

Muramyl dipeptide, a structural analogue of microbial cell wall components, has been extensively employed to modulate the immune responses (23,80,81), the oxygen metabolism of leukocytes (82) and the growth of certain tumor cells (81).

Although the validity of the models, which had employed microbial cell wall components derived by mechanical breakage, is obvious and informative (8,9,20,21,74-77 and reviews 11-15,17), the major problem that remains still unsolved, is to discover and elucidate the nature of the real biological counterpart of mechanical breakage, namely, what are the chemical and biological properties of those fragments, with distinct pathobiological effects, which are generated when bacterial cells proliferate in the tissues and are attacked by both humoral and cellular elements? Which enzymes or combination of enzymes are involved? Can one enhance the biodegradation of bacteria in vivo by releasing the autolytic wall enzymes from their endogenous inhibitors (24,60,61)? Can one employ exogenous enzyme "cocktails" (61,65,83-86) to enhance the biodegradation of cell-wall components in tissue leasions? Can genetic aberration which may be associated with a lack of certain key muralytic enzyme, explain the prevalence of chronic inflammatory sequelae in certain patients? and can one correct such aberrations by genetic engineering? Answers to some of these questions may greatly contribute to our understanding of the complex interrelationships between the host and the parasite.

To avoid the employment of highly artificial methods, to generate fragments of microbial cell walls with phlogistic properties, we have developed a simple model whereby radiolabeled staphylococci were allowed to autolyse in buffers. This was followed by the removal of the large undegraded remnants of the walls and the employment of the solubilized cell wall fragments to induce chemotaxis, both in vitro (87) and in vivo, and the generation of tissue lesions in the rat (29).

Gel filtration on Sephadex G-150 columns of the autolysates revealed 3 distinct peaks with molecular weights of 10^6, 10^5 and 10^4 daltons respectively. Severe arthritis was initiated in the rat knee joint following injection of cell wall materials present in the high-molecular-weight peak.

Preliminary studies have shown that the first peak contained solubilized peptidoglycan-teichoic acid complexes (88). On the other hand, a much milder arthritis was initiated after the injections of the materials eluted in the second and third peaks. These fractions contained peptidoglycan but much less teichoic acid (88). The joint lesions involved both the synovial membrane, the cartilage and the bone (Figs. 1,2) and were essentially similar to those

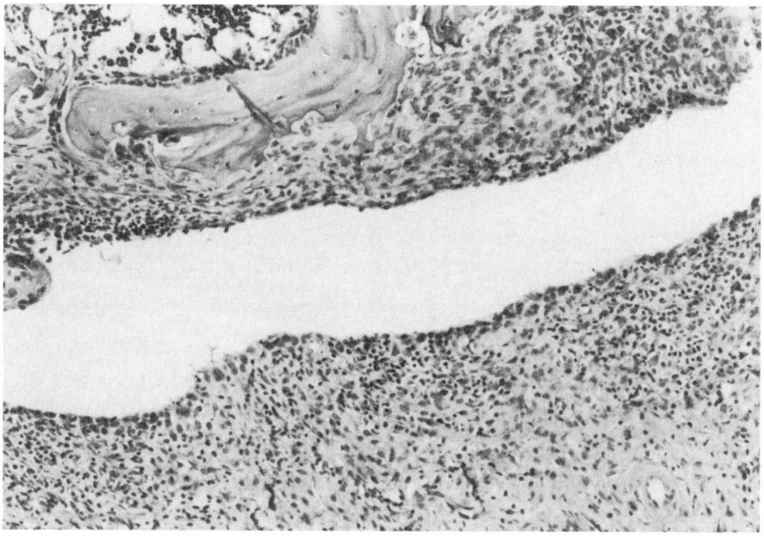

Fig. 1. A section of synovium obtained from a rat 7 days following the intra-articular injection of materials present in the first high-molecular weight peak derived by gel filtration. Note the proliferation of the synovium and the infiltration by numerous inflammatory cells and the attack of the bone by large mononuclear cells (Hematoxylin-eosin x 380).

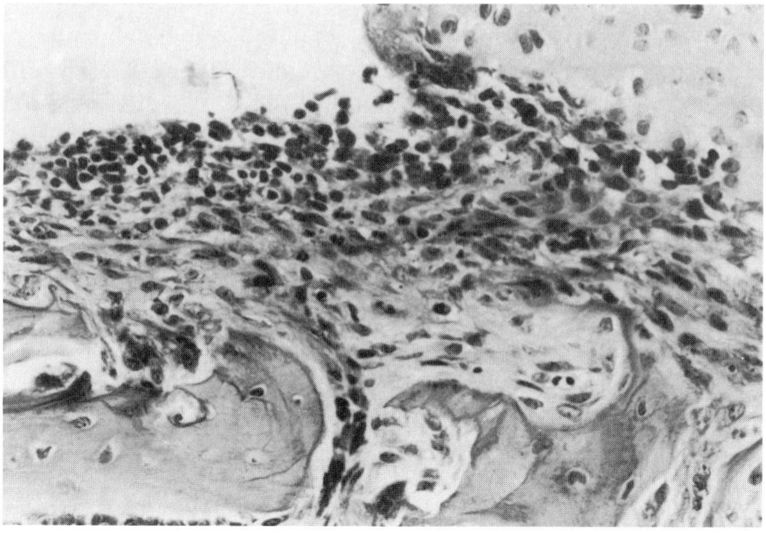

Fig. 2. A higher magnification of a section shown in Fig. 1 (Hemotoxylin-eosin x 780).

induced by cell-free preparations derived from conicated streptococci (89). The first fraction eluted, contained very potent chemotactic agents for neutrophils as determined quantitatively with a Boyden chamber (29,87). Migration of the PMNs took place both in the absence or presence of fresh serum, suggesting that the lysates contained both cytatoxins and cytotaxigens.

Examination of the chemotactic power of the lysates in vivo revealed that Swiss mice, which had been injected intraperitoneally with small amounts of the lysates developed, within 24 hours, a rich exudate which contained approximately 3×10^8 cells, 75% of which were neutrophils and 15-20% were highly vacuolated macrophages. Unlike human blood neutrophils, the mouse cells generated only a very small chemiluminescence response when stimulated with opsonized streptococci, suggesting that the cells were deactivated (Ginsburg to be published). Animals sacrificed 72-96 hours following the injection of the lysates contained a peritoneal exudate rich in macrophages (5×10^7). These were highly phagocytic for streptococci and produced large amounts of superoxide. We have also shown that rats injected intraarticularly with viable staphylococci developed severe arthritis (90,91). Apparently cell walls could be found within large macrophages in the inflamed synovium for long periods (Fig 3). Even penicillin-grown bacterial, which theoretically should have been degraded much faster than untreated cells (60,61) persisted for long periods within macrophages in the inflamed sites (91). Figure 4 shows a section of the synovium of a rat 2 months following the intraarticular injection of viable group A streptococci showing the persistence of cell wall components within a large macrophages. (See 11-19).

Thus, while staphylococcal cell walls could be degraded under in vitro conditions (28), the degradation in vivo is much slower. These simple in vivo models may shed more light on the phlogistic nature of degradation products of bacteria generated under more "physiological" conditions.

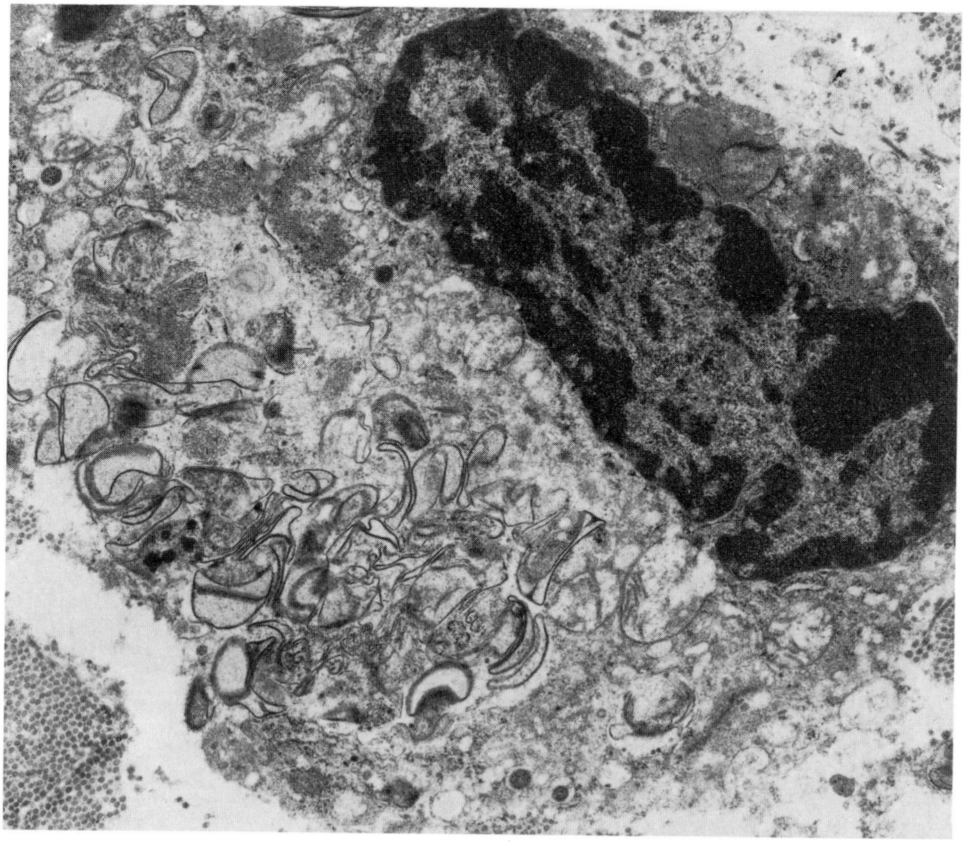

Fig. 3. A section of the rat synovium 5 days following the intra-articular injection of viable Staph.aureus. Note the presence of numerous cell wall structures within a large macrophage (E.M. x 23,500).

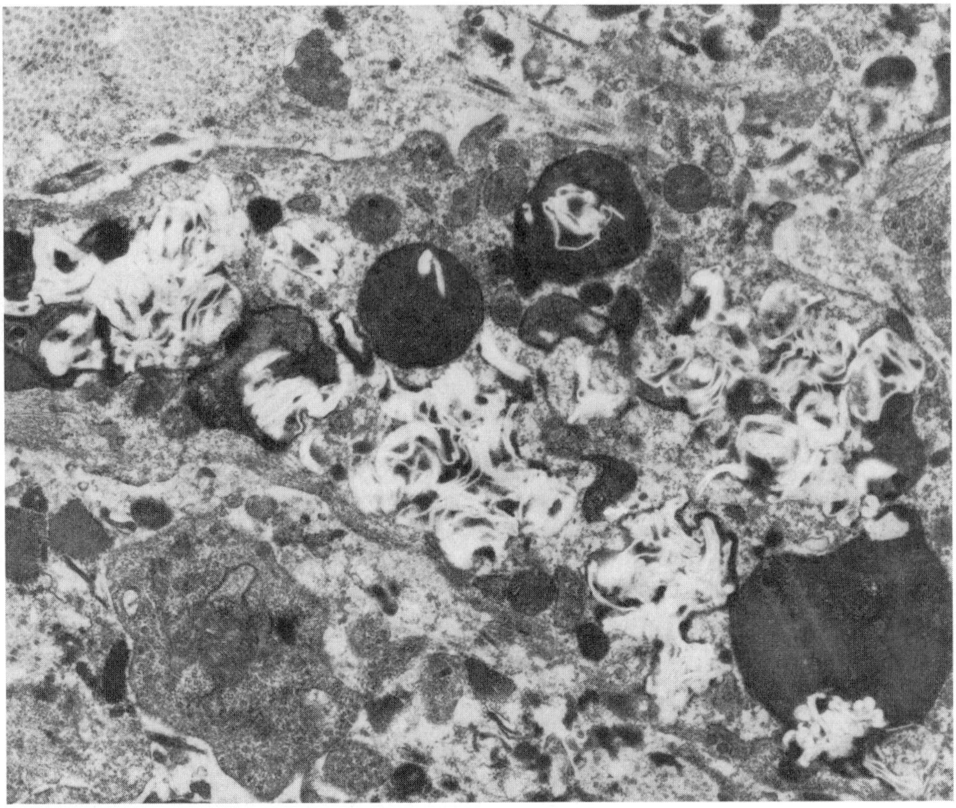

Fig. 4. A section of a rat synovium 64 days following the intra-articular injection of viable group A streptococci. Note the presence of apparently intact cell walls in a macrophage in the inflammatory site (E.M. x 20,235).

Comments

The present communication is a brief overview on the cell and tissue factors, which participate in the biodegradation of microbial cells. Since animals, which recover from bacterial infection, possess circulating antibodies to both peptidoglycan and polysaccharide portions of the cell walls (including antibodies to teichoic acid, lipoteichoic acid and LPS) (92-94) it is postulated that some degree of degradation of the cell walls of bacteria does take place in vivo. On the other hand, the prolonged persistence of both whole cell walls and fragments are demonstrated in many laboratory models.

If enzymatic degradation of microbial cell wall components in tissues fails, for any of the reasons discussed above, can we then postulate that the body of mammalians eliminates microbial components simply by removal by the traffic of phagocytes, which leave the inflamed sites? If so, where do these leukocytes end? If we consider such a possibility, and there are several experimental models which support this contention (See reviews 12,15), we may also have to take into consideration the existence in sites of chronic inflammation (mostly in granulomas) of several obstacles to such one-way traffic of leukocytes. Macrophages laden with microbial cell-wall components, at the deepest sites of the inflamed bed, may be trapped because of the local generation, by activated lymphocytes, of migration inhibition factor (MIF). Furthermore, immune complexes, comprising of PPG antibody and complement, may be trapped either along endothelial cells or under basement membranes where they could generate chemotactic agents for long periods. As long as chemotactic factors and lymphokines will continue to be generated in the lesions, the chances that cell wall components will be readily eliminated from the lesions, are scant. Thus, the question of how microbial cell wall components are disposed of, remains enigmatic, and further research in this controversial field of research is warranted (95).

Table 1

Factors Which May Interfere With the Biodegradation of Pathogenic Bacteria *In Vivo*

A) Microbial Factors

Outer coats	hyaluronate
	acid polysaccharides
	teichoic acid
	poly D glutamate
	lipopolysaccharides
	slime layers
Cell wall structures	
	degree of O-acetylation
	degree of cross linking
	content of free amino acids
	content of N-acetylglucosamine
	deficiency of autolytic wall enzymes
Regulators of autolysins	
	lipoteichoic acid
	phospolipids

B) Host Factors

Serum agents	Cellular agents	
proteinases	Defects in:	chemotaxis
proteinase inhibitors		complement components
lysosomal enzymes inhibitors		
anionic polyelectrolytes		phagocytosis
cationic polyelectrolytes		killing
immunoglobulins		oxygen radicals
fibrin	Overwhelming	phagocytosis
fibronectin		

C) Drugs

steroids
non-steroidal anti-inflammatory agents
antitumor agents
antibiotics

D) Toxins

streptolysins
staphylolysins
coagulase
phospholipases

Inhibitors of lysosome-phagosome fusion
phosphatides
sulfatides

References

1. Klebanoff,S., R.A. Clark. 1978. The neutrophil, function and clinical disorders. Elsvier, New York.

2. Klebanoff , S.P. 1982. In: Advances in Host Defence Mechanisms Vol.I. (J.I. Gallin, A.S. Fauci, eds.) Raven Press, New York p. 111.

3. Babior, B., J.T. Curnutte and B. McMurrich. 1976. J.Clin.Invest. 58, 989.

4. Walters, M.N. and J.M. Papadimitriou. 1978. CRC Critic. Rev. Microbiol. 8, 377.

5. Densen,P., G.L. Mandell. 1980. Rev.Infect.Dis. 2:817.

6. Root, A.K., M.D. Cohen. 1981 Rev.Infect.Dis. 3, 565.

7. Taylor, P.W. 1983. Microbiol. Rev. 47, 46.

8. Schwab, J.H. 1965. J.Bact.90, 1405.

9. Schwab, J.H., W.J. Cromartie, S.M. Ohanian, J.E. Craddock. 1967. J.Bact. 94, 1728.

10. Dannenberg, A.M. 1968. Bacteriol.Rev. 32, 85.

11. Ginsburg, I., S. Mitrani, N. Neeman, M. Lahav. 1975. In: 'Mononuclear Phagocytes in Immunity, Infection and Pathology. (R.Van Furth, ed.) Blackwell Scientific, Oxford, p. 981.

12. Ginsburg, I., M.N. Sela. 1976. CRC Critic.Rev.Microbiol. 4,249

13. Ginsburg, I. 1972. J.Infect.Dis. 126, 294 and 419-456.

14. Ginsburg, I, U. Zor, Y. Floman. 1977. In: Bayer-Symposium VI. Experimental Models of Chronic Inflammatory Diseases. Springer Verlag, Berlin 256.

15. Ginsburg, I. 1979. In: Lysosomes in Applied Biology and Therapeutics. Vol.6, (J.T. Dingle, P.J. Jacques and H. Shaw, eds). North Holland, Amsterdam, p. 327.

16. Page, R.C., P. Davies, A.C. Allison. 1974. Lab.Invest. 30, 568

17. Cromartie, W.J. 1981. In: Arthritis Models and Mechanisms. (H. Deicher and L.C. Schultz, eds) Springer Verlag, Berlin. p. 24.

18. Heymer, B., R. Spanel, O. Hafferkamp. 1982. In: Current Topics in Pathology. (C.L. Berry, ed) Springer Verlag, Berlin. p. 123

19. Schleifer, K.H., B. Heymer. 1975. Z. Immun.-Forsch. 149, 103.

20. Koga, T., S. Kotani, T. Narita, C.M. Pearson. 1976. Int.Archs. Allergy Appl.Immun. 51, 206.

21. Kohashi, O., C.M. Pearson, Y. Watanube, S. Kotani , T. Koga. 1976. J.Immunol. 116, 1635.

22. Schwab, J.H. 1977. In: Microbiology, 1977. (D. Schlesinger, ed) Amer.Soc.Microbiol., p. 366.

23. Chedid, L., F. Audibert. 1984. J.Immunology and Immunopharmacology. 4, 74.

24. Ginsburg, I., M. Lahav. 1983. Clin.Immunol. Newsletter. 4, 147.

25. Zeya, H.I., J.K. Spitznagel. 1966. 91, 755.

26. Elsbach, P. 1980. Rev.Infect.Dis. 2, 106.

27. Rogers, H.L., H.R. Perkins, J.B. Ward. 1980. Microbial cell walls and membranes. Chapman and Hall. London, p. 437.

28. Ginsburg, I., M. Lahav, P. Giesbrecht. 1982. Inflamation. 6,269

29. Ginsburg, I., J. Goultchin, A. Stabholtz, N. Neeman, M. Lahav, L. Landstrom, P. Quie. 1980. In: Trends in Inflamation Research I. Agents and Actions, Vol.7 (Velo, G.P., ed) Birkhauser Verglad, Basel, p. 260.

30. Brumfitt, W., A.C. Wardlaw, J.T. Park. 1958. Nature 181, 1783.

31. Perkins, H.R. 1965. Biochem. J. 95, 876.

32. Strominger, J.L., J.M. Ghuysen. 1968. Science 156, 213.

33. Hayashi, H., Y. Araki, E. Ito. 1973. J.Bacteriol. 113, 592.

34. Avilla, J., J. Convit. 1975. Biochem.J. 152, 57.

35. Holtje, J.V., A. Tomasz. 1975. Proc.Nat.Acad.Sci.U.S.A. 72,1690

36. Hakenbeck, R., J.V. Holtje, H.Labischinski (eds) 1983. The Target of Penicillin. The murein sacculus of bacterial cell walls architecture and growth. Walter de Gruyter & Co. Berlin, New York.

37. Wecke, J., M. Lahav, I. Ginsburg, P. Giesbrecht. 1982. Arch. Microbiol. 131, 116.

38. Wecke, J., M. Lahav, I. Ginsburg, E. Kwa, P. Giesbrecht. 1983. In: Target of Penicillin (R. Hakenbeck, J. Holtje and H. Labischinski, eds.) Walter de Gruyter, Berlin-New York,pp.329-334.

39. Monodane, T., Y. Matsushima, Y. Hirachi, S. Kotani. 1978. Microbiol.Immunol. 22, 57.

40. Cohn, Z.A. 1963. J.Exp.Med. 117, 43.

41. Sela, M., E. Katchalski. 1959. Adv.Prot.Chem. 14, 391.

42. Jones, T.C., J.G. Hirsch. 1972.J.Exp.Med. 136, 1173.

43. Goren, M.B., P. Darcy Hart, M.R. Young, J.A. Armstrong. 1976. Proc.Natl.Acad.Sci. U.S.A. 73, 2510.

44. Edelson, P.J. 1982. Rev.Infect.Dis. 4, 124.

45. Lorian, V., B. Atakinson. 1984. J.Infect.Dis. 149, 719.

45A. Reinicke, B., P. Blümel, H. Labischinski, P. Giesbrecht. 1985. Arch.Microbiol. In Press.

46. Reinicke, B., P. Blümel, P. Giesbrecht. 1983. Arch.Microbiol. 135, 120.

47. Mandell, L.A. 1982. Rev.Infect.Dis. 4, 683.

48. Gallin, J.I. 1981. Rev.Infect.Dis. 3, 1196.

49. Johnston, R.B. 1980. In: The Reticuloendothelial System Vol. 2, (A. Sbarra and R. Strauss, eds.) Plenum Publishing Corp. p.397.

50. Ginsburg, I. 1977. Rheum.Rehabil. 16, 141.

51. Tipper, D.J. 1969. J.Bacteriol. 97, 837.

52. (Rogers, H.L., H.R. Perkins and J.B. Ward, eds.) 1980. In: Microbial Cell Walls and Membranes. Chapman & Hall, London. p. 437

53. Giesbrecht, P., J. Wecke. 1980. In: Electron Microscopy. Seventh European Congress of Electron Microscopy. (P. Brederoo and W. De Priester, eds.) Leiden. p. 446.

54. Wadstrom, T., O. Vesterberg. 1971. Acta Pathol.Microbiol.Scand. Section B 79, 248.

55. Kessler, R.E., I. Van de Rijn, M. McCarty. 1981. In: Chemistry and Biological Activities of Bacterial Surface Amphiphiles. (Shockman, G.D. and A.J. Wicken, eds.) Academic Press, New York p. 239.

56. Tomasz, A. 1979. Ann. Rev. Microbiol. 33, 113.

57. Alkan, M.L., E.H. Beachey. 1978. J.Clin.Invest. 61, 671.

58. Sela, M.N., M. Lahav, I. Ginsburg. 1977. Inflammation 2, 151.

59. Wilkens, T.J., H. Goodman, B.J. MacKay, V.J. Iacono, J.J. Pollock. 1982. Infect.Immun. 38, 1172.

60. Warren, G.H., J. Gray. 1967. Canad.J.Microbiol. 13, 321.

61. Efrati, C., T. Sacks, N. Ne'eman, M. Lahav, I. Ginsburg. 1976. Inflammation 1, 371.

62. Lahav, M., N. Ne'eman, E. Adler, I. Ginsburg. 1974. J.Infect. Dis. 129, 528.

63. Lahav, M., N. Ne'eman, M.N. Sela, I. Ginsburg. 1979. Inflammation 3, 365.

64. Ne'eman, N., M.N. Sela, S. Chanes, L. Bierkenfeld, D. Kutani, M. Lahav, I. Ginsburg. 1979. Inflammation 3, 379.

65. Willet, H.P., H. Thacore. 1967. Canad.J.Microbiol. 13, 581.

66. Ginsburg, I., M.N. Sela, N. Ne'eman, M. Lahav. 1983. In: Staphylococci and Staphylococcal Infections. (J.H. Adlam and C.C. Easmon, eds.) Academic Press, New York. p. 325.

67. Sela, M.N., I. Ofek, M. Lahav, I. Ginsburg. 1978. Proc.Soc.Exp. Biol. & Med. 159, 126.

68. Ginsburg, I., M. Lahav, N. Ne'eman, Z. Duchan, S. Chanes, M.N. Sela. 1976. Agents and Actions 6, 292.

69. Spitznagel, J.K., K.L. Goodrum, D.J. Warejca. 1983. Amer.J. Path. 112, 37.

70. Wilder, R.L., J.B. Allen, L.M. Wahl. 1983. Arth. and Rheumatism 26, 1442.

71. Ginsburg, I., H.A. Gallis, R.M. Cole. 1969. Science 166, 1161.

72. Waksman, B.H., C.M. Pearson, J.T. Sharp. 1960. J.Immunol.85,403

73. Ginsburg, I., R. Trost. 1971. J.Infect.Dis. 123, 292.

74. Eisenberg, R., A. Fox, J.J. Greenblatt, S.K. Anderle, W.J. Cromartie, J.H. Schwab. 1982. Infect.Immun. 38, 127.

75. Cromartie, W.J., J.G. Craddock, J.H. Schwab, S.K. Anderle, C.H. Yang. 1977. J.Exp.Med. 146, 1585.

76. Fox, A., R.R. Brown, S.K. Anderle, C. Chetty, W.J. Cromartie, H. Gooder, J.H. Schwab. 1982. Infect.Immun. 35, 1003.

77. Chetty, C., D.G. Klapper, J.H. Schwab. 1982. Infect.Immun. 38, 1010.

78. Smailowicz, R., J.H. Schwab. 1977. Infect.Immun. 17, 599.

79. Leong, P.A., J.H. Schwab, M.S. Cohen. 1984. Infect.Immun. 45, 160

80. Chedid, L., F. Audibert, A.G. Johnston. Prog.Allergy 25, 63.

81. Lederer, E. 1982. Clin.Immunol.Newsletter 3, 83.

82. Kagawa, K., M. Maku, Y. Ichinose, S. Nagao, A. Tanaka, A. Tomoda. 1984. Infect.Immun. 45, 82.

83. Janusz, M.J., C. Chetty, R.A. Eisenberg, W.J. Cromartie, J.H. Schwab. 1984. J.Exp.Med. 160, 1360.

84. Roberson, B.S., J.H. Schwab, W.J. Cromartie. 1960. J.Exp.Med. 112, 751.

85. Schwab, J.H. 1962. J.Exp.Med. 116, 17.

86. Ginsburg, I. 1975. Infect.Immun. 11, 869.

87 Ginsburg, I., P.G. Quie. 1980. Inflammation 4, 301.

88. Ne'eman, N. 1981. Ph.D. Thesis, Hebrew University, Jerusalem, Israel.

89. Stein, H., R. Yarom, S. Levine, T. Dishon, I. Ginsburg. 1973. Proc.Soc.Exp.Biol. & Med. 143, 1106.

90. Ginsburg, I., M. Lahav, M. Sadovnik, J. Goultchin, J. Wecke, P. Giesbrecht. 1985. Tissue Reaction. In Press.

91. Ginsburg, I., R. Borinski, M. Sadovnik, S. Shauli, J. Wecke, P. Giesbrecht, M. Lahav. 1985. In: The Influence of Antibiotics on the Host-Parasite Relationships. II. (D. Adam, H. Hahn and W. Opferkuch, eds.)

92. Schachenmayer, W., B. Heymer, O. Haferkamp. 1975. Z.Immun-Forsch 149, S 179.

93. Verbrugh, H.A., R. Peters, M. Rozenberg-Arska, P.K. Peterson, J. Verhoef. 1981. J.Infect.Dis. 144, 1.

94. Ginsburg, I., O. Rozengarten, T. Dishon, M. Szabo, S. Bergner-Rabinowitz, I. Ofek. 1972. In: Streptococcal Diseases and the Community. (M.J. Haverkorn, ed.) Excerpta Medica, Amsterdam. p. 97.

95. Ginsburg, I., M. Lahav. 1983. Europ.J.Clin.Microbiol. 2, 186.

96. Krause, R.M. 1958. J.Exp.Med. 108, 803.

97. McCarty, M. 1952. J.Exp.Med. 96, 555.

98. Johannsen, L., H. Labischinski, P. Giesbrecht. This Symposium.

STATE OF PEPTIDOGLYCAN IN SPHEROPLASTS OF PROTEUS MIRABILIS GROWN IN THE PRESENCE OF DIFFERENT ß-LACTAM ANTIBIOTICS

Kora Huber, H.H.Martin

Institut für Mikrobiologie , Technische Hochschule ,
D-6100 Darmstadt ,Federal Republic of Germany

The Gram-negative bacterium *Proteus mirabilis* has specific properties which make it useful for the study of the antibacterial action of ß-lactam antibiotics.Like many other Gram-negative organisms , *Proteus* bacteria loose the shape and mechanical stability of their cell wall during growth in the presence of penicillin,and the cells are converted into fragile spheroplasts.However,the spheroplasts are not killed and lysed in this situation.Instead,they continue to grow and multiply in the spheroplast state.The surface of such spheroplasts is extensively disorganized as is evident from the release of vesicular outer membrane fragments and pili from the spheroplast surface.The fragility and osmotic sensitivity of the spheroplasts also indicates that the cell wall peptidoglycan must have lost its normal function as backbone of the cell wall.

The present study is concerned with the condition of this functionally defective peptidoglycan.Relevant questions are,

i. Is peptidoglycan at all present in spheroplast walls *or* is peptidoglycan synthesis completely inhibited according to the general model of action of ß-lactam antibiotics ?

ii. If peptidoglycan is present in spheroplast walls,is it macromolecular and insoluble *or* is it (water)soluble ? Are the peptide side chains of the peptidoglycan crosslinked *or* uncrosslinked because of specific inhibition of peptidoglycan transpeptidase function by the ß-lactam antibiotics ?

iii. Is peptidoglycan degraded during spheroplast growth,is it excreted into the growth medium ?

The problem has been studied previously in our laboratory with spheroplasts grown in the presence of penicillin G (1).Subsequently,it was found that viable spheroplast growth of *P.mirabilis* can also be obtained with a variety of other structurally different

Biological Properties of Peptidoglycan

ß-lactam antibiotics.For induction of spheroplast growth and further study of peptidoglycan in such spheroplasts a group of more recently developed antibiotics was selected on the basis of their high growth inhibitory activity on Gram-negative bacteria and their typical differences in binding to the seven penicillin-binding proteins of P.mirabilis (2).These included the aminothiazolcephalosporins cefotaxime and ceftizoxim,the 7-methoxicephalosporins cefoxitin and latamoxef,the monobactam aztreonam,and the carbapenem imipenem. A search for peptidoglycan in spheroplasts was carried out according to the procedure outlined in Table 1.

Table 1. Fractionation of Spheroplasts and Growth Medium of P.mirabilis for the Localization of Peptidoglycan

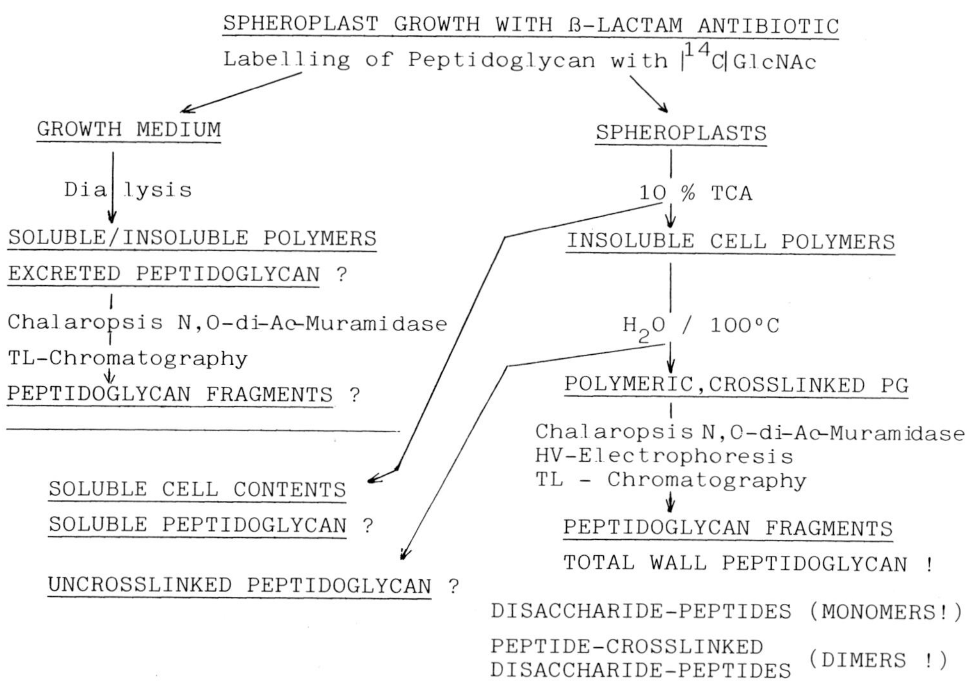

After growth with $|^{14}C|$ N-Acetylglucosamine for the radioactive labeling of peptidoglycan the harvested spheroplasts were extracted with trichloroacetic acid. Soluble peptidoglycan was a possible component of the extracted cell content.In addition,extraction of the TCA-insoluble cell polymers with boiling water could possibly bring polymeric but uncrosslinked peptidoglycan into solution. Water-insoluble,peptide-crosslinked peptidoglycan was identified after specific degradation into disaccharide-peptide fragments with the endo-N,O-diacetylmuramidase from Chalaropsis.The amount of the total radioactive fragments was a measure of total cell wall peptidoglycan.From the ratio of the peptide-crosslinked (Dimers!) and the uncrosslinked (Monomers!) disaccharide-peptides the peptide-crosslinkage of the peptidoglycan was determined (1).

A search was also carried out for peptidoglycan excreted into the growth medium in soluble or insoluble polymeric form.

No evidence for excretion of peptidoglycan during spheroplast growth was found. Similarly,none of the different ß-lactam antibiotics caused the spheroplasts to form abnormal,TCA-or water-soluble uncrosslinked peptidoglycan. However,all spheroplasts contained polymeric,crosslinked ,insoluble peptidoglycan,as shown in Table 2 , together with different affinities of the respective antibiotics for the penicillin-binding proteins (PBPs).

Table 2.Peptidoglycan in Spheroplasts of *P.mirabilis* grown in the Presence of different ß-Lactam Antibiotics

Antibiotic mg/l	Preferred Binding to PBP	Peptidoglycan per mg cell protein % of control	Peptide-crosslinkage molar ratio $\frac{\text{Dimers}}{\text{Monomers}}$
Control no Antibiot.	-	100	0.64
Cefotaxime 50	3,4,1B	84	0.59
Ceftizoxime 10	3,4,1B	46	0.72
Cefoxitin 25	3,5,6,1A, 1B	54	0.65
Latamoxef 25	3,5,6,1A 1B	34	0.66
Aztreonam 25	3,4,1A	113	0.57
Imipenem			
25	1A,1B,2,3,4	90	0.54
100	1A,1B,2,3,4, 5,6	3	1.03

With some antibiotics the amount of peptidoglycan in the spheroplasts was considerably lower than in normal control cells,with other antibiotics there was little decrease of peptidoglycan content or none at all.However,with no antibiotic was there an extensive decrease in peptide-crosslinkage.
Imipenem presented a remarkable exception as the only antibiotic where higher concentrations caused an almost complete inhibition of peptidoglycan synthesis,paralleled by the saturation of all penicillin-binding proteins.
In conclusion,the action of ß-lactam antibiotics on a typical Gram-negative bacterium leads to the loss of the specific function of the peptidoglycan as the rigid supporting structure of the cell wall, but not to the large scale inhibition of the synthesis or peptide-crosslinkage of peptidoglycan,or to the excretion of soluble peptidoglycan. The precise cause of the functional damage to the peptidoglycan by ß-lactam antibiotics is still not known in chemical terms. It may have to do with the disorganization of the macromolecular architecture of the peptidoglycan network due to the inhibition of individual,essential , but quantiatively minor peptidoglycan transpeptidase functions among the multiple penicillin-binding proteins.

Acknowledgement

This study was supported by grants from the Deutsche Forschungemeinschaft and the Fonds der Chemischen Industrie

References

1. Martin,H.H.,J.Gmeiner.1979 . Eur.J.Biochem.95, 487-495

2. Schilf,W., H.H.Martin.1980 . Eur.J.Biochem.105 , 361-370

DEGRADABILITY BY LYSOZYME OF STAPHYLOCOCCAL CELL WALLS AS A FUNCTION OF O-ACETYL GROUPS

L. Johannsen, H. Labischinski, P. Giesbrecht

Robert Koch-Institut des Bundesgesundheitamtes
Nordufer 20, D-1000 Berlin 65

Introduction

The degradability by lysozyme of bacterial cell walls is of special interest because it is well known that cell walls that cannot be degraded within the phagocytes might play a crucial role in the induction of chronic inflammatoric processes in mammalian joints. Since staphylococcal cell wall material is known to be very resistant against the action of lysozyme, we used staphylococci and staphylococcal cell walls in order to discover the factors responsible for this resistance. It has already been demonstrated that this "intrinsic" resistance was even enhanced in cell walls grown under influence of Chloramphenicol (1). The increase in resistance was accompanied by an enhanced level of O-acetyl groups (2) which are substituents located on the C-6-position of the muramic acid of cell walls. Two conditions are known to generate an elevated O-acetylation: This occurs when cells enter stationary phase of growth or when cells are treated with certain bacteriostatic antibiotics (2,3,4). This study focusses on the dependence of O-acetylation and degradability by lysozyme of staphylococcal cell walls.

Materials and Methods

In all experiments Staphylococcus aureus, strain SG 511 from the culture collection of the Robert Koch-Institut was employed. Radioactive labelling, isolation of cell walls, and determination of O-acetylation was performed as described previously (2,5).
To remove O-acetyl groups from cell walls, whole cells or their isolated cell walls were suspended in 100 µl 0.05 M NaOH at 37°C for 2 hours.
Enzymatic degradation of radioactively labelled cell walls was performed with isolated cell walls or with complete cells as

Biological Properties of Peptidoglycan

substrate using 1 mg/ml of Hen-Egg-White-Lysozyme in a 150 mmol pH 5 sodium-acetate buffer at 37 °C. After different periods of incubation the amount of radioactive cell wall material released into the buffer was separated by centrifugation at 10,000 x g for 4 min. and the radioactivity in the supernatant was measured.

Results and Discussion

Degree of O-Acetylation.

Cell wall qualities with different levels of O-acetylation were maintained by using cell walls from the logarithmic phase of growth or from the stationary phase or by removing O-acetyl groups from these cell walls (Table 1). Cell walls from the log. phase expressed a medium level of about 50 percent O-acetylation, whereas walls from the stat. phase of growth revealed an enhanced level of about 70 percent. In both cases O-acetyl groups could be completely removed by treatment with a mild alkali solution.

Degradation of Isolated Cell Walls.

Isolated cell walls with the different degrees of O-acetylation mentioned above were subjected to degradation by lysozyme. Fig. 1 shows the cell wall material released by lysozyme as a percentage of the total cell wall material which was inserted, drawn against the time. These experiments revealed some unexpected findings: Both isolated cell walls from log. phase as well as from stat. phase of growth could not be degraded by lysozyme at all. The substitution of 50 to 70 % of muramic acid with O-acetyl groups apparently seems sufficient to completely prevent the cell walls from being released by lysozyme.

Table 1: O-Acetylation of Staphylococcus aureus Cell Walls (Typical Values)

cell walls from	O-acetyl groups as nmol per mg of cell wall	O-acetyl groups as % of muramic acid
log. phase	180	50
stat. phase	240	70
after mild alkali	0	0

Fig. 1:

Degradation of Isolated Cell Walls by Lysozyme

- ◆ log. phase
- ■ log. phase, O-acetyl groups removed
- × stat. phase
- ● stat. phase, O-acetyl groups removed

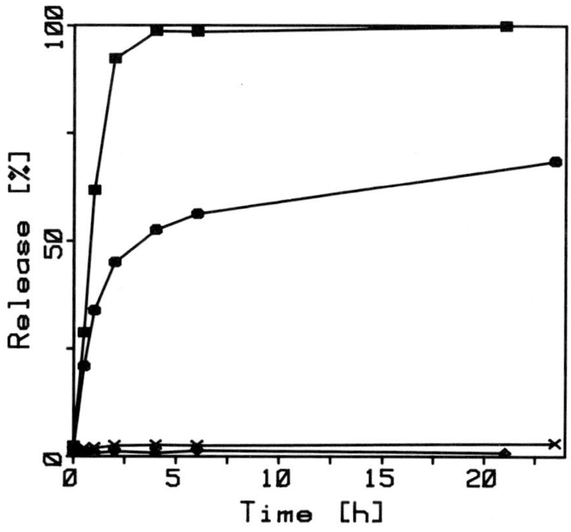

After removal of O-acetyl groups the wall material could be released by the action of lysozyme, however, to a different degree depending on the growth phase they were isolated from.
After removal of the O-acetyl groups from the walls stemming from log. phase of growth this cell wall was nearly completely degradable (>90%). For these walls from log. phase the O-acetylation was the only factor hindering the action of lysozyme.
The degradation behavior of de-O-acetylated walls from stationary phase was different: Indeed the cell wall was clearly better degraded by lysozyme as compared to the walls still wearing their O-acetylation, but this degradation was not complete. There remained a considerable amount which was apparently resistant against lysozyme. Here one can draw the conclusion that there are at least two different qualities of cell wall material, of which one is susceptible to lysozyme and one is not, and this difference is not due to different degrees of O-acetylation. This also means that during the cells' change from the log. phase of growth to the stationary phase not only an enhanced degree of O-acetylation of cell walls is generated but besides that also a somehow different quality of cell wall is constructed. It might be that these qualities of wall material with different lysozyme-sensitivity independent of O-acetylation are due to different packing-types of murein (6).

Fig. 2:

Degradation of Cell Walls from Intact Cells by Lysozyme

◆ log. phase
■ log. phase, O-acetyl groups removed
× stat. phase
● stat. phase, O-acetyl groups removed

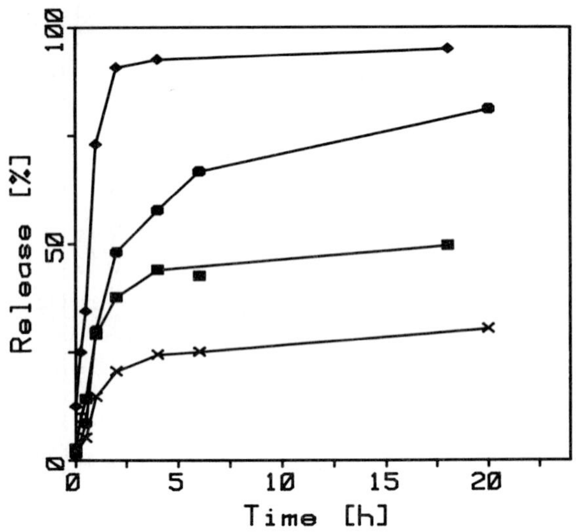

Degradation of Cell Walls from Intact Cells.
To approach more closely the conditions found in phagolysosomes the same experiments as those performed with isolated cell walls were performed using complete intact cells as substrate, having labelled only cell walls (see also 8). The results obtained in these experiments (Fig. 2) showed some remarkable differences as compared to those with isolated walls.
Both cell wall materials from complete log. phase and from complete stat. phase cells could be degraded by lysozyme. Because lysozyme did not cause a release of cell wall material on isolated cell walls the release found using intact cells can apparently be induced by the action of the cells' own cell wall autolytic systems.
In case of log. phase cells the combination of lysozyme and autolysins was capable of nearly completely destroying the log. phase cell wall. Using stat. phase cells lysozyme and autolysins together could only release a certain amount (about 30 %) of cell wall material. This degree of release could also be achieved by using autolysins alone (data not shown).
There was an extreme difference between the degradation of log phase walls having a "normal" degree of O-acetylation as compared to those from stationary phase with an elevated level, by the lysozyme and the autolysins acting together.
After treatment of the complete cells with alkali the degradation

of those cells revealed some unexpected results. First, it must be mentioned that the removal of O-acetyl groups with the procedure used in this experiments completely inactivated the cell wall autolytic enzymes as judged from corresponding release measurements (data not shown).

The degree of released wall material from stationary phase cells could be enhanced when we removed the O-acetyl groups. Again the cell wall material was only partially released to the supernatant, as has been shown with the isolated stationary phase cell walls. Unexpected was the behaviour of alkali treated log phase cells. After removal of O-acetyl groups and thereby inactivating autolytic enzymes there was a certain reduction in degradability. The respective de-O-acetylated isolated cell walls were completely degraded (Fig. 1). At the present point of investigations this might be explained by problems of accessing all parts of the wall in complete cells or by the loss of special parts of the cell wall during isolation procedure.

From the results shown it seems clear that there is a strong influence of O-acetylation on the resistance of staphylococcal cell walls against lysozyme, especially, since in the meantime we have gathered some evidence that lysozyme can cut susceptible chemical bonds in the wall even if this will not result in release of cell wall material (see also 9), but, of course, should support the action of other cell wall lytic enzymes. So the O-acetylation is indeed the crucial factor hindering the action of lysozyme, though apparently not the only one.

This is of special meaning when considering the fact that cell wall material with a low degree of O-acetylation can be completely degraded by lysozyme and autolysins together, while this is not true in stat. phase cells with a high O-acetylation in their walls.

It is tempting to speculate that it is this highly O-acetylated material which will persist in the phagocytes and which might contribute to the medical problems mentioned in the introduction. This should be considered when applying antibiotics, since we could demonstrate that several bacteriostatic drugs do effect an enhanced level of O-acetyl groups in staphylococcal cell walls (2,7).

The results can be summarized in the following fashion:

1) Lysozyme alone is not able to degrade O-acetylated staphylococcal cell walls.
2) The O-acetylation of the cell wall is the decisive factor - in the case of log. phase cell walls seemingly the only one - responsible for hindering the degradation of the walls by lysozyme.
3) In the case of stationary phase cell walls there remains - even after complete removal of O-acetyl groups - an unsoluble residue revealing an "intrinsic" resistance against lysozyme action.
4) This leads to the certainty that there must be different cell wall qualities in staphylococcal cells, besides the obvious differences in O-acetylation.

References

1. Reinicke, B., P. Blümel, P. Giesbrecht. 1983. Arch. Microbiol. 135, 120-124.

2. Johannsen, L., H. Labischinski, B. Reinicke, P. Giesbrecht. 1983. FEMS Letters 16, 313-316.

3. Johannsen, L., H. Labischinski, P. Burghaus, P. Giesbrecht. 1983. In: The Target of Penicillin (R. Hakenbeck, J.-V. Höltje and H. Labischinski, eds.). Walter de Gruyter, Berlin - New York. pp. 261-266.

4. Rosenthal, R.S., M.A. Gfell, W.J. Folkening. 1985. Infect. Immun. 49, 7-13.

5. Wecke, J., M. Lahav, I. Ginsburg, P. Giesbrecht. 1982. Arch. Microbiol. 131, 116-123.

6. Labischinski, H, L. Johannsen. 1985. Contribution to this volume.

7. Burghaus, P., L. Johannsen, D. Naumann, H. Labischinski, H. Bradaczek, P. Giesbrecht. 1983. In: The Target of Penicillin (R. Hakenbeck, J.-V. Höltje and H. Labischinski, eds.). Walter de Gruyter, Berlin - New York. pp. 261-266.

8. Wecke, J., E. Kwa, L. Johannsen, P. Giesbrecht, M. Lahav, I. Ginsburg. 1985. Contribution to this volume.

9. Westmacott, P., H.R. Perkins. 1979. J. Gen. Microbiol. 115, 1-11.

O-ACETYLATION OF STAPHYLOCOCCAL CELL WALLS AS AN IMPORTANT FACTOR FOR THEIR DEGRADABILITY WITHIN BONE MARROW-DERIVED MACROPHAGES

Jörg Wecke, Elly Kwa, Lars Johannsen, Peter Giesbrecht
Robert Koch-Institute of the Federal Health Office, D-1000 Berlin 65

Meir Lahav, Isaac Ginsburg
The Hebrew University, Hadassah Medical Center, Jerusalem

Introduction

The cellular immune system is known to be rather ineffective in degrading bacterial cell walls of a number of pathogenic bacteria, such as streptococci and staphylococci (1).
Consequently, undegraded bacterial wall material is still detectable in the infected host for considerable periods of time after the infection is overcome. These cell walls may induce chronic inflammatory processes (5,6). The reason for the low degradability of bacterial cell walls by wall-degrading enzymes of the host is largely unknown. It could be demonstrated by in-vitro experiments that the degree of O-acetylation of murein may influence the resistance of murein against degradation by lysozyme (2,3). However, the role of O-acetylation during the degradation of bacterial cell walls within the cellular immune system is still unknown.

Material and Methods

Bacterial strain and cultivation: Staphylococcus aureus strain SG 511 from the culture collection of the Robert Koch-Institute, Berlin was used. Bacteria kept on agar plates were transferred to flaskes containing an aqueous solution of 2,5% pepton + 0,5% NaCL and cultivated with shaking for 22h at 37°C. Such bacteria from the stationary phase of growth were used for the NaOH-treatment and the phagocytosis experiments.

Removal of O-acetyl groups

O-acetyl groups were removed from the murein, as previously described (3).

Biological Properties of Peptidoglycan

Assay of viable counts

Viable counts were determined on agar plates using a spiral plater (Spiral Plater Model CU Spiral systems Instruments Inc., Bethesda Maryland, USA)

Radiochemical experiments

Cell wall labeling of staphylococci with ^{14}C-N-acetylglucosamine and radioactivity measurements were performed as described in a previous study (7). Cell wall degradation was determined in terms of the amounts of radioactive material released into the surrounding medium from the labeled staphylococci within the macrophages.

Phagocytosis experiments

Marrow cells were harvested from femur and tibia of young, adult BALB/C mice (Zentralinstitut für Versuchstierzucht, Hannover,FRG) and grown in Dulbecco's minimum essential medium as previously described (4). For the phagocytosis experiments petriperm dishes (Heraeus, FRG) with Teflon film bottoms were used. Macrophages (9-11 days old) were fed with antibody-coated staphylococci or staphylococcal cell walls (30 min ingestion time). After different digestion times (3h, 1 day, 4 days) the macrophages were fixed (7). All experiments were performed with adherent macrophages.

Electron microscopy

The electron microscopical methods were performed as previously described (7).

Results and Discussion

In order to determine the degradation of bacterial cell walls within macrophages quantitatively, the release of their specific wall label was measured after different digestion times. The amount of soluble wall label increased with the digestion time and the release from staphylococci containing de-O-acetylated murein proved to be in any case higher than that from normal cells (Fig. 1). Astonishingly, the release of label from isolated normal cell walls (Fig. 2) was largely reduced as compared with that of the intact bacteria (Fig. 1). But after removal of O-acetyl groups from the isolated cell walls the release turned out to be approximately the same as it was the case using intact de-O-acetylated bacteria. One could speculate that the higher degree of wall degradation of intact staphylococci (as compared with isolated autolysin-free cell walls) was induced by the joint activities of autolytic enzymes of intact bacteria and of lytic enzymes of the phagolysosome.

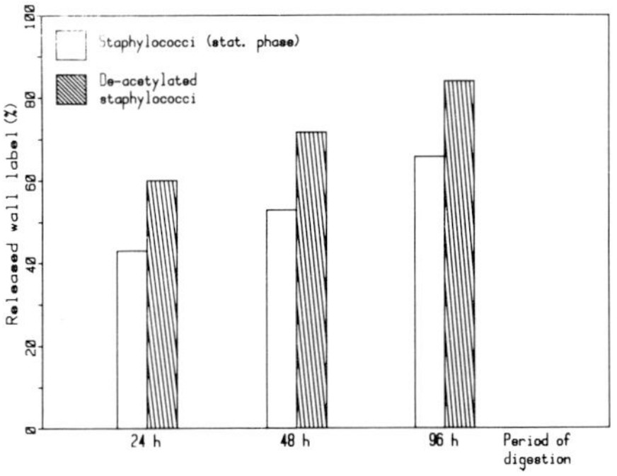

Fig. 1 Cell wall degradation of staphylococci in bone marrow-derived macrophages as measured by released wall label

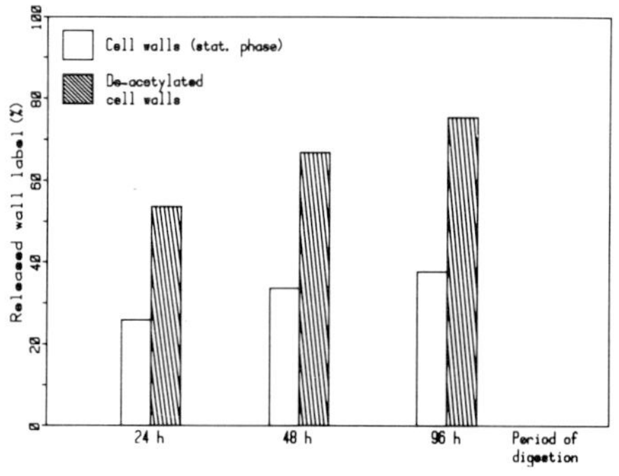

Fig. 2 Degradation of isolated cell walls of staphylococci in bone marrow-derived macrophages as measured by released wall label

While the release measurements revealed the quantitative aspects of wall degradation, the electron microscopical techniques gave a more detailed picture of its mechanisms. After one day of digestion the cell wall of the ingested normal bacteria seemed to be more or less intact (Fig. 3), although the release measurements revealed at least a certain degree of wall degradation. But also in this case a removal of O-acetyl groups from the murein resulted in a considerable degradation of the cell walls (Fig. 4). Furthermore, the bacterial cytoplasm was more or less completely degraded. Very typical of the degradation process of de-O-acetylated bacteria was a sponge-like appearance of the bacterial wall (Fig. 5), while cell walls from normal bacteria showed a more compact appearance. Only after 4 days of digestion, normal staphylococci cell walls also revealed different stages of degradation. Contrary to these experiments with normal staphylococci, it was rather difficult to detect bacteria or remnants of bacteria after a 4 days digestion of de-O-acetylated staphylococci within macrophages.

While the activity of autolytic wall enzymes resulted in wall degradation starting from the inside (7) the crenellated surface of the cell wall of Fig. 6 might be taken as an indication of an

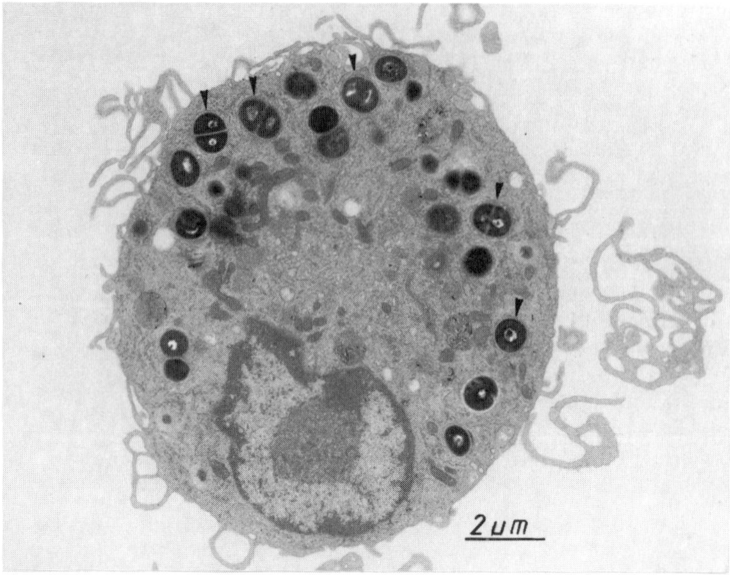

Fig. 3 Bone marrow derived macrophage after 1 day of digestion of staphylococci (arrows) : only minor changes of bacteria can be detected

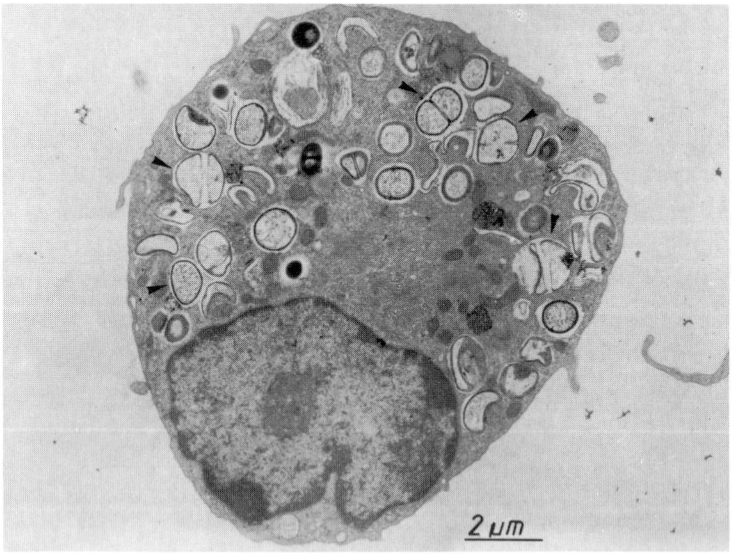

Fig. 4 Bone marrow derived macrophage after 1 day of digestion of de-O-acetylated staphylococci (arrows) : numerous changes of bacteria can be seen

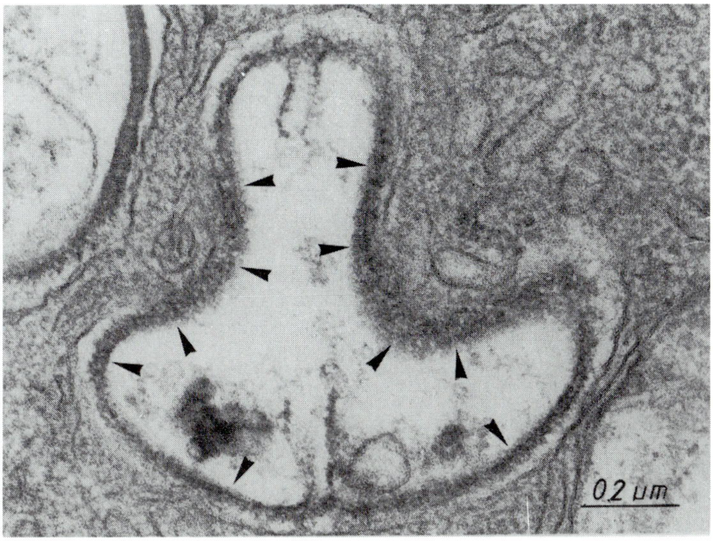

Fig. 5 De-O-acetylated staphylococcal cell within macrophage (1 day digestion) : Sponge-like appearance of bacterial wall (arrows)

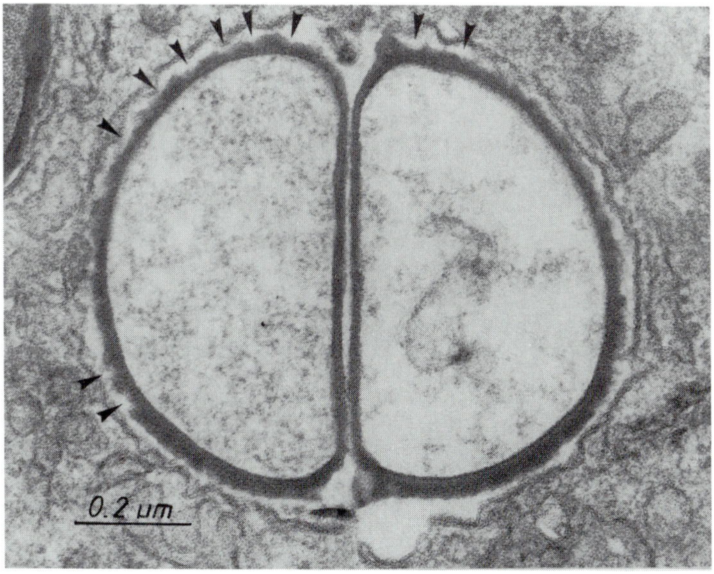

Fig. 6 Staphylococcal cell after 4 days of digestion within macrophage : Crenellated appearance of the bacterial surface (arrows)

enzymatic attack from the outside, involving mainly the activity of enzymes of the macrophage and only to a limited extent that of the autolytic wall enzymes. Furthermore, the electron microscopical results indicated the presence of different qualities of cell wall material: a rather high resistant material at their basis and a wall material, located at the surface, which was rather easy to be degraded. These findings are in good agreement with results of some of our recent biochemical studies (3).

Summary

The presence of O-acetyl groups in the murein proved to be an important factor for the resistance of staphylococcal cell walls against the lytic attack of phagocytes. In contrast to our in-vitro results, the autolytic wall activity was shown to be only of minor importance for the degradation of bacterial cell walls within the cellular immune systems. The cell wall of staphylococci consisted of different types of wall material with different degradability.

References

1. Ginsburg, I., Sela, M. : Critical Rev. Microbiol. 1976, 4, 249-332

2. Johannsen L., Labischinski, H., Reinicke, B., Giesbrecht, P. : FEMS Microbiol. Lett. 1983, 16, 313-316

3. Johannsen, L., Labischinski, H., Giesbrecht, P. : Contribution this volume

4. Neumann, C., Sorg, C. : Eur. J. Immunol. 1980, 10, 834-840

5. Rosenthal, R.S., Fleming, T.J. : Contribution this volume

6. Schwab, J.H., Cromartie, W.J., Ohanian, S.H., Craddock, J.G. : J. Bacteriol. 1967, 95, 1728-1735

7. Wecke, J., Lahav, M., Ginsburg, J., Giesbrecht, P. : Arch. Microbiol. 1982, 131, 116-123

METABOLIC FATE OF PEPTIDOGLYCAN MONOMER FROM Brevibacterium divaricatum AND BIOLOGICAL ACTIVITY OF ITS METABOLITES

J. Tomašić, Z. Valinger
Institute of Immunology, Rockefellerova 2, 41000 Zagreb, Yugoslavia

I. Hršak, B. Ladešić
Institute "Rudjer Bošković", P.O.Box 1016, 41000 Zagreb, Yugoslavia

Introduction

In the course of our studies on the action of penicillin on Brevibacterium divaricatum we found that the culture fluid of treated bacteria contained uncrosslinked peptidoglycan complex (1). Lysozyme hydrolysis of this complex and subsequent gel chromatography of the products yielded uncrosslinked peptidoglycan monomer (PGM), the repeating unit of peptidoglycan chain. This unit was further characterized by chemical modifications and NMR spectroscopy as GlcNAc-MurNAc-L-Ala-D-isoglutaminyl-meso(ε NH_2)-diaminopimelyl-D-Ala-D-Ala (2,3).

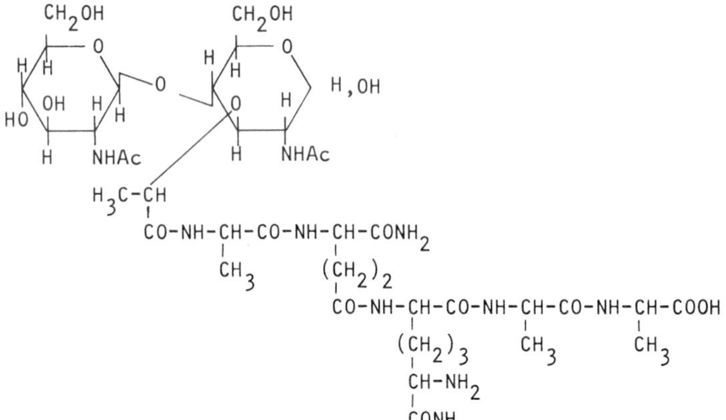

Fig.1.

A very recent study of Martin et al. (4) by fast atom bombardment mass spectrometry confirmed the proposed structure (Fig.1) and revealed the presence of

Biological Properties of Peptidoglycan

anhydro-component in the PGM preparation in very low amounts (approx. 5%). The anhydro-linkage is between C_1 and C_6 of the muramyl entity.

In our previous experiments we have shown that PGM stimulates humoral immune response to particulate antigens in mice (5,6) and exhibits antitumor and antimetastatic activity in mice bearing B-16 melanoma (7), Lewis lung carcinoma (8,9) and mammary carcinoma (10).

Metabolic Fate of Peptidoglycan Monomer

In vivo studies

In view of possible application of PGM as the adjuvant or in therapy of tumors, it was of particular interest to carry out preliminary studies of its metabolic fate in experimental animals. $[^{14}C]$ PGM was used throughout. It was obtained using $[1\text{-}^{14}C]$ sodium acetate as the precursor, which was added to the cultures simultaneously with penicillin. $[^{14}C]$ PGM isolated subsequently from the culture fluid contained radioactivity in the peptide chain (53%) and N-acetylamino groups (47%) of the glycan portion (2).

The metabolic fate of PGM was investigated following various routes of administration. Thus, after i.v. administration of $[^{14}C]$ PGM to mice (11), the largest part of applied radioactivity was found excreted in urine as soon as 3 hrs after injection (Table I). Gel chromatography of pooled mice urines resulted in the separation of radioactive products, which were characterized as the unchanged PGM and the corresponding pentapeptide, L-Ala-D-isoglutaminyl-*meso* (ε NH_2)-diaminopimelyl-D-Ala-D-Ala. The pentapeptide, which could be considered as the metabolite of PGM, was apparently formed by hydrolysis with the specific enzyme N-acetylmuramyl-L-alanine amidase. Due to the presence of D-amino acids in the molecule of the pentapeptide, it is probably not the substrate for non specific peptidases from plasma and, therefore, is not further hydrolyzed to smaller peptides.

Different kinetics of excretion of radioactivity was observed following subcutaneous administration of PGM to mice as compared to i.v. administration (Table I), although total amounts of radioactivity excreted in 6 hrs were practically the same (12). Radioactive products isolated from pooled urines were again characterized as the unchanged PGM and the corresponding pentapeptide. It should be noted that the ratio of radioactivity in two isolated peaks was 3:1, as compared to the ratio of 2:1 for products found in urine after i.v. administration. Such results indicate that a lower amount of PGM is metabolized if administered s.c. than in the case of i.v. injection.

Table I

PGM administered	Percent of radioactivity recovered from urine		
	First 3 hrs	3-6 hrs	6-24 hrs
i.v.	60.1	7.2	7.4
s.c.	34.0	22.3	3.3
p.o.	2.7	2.5	4.2

PGM was administered in saline

Following oral administration of $[^{14}C]$ PGM 70% of applied radioactivity was found accumulated in stomach and intestine (12) in the first 3 hrs. Total cumulative excretion of radioactivity in urine even after 72 hrs did not exceed 15%, but exhaled $[^{14}C]$ CO_2 comprised approx. 30% of administered radioactivity. Relatively large amount of detected ^{14}C CO_2 is an indication of extensive degradation of PGM molecule most probably in the intestinal tract. Gel chromatography of stomach and intestines extracts yielded several products, probably smaller peptides resulting from the original molecule.

Parallel experiments were also carried out in order to determine whether the route of administration affects the immunostimulating activity of PGM. As expected from the results of studies of metabolic fate, it was found that PGM stimulates immune response to sheep red blood cells (SRBC) in mice if administered i.v. or s.c. but is not active if administered per os, since the molecule undergoes degradation in intestinal tract (12).

In vitro studies

The occurrence of the pentapeptide in urines of mice treated i.v. or s.c. with PGM was an indication that the enzyme N-acetylmuramyl-L-alanine amidase is involved in the metabolic processing of PGM. The action of this amidase is expressed as the hydrolysis of the bond between N-acetylmuramic acid and L-alanine. A study was undertaken to establish in which organ or tissue enzymatic hydrolysis of PGM takes place. Contratry to expectations, PGM was not hydrolyzed in the liver, but incubation with plasma or serum resulted in rapid, practically complete hydrolysis of PGM to the disaccharide GlcNAc-MurNAc and the corresponding pentapeptide (13). Such results indicate that blood is the site where biotransformation of PGM takes place. $[^{14}C]$ PGM labeled both in the glycan and peptide portions of the molecule was shown to be a very convenient substrate for the amidase in various mammalian and in the human sera (14). The specificity of the enzyme action and the affinity of this enzyme for low molecular weight substrates (peptidoglycan monomer and dimer) has been confirmed.

N-Acetylmuramyl-L-alanine amidase has not been detected in mammalian blood so far, although the presence of lysozyme (15) and β-N-acetylglucosaminidase (16) has been reported. It is possible that the amidase complements the action of other bacteriolytical enzymes in serum involved in the digestion of bacteria, and their processing in the organism. Partially purified amidase was used for hydrolysis of PGM on a preparative scale (14). Purified metabolites were further used for testing of biological activity.

Biological Activity of Peptidoglycan Monomer and its Metabolites

Stimulation of humoral immune response

Humoral immune response to SRBC in mice was assayed by means of Jerne´s technique for plaque forming cells (PFC) in the spleen. In our previous work (5) we have shown that PGM stimulates immune response to SRBC in mice even if administered before the antigen. For comparison purposes, PGM, the disaccharide and the pentapeptide were tested parallely in the same experimental model. Both PGM and the pentapeptide stimulated the immune response to the same extent (up to 200% over 100% for the control mice) (17). The disaccharide, however, had no effect on PFC response.

Induction of delayed hypersensitivity

Delayed type hypersensitivity to SRBC in the foot pad of mice showed the same results as in testing of humoral immune response. Significant enlargement of foot pad was observed in animals treated either with PGM or the pentapeptide, while the disaccharide did not exhibit any effect on foot pad enlargement (17).

Antimetastatic activity

Antineoplastic effects of PGM could be evidenced in the retardation of the tumor growth rate, prolongation of the survival time of tumor bearing mice, and particularly in significant reduction of spontaneous and induced pulmonar metastases (7-10). Comparison of antimetastatic properties of PGM and its metabolites was carried out in mice bearing B-16 melanoma. Following one i.v. injection of PGM or the pentapeptide in saline, the number of induced metastases was reduced up to 50% as compared to the controls. The disaccharide had no significant antimetastatic effect (17).

Effect on microsomal cytochrome P-450

In experimental models in which the biological activity of PGM was tested so far, no apparent toxicity for the treated host has been observed. However, undesirable effects, such as the inhibition of microsomal enzymes in liver and the decrease of cytochrome P-450 content have been found in mice treated with PGM (18). Studies were also carried out to determine whether the decrease in cytochrome P-450 level could be attributed to PGM as an integral molecule, or to its disaccharide or pentapeptide portion, respectively. It was found that only pentapeptide affects significantly cytochrome P-450 levels, while the disaccharide is completely inactive (18).

Conclusions

Experimental data collected so far on the biological activity of PGM originating from Brevibacterium divaricatum indicate that this low molecular weight, non-pyrogenic and non-toxic compound might be applicable for stimulation of humoral and cellular immune response and in the antitumor therapy.

When administered to mice i.v. or s.c., PGM is hydrolyzed by the enzyme N-acetylmuramyl-L-alanine amidase to the disaccharide and the corresponding pentapeptide. Unchanged PGM and the pentapeptide are rapidly excreted in urine. It was also shown, that in several experimental models integral PGM and its pentapeptide portion exhibit practically the same biological activity, while the disaccharide is completely inactive. When administered per os, PGM undergoes extensive degradation in stomach and intestines, resulting in a complete loss of biological activity.

Acknowledgements

The authors thank Mrs. R. Naumski, B.Sc., from "Pliva" Chemical and Pharmaceutical Works, Zagreb, for the donation of raw preparation of peptidoglycan monomer. We also wish to thank Prof. Edgar Lederer, of CNRS, Gif-sur-Yvette, for his support and helpful discussions throughout our research on metabolism of PGM.

References

1. Keglević, D., Ladešić, B., Hadžija, O., Tomašić, J., Valinger, Z., Pokorny, M. 1974. Eur. J. Biochem 42, 389.

2. Keglević, D., Ladešić, B., Tomašić, J., Valinger, Z., Naumski, R. 1979. Biochim. Biophys. Acta 585, 273.

3. Klaić, B., Ljubić, B., Metelko, B., Pongračić, M. 1983. Carbohydr. Res. 123, 168.

4. Martin, S.A., Karnovsky, M.L., Krueger, J.M., Pappenheimer, J.R., Biemann, K. 1984. J. Biol. Chem. 259, 12652.

5. Hršak, I., Tomašić, J., Pavelić, K., Valinger, Z. 1979. Z. Immun. Forsch. 155, 312.

6. Hršak, I., Novak, Dj., Tomašić, J. 1980. Period. Biol. 82, 147.

7. Hršak, I., Tomašić, J., Osmak, M. 1983. Eur. J. Cancer Clin. Oncol. 19, 681.

8. Sava, G., Giraldi, T., Tomašić, J., Hršak, I. 1983. Cancer Immunol. Immunother. 15, 84.

9. Sava, G., Perissin, L., Zorzet, S., Tomašić, J. 1985. Anticancer Res. 5, 301.

10. Sava, G., Tomašić, J., Hršak, I. 1984. Cancer Immunol. Immunother. 18, 49.

11. Tomašić, J., Ladešić, B., Valinger, Z., Hršak, I. 1980. Biochim. Biophys. Acta 629, 77.

12. Valinger, Z., Ladešić, B., Hršak, I., Tomašić, J.: To be published.

13. Ladešić, B., Tomašić, J., Kveder, S., Hršak, I. 1981. Biochim. Biophys. Acta 678, 12.

14. Valinger, Z., Ladešić, B., Tomašić, J. 1982. Biochim. Biophys. Acta 701, 63.

15. Selsted, M.E., Matinez, R.J. 1978. Infect. Immun. 20, 782.

16. Walker, P.G., Pugh, G. 1960. J. Clin Pathol. 13, 353.

17. Hršak, I., Marotti, T., Tomašić, J.: To be published.

18. Trešćec, A., Iskrić, S., Hršak, I., Tomašić, J. 1983. Biochem. Pharmacol. 32, 2354.

DETERMINATION OF THE SIZE DISTRIBUTION OF THE GLYCAN STRANDS RELEASED FROM MUREIN OF E. COLI BY HUMAN SERUM AMIDASE

H. Ludewitz, J.-V. Höltje

Max-Planck-Institut für Entwicklungsbiologie,
Abteilung Biochemie, Spemannstrasse 35, D-7400 Tübingen

Introduction

Two morphogenetic events occur during the propagation of a rod-shaped bacterium, like *E. coli*: Longitudinal growth of the sacculus and formation of a septum that after completion of cell division forms the polar caps of the new cells (1,2). These two processes yield murein with different geometries: A cylindrical middle part and a hemispherical end piece. Since isolated murein sacculi still reflect the shape of the bacterium from which they have been isolated (2-4), the specific morphology of a bacterium is conserved by the murein itself. We would like to understand the structural parameters which determine the three-dimensional arrangement of the murein, a network of glycanstrands crosslinked by short peptide bridges (1). Variations in the amount and chemistry of the peptide bridges as well as in the average length of the sugar chains which parallel changes in the shape of the murein sacculus have been reported (5). However, another important structural aspect of murein has never been determined before. The actual size of the individual glycan strands as well as their length distribution are not known yet.

Biological Properties of Peptidoglycan

Results and Discussion

The experimental strategy we employed to investigate the size distribution of the glycan strands was as follows: 1) Isolation of the intact glycan strands from the murein network. 2) Fractionation of the sugar chains according to size (i.e. length). 3) Calculation of the chain lengths by determination of the ratio of terminating to total aminosugars. Mainly two enzymes would be useful to isolate the sugar chains: Either an endopeptidase cleaving the peptide bridges or an amidase removing the peptides by splitting the amide bond between the muramic acid and the L-alanine residue. The enzyme of choice necessarily has to be absolutely devoid of contaminating glucosaminidases or muramidases. Since earlier experiments with endopeptidase of *E. coli* failed to release the glycan strands from the murein network (W. Keck, unpublished results) we decided to employ human serum amidase which can easily be prepared and well be separated from serum lysozyme (6,7). To be absolutely sure to isolate uncleaved, native glycan strands amidase digestion of murein was carried out in the presence of antibodies against human serum lysozyme. The average chain length as calculated from our data for the isolated sugar chains is the same as what has been determined for whole sacculi (1); thus our procedure left the sugar chains intact.
The most successful method to fractionate the isolated glycan strands according to size turned out to be gel filtration on a hydrophilic vinyl polymer gel (Fractogel TSK; Merck, Darmstadt, W-Germany). The reliability of the gel permeation method for a separation according to length is indicated by a straight line in a plot of the elution volume as a function of the logarithm of the chain length. A typical fractionation profile of the glycan strands isolated from murein of cells from the logarithmic growth phase is shown in Figure 1. As can be seen, a clear separation into different peaks at least in the low molecular weight region was achieved.

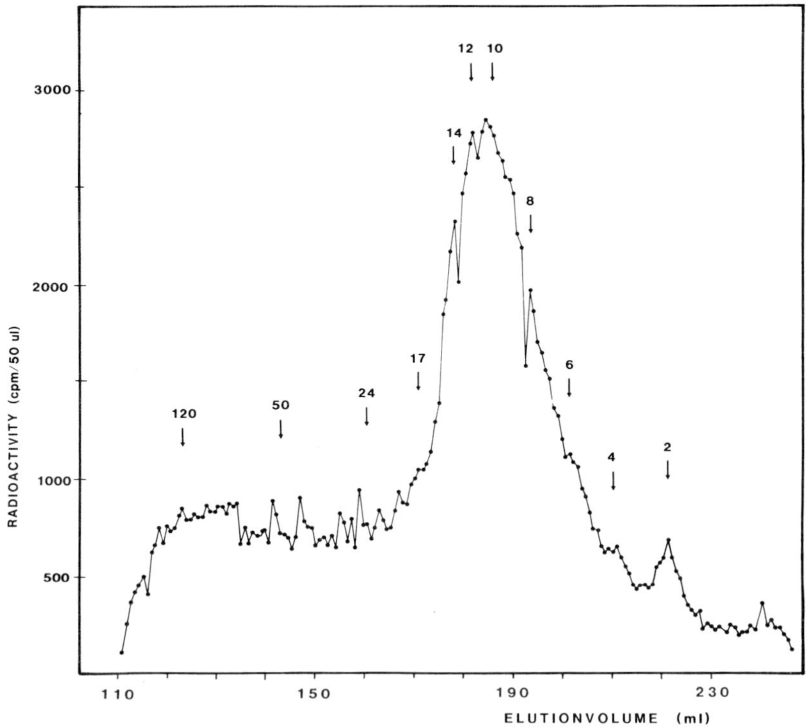

Fig. 1: Fractionation of the glycan strands of murein of E. coli W7 by gel filtration on a Fractogel TSK column. N-acetyl-D-(^{3}H)glucosamine-labelled murein sacculi (0.5 mg/ml; 2x10^6 cpm/ml) were incubated in 0.05 M Tris/HCl-buffer, pH 7.9, containing 0.005 M $MgCl_2$ and 0.05 % NaN_3 with human serum amidase (3 mg protein/ml; 2 x 10^{-3} units/mg) in the presence of anti-human lysozyme IgG for 12 hours at 37°C. The sample was applied to a Fractogel-TSK-HW-55 (S) (Merck, Darmstadt) column (1.5 x 200 cm) and eluted (0.8 ml/6 min) with 0.05 M sodium phosphate buffer, pH 6.6, containing 0.05 M LiCl.

In murein of E. coli there exists a natural tag for one of the chain ends. The glycan strands have a normal N-acetyl-glucosamine at the non-reducing end of the chain but an altered muramic acid at the reducing end. The C-1-hydroxyl group of the terminating muramic acid is blocked by the formation of an 1,6-

anhydro group (8,9). It follows then that the ratio of terminating 1,6-anhydro sugars to total sugars reveals the length of the glycan strand. In order to utilize this natural marker, a cleavage method was needed that preserves this structure. Thus, we had to employ an enzymic digestion. A combination of an endo- and an exoglycosidase, egg white lysozyme and chitinase, resulted in the formation of N-acetylglucosamine, N-acetylmuramic acid as well as 1,6-anhydro muramic acid.
Using this approach the lengths of the sugar chains in the various fractions obtained after TSK gel permeation chromatoraphy was determined. Lengths of 2,4,6,8,10,12, and 14 disaccharide units were obtained for the separated peak fractions Fig. 1). The longest glycan strands reached 120 disaccharide units. It is interesting to note that the chains seem to exist in certain size classes differing by 2 disaccharide units. Furthermore, to our surprise the shortest chain was not a disaccharide but a tetrasaccharide although during murein biosynthesis disaccharide precursors are polymerized. Very likely, newly synthesized glycan strands are cleaved by the action of the lytic transglycosylase (8) during cell growth. The predominant length in the murein of a logarithmically growing culture of *E. coli* was about 10 disaccharide units which in terms of the dimension of an *E. coli* cell corresponds to about one threehundredth of the periphery of an *E. coli* cell. Quite a lot of chains were extremely short comprising only 4 disaccharide units. One may speculate that this represents mostly material from the spherical polar caps of the cell, since stereochemical calculations demand for such short chains at the ends of the rod (10).
An analysis of murein of cells from different growth phases showed dramatic changes in the length distribution when *E. coli* enters the stationary phase of growth (Fig. 2).
Two changes were evident: First, the disappearance of the very long chains and second, a dramatic increase in the amount of

tetrasaccharides. About 40 % of all chains were tetrasaccharides which of course represents only a very small amount of total murein. The changes sum up to an overall decrease of the average chain length from 16 in murein from the logarithmic phase of growth to 10 disaccharide units in murein from stationary cells. We think that these changes in the length distribution of the glycan strands reflect the morphological change that occurs when E. coli becomes stationary, which is a change from a rod-shaped cell to a very short, sphere-like cell.

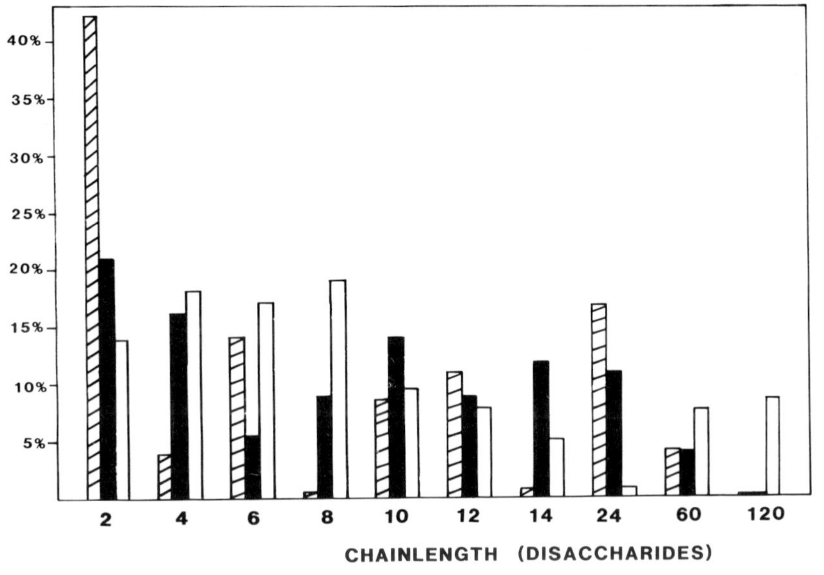

Fig. 2: Length distribution of the glycan strands of murein of E. coli EH 3247 from different growth phases (▨: stationary; ■: late logarithmic; □: logarithmic).

Acknowledgement

We thank Uli Schwarz for his support and valuable advice.

References

1. Höltje, J.-V., U. Schwarz. 1985. In: Molecular Cytology of Escherichia coli (N. Nanninga, ed.). Academic Press Inc. (London) Ltd., p 77.

2. Schwarz, U., A. Asmus, H. Frank. 1969 J.Mol.Biol. 41, 419.

3. Schwarz, U., W. Leutgeb. 1971. J.Bacteriol. 106, 588.

4. Henning, U., K. Rehn, V. Braun, B. Höhn, U. Schwarz. 1972. Eur.J.Biochem. 26, 570.

5. Essig, P., H.H. Martin, J. Gmeiner. 1982. Arch.Microbiol. 132, 245.

6. Valinger, Z., B. Ladesic, J. Tomasic. 1982. Biochem.Biophys. Acta 701, 63.

7. Mollner, S., V. Braun. 1984. Arch.Microbiol. 140, 171.

8. Höltje, J.-V., D. Mirelman, N. Sharon, U. Schwarz. 1975. J.Bacteriol. 124, 1067.

9. Taylor, A., C.B. Das, J. Van Heijenoort. 1975. Eur.J.Biochem. 53, 47.

10. Labischinski, H., G. Barnickel, D. Naumann, P. Keller. 1985. Ann.Inst.Pasteur/Microbiol. 136A, 45.

Endogenous Induction of Bacterial Lysis by Cloned PhiX174 Gene E Product

U. Bläsi, R.E. Harkness, A. Witte, G. Halfmann and W. Lubitz*
Institute of Genetics and Microbiology, University of Munich,
Maria Ward Str. 1a, D-8000 Munich 19, FRG.

Introduction

Lysis of bacteria following infection with bacteriophages is, with the exception of only a few phage species, the general mechanism for release of mature progeny phages. Most of the larger phages (e.g., Lambda), code for several enzymes which alter the permeability properties of the inner membrane and which actively degrade the rigid peptidoglycan layer of the host (1). An alternate method used by some phages is to inhibit peptidoglycan synthesis of the host bacterium (2). As a consequence of this inhibition, host cell autolytic enzymes may be activated resulting in lysis of the bacterial host (3).
In this communication, we provide evidence that activation of cellular autolytic enzymes also occurs with the cloned lysis gene of phage PhiX174 as a result of membrane perturbation. This mechanism of lysis induction, which may also be used by phage MS2 (4), is an example of a novel method of endogenous triggering of autolysis.

Results and Discussion

The genome of phage PhiX174 codes for only one lysis protein, the gene E product (gpE). This gene product is able to induce lysis when expressed from a cloned state. A list of gene E carrying plasmids from our collection, including chimeric gene E hybrids and gene E deletions, is given in table 1. Since uncontrolled expression of gene E is lethal to the cells, a stringent translational control had to be imposed on gene E or its derivatives. Two control systems have been employed: i) lactose

Biological Properties of Peptidoglycan

Table 1: Plasmids carrying lysis-inducing PhiX174 gene E

plasmid	transcriptional control	resistance determinant	state of gene E	reference
pUH12	lac iq1, lac po	am	wt	(5)
pUH51	lac iq, lac po	tc	wt	(6)
pUH123	lac iq, lac po	am, tc	wt	(6)
pSB4	cI857, lambda pL	km	hyb	(7)
pSB12	cI857, lambda pL	am	wt	(8)
pSU815	cI857, lambda pL	am	del	(9)

promoter-operator (lac po) region, and ii) leftward lambda promoter (pL). In the first case expression of gene E could be achieved by inactivation of the lac repressor with the lactose analogue IPTG. In the second control system, lysis of bacteria could be induced by heat inactivation of the thermosensitive lambda cI857 repressor.

Following induction of gene E expression from plasmid pSB12 in E. coli (fig. 1a), lysis onset occurred approximately 10 min after temperature upshift in samples taken at any time point during exponential growth of the culture. Expression of gene E during the stationary growth phase, however, was not followed by lysis (fig. 1a). That cell lysis was specific for expression of gene E, and not for any other sequence carried on plasmid pSB12, can be seen from the normal growth behaviour of cells carrying plasmid pSB22 (fig. 1b). This latter plasmid is identical to pSB12 except for an amber mutation in gene E (Eam3, lysis negative allele) (fig. 1c). Lysis inducing properties of the gene Eam3 allele of pSB22 were re-established when expressed in an E. coli amber suppressor strain (fig. 1b).

In addition to the requirement for actively growing cells, other cellular and environmental parameters also influenced the lysis inducing properties of gpE. These are listed in table 2.

The requirements for a functional autolytic enzyme system and energized inner membrane are the most crucial cellular prerequisites for gpE induced lysis. These features give a clue to the mechanism of gpE induced lysis. The first describes the

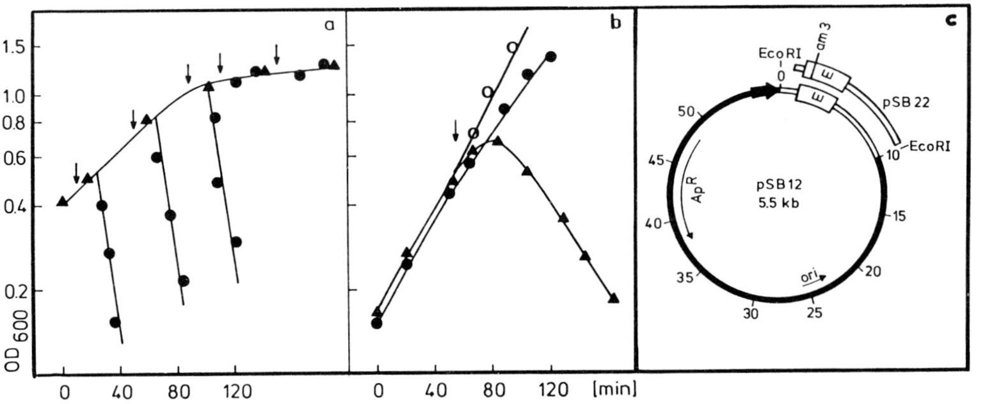

Fig.1. Expression of PhiX174 gene E from plasmids pSB12 or pSB22 a) At times indicated by the arrows, samples of E. coli strain K12ΔH1Δtrp(pSB12) (8) grown at 30°C (▲ - ▲) were shifted to 42°C (● - ●). b) Growth of E. coli K12ΔH1Δtrp(pSB22) at 30°C (● - ●) or at 42°C (o - o). At time indicated by arrow the amber-suppressor strain E. coli HF4714 (8) was shifted from 30°C to 42°C (▲ - ▲). c) Plasmids pSB12 and pSB22, direction of transcription from lambda pL is indicated by heavy arrow.

cellular system which is activated by gpE, and the latter, necessary conditions for the mode of induction. In this process, the presence of at least one regulatory gene function of the autolytic system is essential for gpE- and drug-induced autolysis (11). Additional evidence suggests, however, that exogenously added penicillin and endogenously expresssed gpE use largely alternate pathways for lysis induction in E. coli (13, 14).

Table 2: Cellular and environmental factors influencing gene E product induced lysis

characteristic features	reference
- dependence on growth phase of the cells	(10)
- pH of the culture medium	(10)
- lysis can be inhibited by Mg++	(10)
- requirement of a functional autolytic system	(11)
- requirement of an energized inner membrane (pmf)	(12)

Table 3: Molecular characteristics of the gene E product

characteristic features	reference(s)
- consists of 91 amino acids (10 kd)	(15)
- very hydrophobic polypeptide (pI 9.8)	(15, 16)
- represents a membrane protein (im and om)	(16, 17)
- lysis inducing function within N-terminal part of the protein	(7)
- C-terminal part of the molecule influences conformation and/or regulation of functional site	(7,9)

gpE represents a membrane protein (Tab. 3), and this feature may explain some of the requirements for lysis induction. Other molecular characteristics of gpE (Tab. 3) suggest that it integrates into the inner and outer membranes of E. coli in a highly ordered state. On the basis of biochemical characteristics and theoretical computer analysis of the structure, it may well be possible that a complex of gpE molecules, either in multimeric form or with other membrane proteins, is built up prior to the initiation of lysis. Because lysis onset, and rate of lysis, depend on the efficiency of gene E expression (8), formation of a functional gpE complex(es) may determine triggering of lysis. The energetic parameters which have been determined during gpE induced lysis (12), showed that ion fluxes and breakdown of membrane potential were coincident with onset of lysis, and that

Table 4: Cellular effects associated with the activation of the autolytic system by gene E expression

characteristic features	reference(s)
- degradation of peptidoglycan	(18)
- membrane protein composition altered	(16,19)
- induction of phospholipase activity	(20)
- ion leakage and drop of membrane potential	(12)

all these changes occurred simultaneously. Furthermore, other cellular alterations, such as the degradation of peptidoglycan and changes of membrane protein composition have been observed in PhiX174 infected cells (Tab. 4).

It should be emphasized at this point, that there appear to be some differences between the extent of peptidoglycan degradation by gpE alone and cells infected with PhiX174. In the case of gene E expression from a plasmid, the damage to peptidoglycan seems to be a secondary effect caused by the membrane perturbing effects of gpE. In phage infected cells, peptidoglycan degradation may be the consequence of a more complex interaction of gpE with other phage gene products. Studies to determine this difference are in progress.

One hypothetical model for gpE induced lysis of E. coli, could be the formation of a transmembrane tunnel or unspecific pore. Consequently, as soon as contact between the external environment and the cytoplasmic content of the cells is established, cellular material is extruded and membrane bound enzymes, normally summarized as autolysins, are induced. Whether this model holds to be true or not remains to be revealed by further studies.

Acknowledgements

This research was supported by a grant from the Deutsche Forschungsgemeinschaft (Lu 213/4-5), R.E.H. was the recipient of a Natural Sciences and Engineering Research Council of Canada post-doctoral fellowship.

References

1. Garrett, J., R. Fusselman, J. Hise, L. Chiou, D. Smith-Grillo, J. Schulz, R. Young. 1981. Mol. Gen. Genet. 182, 326.

2. Ozaki, M., R.C. Valentine. 1973. Biochim. Biophys. Acta 304, 707.

3. Tomasz, A. 1984. In: Microbial Cell Wall Synthesis and Autolysis (C. Nombela, ed.). Elsevier Science Publishers, p. 3.

4. Höltje, J.-V., J. Van Duin. 1984. In: Microbial Cell Wall Synthesis and Autolysis (C. Nombela, ed). Elsevier Science Publishers, p. 195.

5. Henrich, B., W. Lubitz, R. Plapp. 1982. Mol. Gen. Genet. 185, 493.

6. Henrich, B., W. Lubitz, R. Plapp. 1983. In: The Target of Penicillin (R. Hakenbeck, J.-V. Höltje and H. Labischinski, eds.), de Gruyter, Berlin, p. 197.

7. Bläsi, U., W. Lubitz. 1985. J. Gen. Virol. 66, 1209.

8. Bläsi, U., B. Henrich, W. Lubitz. 1985. J. Gen. Microbiol. 131, 1107.

9. Schüller, A., R.E. Harkness, U. Rüther, W. Lubitz. 1985. Nucl. Acids Res. 13, 4143.

10. Lubitz, W., G. Halfmann, R. Plapp. 1984. J. Gen. Microbiol. 130, 1079.

11. Lubitz. W., R.E. Harkness, E.E. Ishiguro. 1984. J. Bacteriol. 159, 385.

12. Witte, A., E.P. Bakker, W. Lubitz. 1986. (in preparation)

13. Halfmann, G., M. Leduc, W. Lubitz. 1984. FEMS Microbiol. L. 24, 205.

14. Halfmann, G., W. Lubitz. 1986. J. Bacteriol. 166 (in press)

15. Barrell, B.G., G.M. Air, C.A. Hutchison. 1976. Nature 264, 34.

16. Bläsi, U., R. Geisen, W. Lubitz, B. Henrich, R. Plapp. 1983. In: The Target of Penicillin (R. Hakenbeck, J.-V. Höltje and H. Labischinski, eds.), de Gruyters, Berlin, p. 205.

17. Altman, E., K. Young, J. Garrett, R. Altman, R. Young. 1985. J. Virol. 53, 1008.

18. Lubitz, W., R. Plapp. 1980. Curr. Microbiol. 4, 301.

19. Lubitz, W., R. Schmid, R. Plapp. 1981. Curr. Microbiol. 5, 45.

20. Lubitz, W., A.P. Pugsley. 1985. FEMS Microbiol. L. 30, 171.

RADIOIMMUNOASSAYS FOR THE SELECTIVE DETECTION OF THE GLYCAN MOIETY AND THE PENTAPEPTIDE SUBUNIT IN SECRETED PEPTIDOGLYCANS

P. Zwerenz, F. Hagen, G. Leitherer, P.H. Seidl, and K.H. Schleifer
Lehrstuhl für Mikrobiologie, Technische Universität München,
Arcisstraße 21, D-8000 München 2, FRG

Introduction

Secretion of peptidoglycan fragments into the culture filtrate was reported by several groups (1-6; Zeiger, this book; Rosenthal and Fleming, this book) and became a point of interest with respect to biosynthesis of peptidoglycan (3,6) and with respect to the biological activities of peptidoglycan (4,5; Zeiger, this book; Rosenthal and Fleming, this book). Most studies on peptidoglycan turnover products (1-4,6) or 'soluble peptidoglycan' (5) secreted into the culture filtrate employ the measurement of radioactive *in vivo* labelled peptidoglycan. The present paper describes sensitive radioimmunoassays for selective detecting the glycan moiety and the pentapeptide subunit in unlabelled peptidoglycans secreted by bacterial cells.

Materials and Methods

The principle of the test system was recently described in detail (7) and is summarized in Fig. 1. Specific antibodies to R-D-Ala_2-OH were previously reported (8,9), glycan strand specific antibodies were obtained from H. Mett, Ciba-Geigy, Basel, Switzerland. Specificity for the glycan and the absence of antibodies to R-D-Ala_2-OH were proved in specific ELISAs for measuring antibodies to the glycan (Zauner et al., this book) or to R-D-Ala_2-OH (9). As haptens, synthetic H-L-Ala-D-Glu(L-Lys-D-Ala-D-Ala-OH)-OH and glycopeptides lacking C-terminal R-D-Ala_2-OH, labelled with ^{125}iodine (10), were employed. The glycopeptides were isolated from highly purified peptidoglycan of *Lactobacillus acidophilus* and radiolabelled according to the methodology recently described (11). The specificity of the assays for detecting R-D-Ala_2-OH or glycan containing material was proved in detail by appropriate binding studies and binding inhibition experiments (7,10,11). All culture filtrates tested as inhibitors (compare Fig. 1) were taken from the late logarithmic phase. For removal of low molecular weight mate-

Biological Properties of Peptidoglycan

rial, aliquots of the culture filtrates were dialysed against twice distilled water (molecular weight cut-off 10.000). The content of glycan or R-D-Ala_2-OH containing material in the culture filtrates was calculated from their inhibitory capacity. Quantification of soluble peptidoglycan secreted into the medium was achieved employing standard curves with known concentrations of soluble peptidoglycan in the inhibition studies. Soluble peptidoglycans were prepared from highly purified peptidoglycans of <u>Staphylococcus epidermidis</u> ATCC 14990 and <u>Staphylococcus aureus</u> 52A5 (strain obtained from D. Mirelman, The Weizman Institute, Rehovot, Israel) by digestion with lysozyme, dialysis of the soluble material and lyophilization. The influence of sublethal concentrations of penicillin G on secreted high molecular weight peptidoglycan was studied by adding 0.2 mg/l penicillin G. Cells from the logarithmic phase were harvested and washed four times with saline for complete removal of secreted material. The cells were divided into two samples; an aliquot was grown for 60 minutes in nutrient broth containing penicillin G, cells grown without penicillin were used as a control.

Results and Discussion

The principle of the test systems employed for specific detection of a) peptides R-D-Ala_2-OH or b) glycan framents in material secreted into the medium by bacterial cells is shown in Fig. 1.

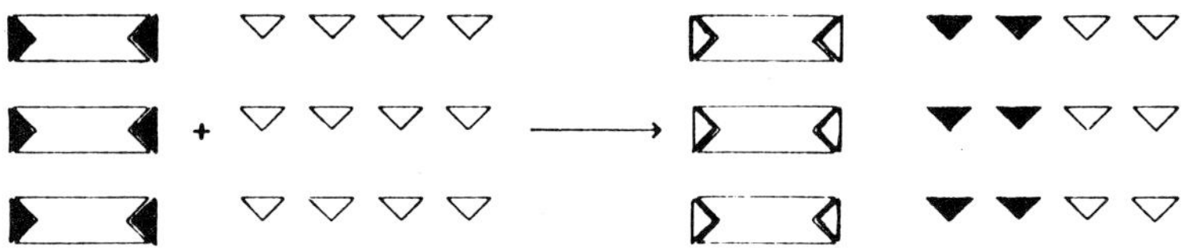

Antibody with specificity a) for peptides R-D-Ala_2-OH
b) for the glycan moiety

Secreted peptidoglycans (culture filtrates) as an inhibitor

a) ^{125}I-labelled synthetic pentapeptide subunits with C-terminal R-D-Ala_2-OH

b) ^{125}I-labelled glycopeptides lacking C-terminal R-D-Ala_2-OH isolated from peptidoglycan

Fig. 1. Scheme of the radioimmuno-assays for detecting secreted material containing a) peptides R-D-Ala_2-OH; b) glycan fragments.

Selective Detection of the Pentapeptide Subunit

Typical inhibition curves obtained in the R-D-Ala_2-OH specific RIA with culture filtrates from Gram-positive cocci are given in Fig. 2a. The specificity of the test system for the sequence R-D-Ala-D-Ala-OH should be pointed out, peptides or peptidoglycan fragments lacking the penultimate D-alanine (e.g. R-L-Lys-D-Ala-OH) showed no inhibition (7-10, Seidl and Schleifer, this book). The results obtained with the culture filtrates of numerous bacterial strains are summarized in Fig. 2b. Culture filtrates from yeasts as well as uninoculated nutrient broth did not inhibit. It is important to know that most of the staphylococci and streptococci listed in Fig. 2b inhibited up to 80%. The percentages given in the Figure represent the minimum inhibitory capacity obtained for the respective genera. All Gram-positive genera investigated secreted material containing peptides R-D-Ala_2-OH into the medium. The fact that we have found secreted material among streptococci in contrast to Mychajlonka et al. (4) may be due to the sensitivity of our test system. Bacilli, because of their strong

Fig. 2a

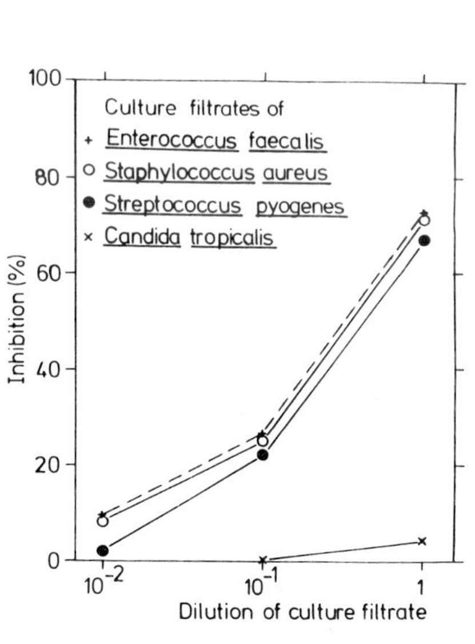

Fig. 2b

Genus	Number of species/strains investigated	Inhibition (%)
Staphylococcus	8/16	> 50
Streptococcus	8/10	> 60
Peptococcus	2/2	> 60
Bifidobacterium	2/2	> 60
Micrococcus	3/3	> 60
Corynebacterium	1	> 60
Clostridium	2/2	> 50
Lactobacillus	6/9	> 40
Bacillus	2/3	20-30
Escherichia	1	< 10
Salmonella	1	< 10
Proteus	1	< 10
Pseudomonas	1	< 10
Enterobacter	1	< 10
Candida	2/2	< 10
Saccharomyces	1	< 10

Fig. 2. Inhibition in binding of ^{125}I-L-Ala-D-Glu(L-Lys-D-Ala_2-OH)-OH to specific antiserum against the sequence R-D-Ala_2-OH (8) by culture filtrates.

carboxypeptidase activity, are known to exhibit only very small amounts of R-D-Ala_2-OH peptides in their peptidoglycan. We consequently detected only minor secretion of R-D-Ala_2-OH peptide-containing fragments among bacilli. In the case of all Gram-negative organisms studied, in full agreement with the literature (4), no secretion of material revealing the sequence R-D-Ala-D-Ala-OH was detected. Fragmentation of the culture filtrates by dialysis or ultrafiltration showed for most Gram-positive organisms that the bulk of secreted material containing R-D-Ala_2-OH peptides had a molecular weight between 10.000 and 30.000. The dialysable materials (presumably nucleotide activated precursors) contributed less but distinct to the inhibitory effect. However, in the case of some strains, the inhibition obtained was mostly due to secreted R-D-Ala_2-OH revealing material of molecular weight <10.000. This was clear comparing the inhibitory effect of undialysed culture filtrates from Lactobacillus acidophilus DSM 20242 (60%) with the negligible effect caused by dialysed material (compare Fig. 3a).

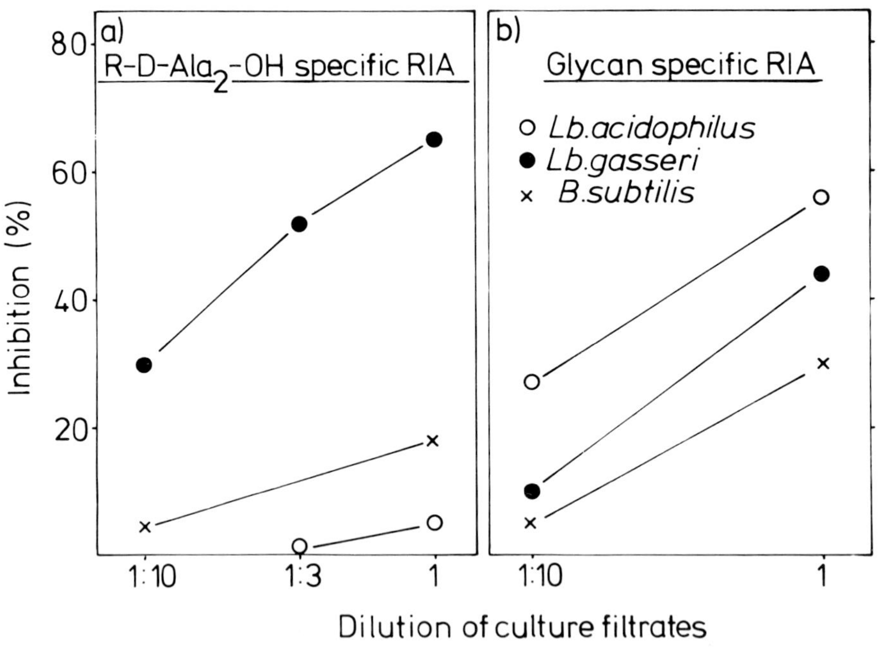

Fig. 3. Inhibition of binding by dialysed culture filtrates from (●) Lactobacillus gasseri DSM 20243, (o) Lactobacillus acidophilus DSM 20242, Bacillus subtilis W23 a) in the R-D-Ala_2-OH specific RIA; b) in the glycan specific RIA.

Selective Detection of the Glycan Moiety

The inhibition obtained with dialysed culture filtrates in the R-D-Ala_2-OH specific RIA or in a glycan specific RIA was compared. It is obvious from Fig. 3 that the soluble peptidoglycan material (MW > 10.000) secreted into the medium by Lb. gasseri DSM 20243 revealed pentapeptide subunits R-D-Ala_2-OH, whereas the peptidoglycan material of MW >10.000 secreted by Lactobacillus acidophilus DSM 20242 (Fig. 3b) lacked such pentapeptides (Fig. 3a).

Quantification of the Assays

Quantification of soluble peptidoglycan secreted into the medium was achieved by employing standard curves with known concentrations of soluble peptidoglycan. Soluble peptidoglycans for standardization were obtained from peptidoglycans by digestion with lysozyme and dialysis of the soluble material. The quantification of secreted peptidoglycan material (culture filtrate dialysed prior to determination) is depicted for the R-D-Ala_2-OH specific RIA in Fig. 4. The culture filtrates of staphylococci and of clinically

Fig. 4

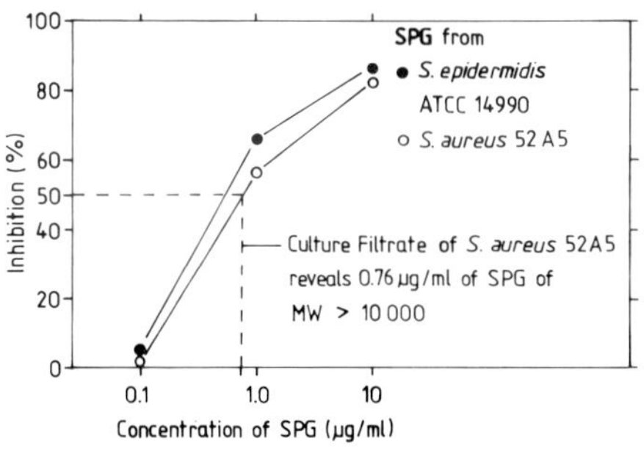

Fig. 5

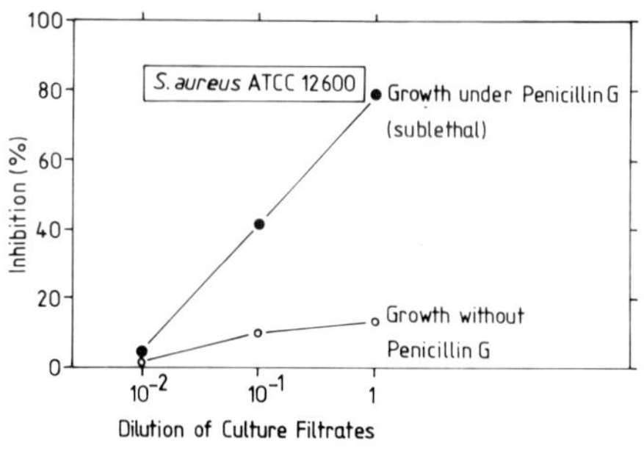

Fig. 4. Quantification of secreted peptidoglycan in dialysed culture filtrates. Standard curves obtained with soluble peptidoglycans prepared from highly purified peptidoglycans of S. aureus 52A5 or S. epidermidis ATCC 14990 by digestion with lysozyme and dialysis of the soluble material.

Fig. 5. Secretion of R-D-Ala_2-OH revealing peptidoglycan material of MW >10.000 by S. aureus ATCC 12600 grown under sublethal concentrations of penicillin G.

important streptococci (Str. pyogenes, Str. agalactiae, Str. equisimilis, Str. dysgalactiae, Str. group G biotype II (human), Enterococcus faecalis, E. faecium, Str. sanguis) contained peptidoglycan material of molecular weight > 10.000 in the range of 1 µg/ml (cells taken from the late logarithmic phase of growth).

Influence of Penicillin G on Secretion of Peptidoglycan

The influence of sublethal concentrations of penicillin G on soluble R-D-Ala_2-OH revealing peptidoglycan of molecular weight > 10.000 secreted by Staphylococcus aureus is depicted in Fig. 5. Cells from the logarithmic phase of growth were harvested, washed for complete removal of secreted material and divided into two samples. An aliquot was grown for 60 minutes under sublethal concentrations of penicillin G, cells grown without penicillin were used as a control.

Conclusion

Loss of peptidoglycan to the surrounding medium has both physiological and medical implications (1-6; Zeiger, this book). The presented test systems allow the determination of secreted peptidoglycans just by employing 10 µl of the culture filtrate as an inhibitor. The test systems are highly specific due to specificity of antibody source employed. Techniques hitherto used to study secretion of peptidoglycan fragments were based on radiolabelling cell walls in vivo and subsequent detection of radiolabelled material in the culture filtrate (1-6). These techniques necessitate incubation of cells in a "wall medium" and further unphysiological treatments. The test systems described in the present report allow studying the secretion of peptidoglycan under fully physiological conditions.

Acknowledgement

Part of this work was supported by grant 01 ZR 112 from the Bundesminister für Forschung und Technologie (BMFT).

References

1. Mauck, J., L. Chan, L. Glaser. 1971. J. Biol. Chem. 246, 1820-1827.

2. Glaser, L., B. Lindsay. 1977. J. Bacteriol. 130, 610-619.

3. Sinha, R.K., R.S. Rosenthal. 1980. Infect. Immun. 29, 914-925.

4. Mychajlonka, M., T.D. McDowell, G.D. Shockman. 1980. Infect. Immun. 28, 65-73.

5. Zeiger, A.R., W. Wong, A.N. Chatterjee, F.E. Young, C.U. Tuazon. 1982. Infect. Immun. 37, 1112-1118.

6. Doyle, R.J., M.A. Motley, P.H.B. Carstens. 1982. Carbohydrate Res. 104, 147-152.

7. Seidl, P.H., K.H. Schleifer. 1985 In: Environmental Regulation of Microbial Metabolism (I.S. Kualev, A.T. Severin, and E.A. Dawes, eds.). Academic Press, London, pp. 443-450.

8. Schleifer, K.H., P.H. Seidl. 1974. Eur. J. Biochem. 43, 509-519.

9. Franken, N., P.H. Seidl, E. Zauner, H.J. Kolb, K.H. Schleifer, L. Weiss. 1985. Mol. Immunol. 22, 573-579.

10. Heymer, B., D. Bernstein, K.H. Schleifer, R.M. Krause. 1975. Z. Immun. Forsch. exp. Ther. 149, 168-178.

11. Seidl, P.H., J.R. Golecki, N. Franken, P. Zwerenz, K.H. Schleifer. FEMS Microbiol. Lett. (in press).

12. Schleifer, K.H., L. Huss, O. Kandler. 1969. Arch. Microbiol. 68, 387-404.

EFFECTS OF PEPTIDOGLYCAN ON THE CELLULAR COMPONENTS OF THE IMMUNE SYSTEM

Roman Dziarski
Northwest Center for Medical Education, Indiana University School of Medicine, Gary, Indiana 46408

This article reviews the effects of peptidoglycan (PG) on lymphocytes and macrophages. The target cells, possible mechanisms of action and the biological and pathological consequences of PG interactions with these cells are discussed. The effects of highly polymerized PG and low molecular weight (MW) glycopeptides are considered.

Interactions of PG with Macrophages

One of the major results of interaction of PG with macrophages is the induction of secretion of various monokines (Table 1), which occurs both *in vivo* and *in vitro*. These monokines have a plethora of immunologic and biologic effects that can regulate differentiation and functions of various cells and affect numerous metabolic processes. Activation of macrophages by PG also results in increased metabolic activity of these cells, which has been demonstrated in a variety of systems (Table 1). Moreover, PG is directly chemotactic for macrophages and inhibits macrophage migration and phagocytosis. Most of these effects can be produced by both high MW PG and low MW glycopeptides, although some of these effects have only been studied with muramyl dipeptide (MDP), the synthetic analog of PG fragment (Table 1).

The exact mechanism of action of PG on macrophages at the molecular level is not known. It has been observed that PG is taken up by macrophages and digested into biologically active low MW glycopeptides, which are then released from the cells (72). It is still not clear, however, if such "processing" of PG by macrophages is required for the induction of the various biologic effects of PG (listed in Table 1), or even if such products of PG digestion play any major biologic role *in vivo*.

Biological Properties of Peptidoglycan

Table 1. Effects of Interaction of PG with Macrophages

Biologic effects	References
Monokine secretion	
Interleukin 1 (IL-1)	1-5
Endogenous pyrogen (EP)	6,7
Slow wave sleep promoting activity (SWS-PA)	8,9
Colony stimulating factor (CSF)	2
Colony stimulating activity (CSA)*	10-13
Synovial activator activity (SAA)	14
Chemotactic factor	15
Fibroblast-activating factor*	16,17
Thymocyte activating factor (TAF, not IL-1 or IL-2)*	18,19
Immunomodulating activities	1,16,20,21
Increased adherence and spreading	22-24
Increased synthesis of prostaglandins and cAMP	16,17
Increased glucose oxidation and glucosamine incorporation	25,26
Increased beta-glucosaminidase, lactate dehydrogenase and total protein synthesis	22,24
Collagenase release	16,17
Enhanced O_2^- generation*	24,27-33
Cytotoxic and cytostatic activity for tumors and virus-infected cells	31,32,34-49
Increased phagocytic and microbicidal activities*	24,50
Enhancement of virus uptake	51
Reticuloendothelial system stimulation	52-55
Aggregation *in vitro*	15
Granuloma formation	56-61
Chemotaxis	62,63
Inhibition of phagocytosis	64
Inhibition of migration	55,58,65-71
Toxicity	52
Degradation of PG and secretion of low MW glycopeptides	72

* Studied only with MDP

PG stimulates macrophages to release a group of closely related (if not identical) monokines; they are: interleukin 1 (IL-1), endogenous pyrogen (EP), leukocyte endogenous mediator (LEM) and slow wave sleep promoting activity (SWS-PA) (Table 1). It is striking that the major biologic effects of these IL-1-like monokines and PG are very similar (Table 2). Based on these observations, three hypotheses of the mechanism of PG activity can be proposed (Table 3). They are not mutually exclusive and there is some evidence to support each of these hypotheses (9,125-127, and Tables 1 and 2), but it should be stressed that they are far from proven. Since they can be tested experimentally, however, they should be useful in the future studies on the

Table 2. Comparison of Biological Effects of IL-1, EP, LEM, SWS-PA and PG

Effects of IL-1, EP, LEM and SWS-PA	PG*	References
Enhancement of thymocyte mitogenic responses	+	73-76
Fever	+	6,7,73-78
Slow wave sleep	+	8,9,75,79,80
Lekopenia followed by leukocytosis	+	73,75,77,78
Stimulation of acute-phase proteins	+	73,75,77,81
Release of granule content from neutrophils	+	73,77,78
Stimulation of oxygen metabolism	+	24,27-33,73,77,82
Chemotaxis	+	62,63,73,77,83
Bone resorption	+	73,75,77,84-87
T cell activation	+	73-75,88,89
B cell activation	+	75,78,88-95
Adjuvant activity	+	1,20,78,94,96-108
T cell replacing (helper) activity	+	103,104,109
Increased prostaglandin synthesis	+	16,17,73,75,77
Collagenase release	+	16,17,73,75,77
Increased cell chemiluminescence	+	77,110,111
Fibroblast proliferation	+	16,17,73-75,77
NK cell activation	+	73,75,112-114
Binding to monoclonal anti-MDP antibodies	+	115
Activity present in urine	+	9,79,116
Defective production and/or responses in humans and mice with autoimmune diseases	+	117-119
Stimulation of IL-2 production	?	73-75,77
Proteolysis in muscles	?	73,75,77
Decrease in plasma Fe and Zn	?	73,75,77
Lactoferrin release from neutrophils	?	77
Synergistic effects with anti-u antibodies and BCGF in B cell activation	?	75,90,120
Generation of stable E-rosette-forming T cells	?	121
Induction of procoagulant activity in macrophages and vascular endothelium	?	122,123
Stimulation of collagen production by mouse epithelial cells	?	124

* Effects exhibited (+) usually by both high MW PG and low MW glycopeptides; ?, not studied with PG and glycopeptides

Table 3. Hypotheses of PG Activity

1. IL-1, EP, LEM and SWS-PA are induced by PG and mediate its effects
2. PG can mimic or reproduce the effects of IL-1, EP, LEM and SWS-PA
3. PG fragments may serve as parts of the active sites of IL-1, EP, LEM and SWS-PA

Table 4. Effects of PG not Exhibited by IL-1, EP, LEM and SWS-PA*

Mitogenicity
Activation of polyclonal antibodies
Arthritogenicity
Lesions in the skin, heart and liver
Complement activation

* Reviewed in reference 78; for further discussion and references see section on interactions of PG with lymphocytes

mechanism of action of both PG and these monokines.

It should also be noted that there are several biologic effects of PG that are not exhibited by IL-1-like monokines (Table 4). It is interesting that these effects are most effectively induced by high MW PG, whereas the effects that are common to PG and IL-1-like monokines (Table 2) are induced by both high and low MW PG. Since the biological effects of polymerized (high MW) monokines have not been studied so far, it is difficult to assess if this difference in the activity of PG and IL-1-like monokines is merely due to the difference in their MWs. This latter possibility seems unlikely at least in some cases, such as arthritogenicity and production of chronic lesions in tissues, because poor biodegradability of PG seems to play a major role in the development of these lesions (56,57,66,78,128,129).

Interactions of PG with Lymphocytes

PG serves as a nonspecific polyclonal activator of lymphocytes (2,43,78,88,89, 92,117,118,130-149), i.e. it activates multiple clones of lymphocytes with various antigen-specific receptors. Lymphocytes of several species (mice, rats, rabbits, guinea pigs, humans) can be activated by PG.

Polyclonal activation of lymphocytes involves three to four separately controlled steps, in which cell proliferation is followed by differentiation and then expression of effector functions (150-155). These steps are: (i) stimulation of small resting lymphocytes to enlarge and enter early G_1 phase of the cell cycle; (ii) commitment of these cells to proliferate (late G_1 phase); (iii) cell proliferation (S, G_2 and M phases); and (iv) cell differentiation and maturation into effector cells, e.g. immunoglobulin (Ig)-secreting cells or cytotoxic T cells. The last two steps of this sequence may be actually overlapping, since some Ig-secreting cells may be still actively proliferating blasts, rather than fully mature non-proliferating plasma cells (149,156). Some polyclonal activators activate all the steps and some only certain steps of this sequence (149-154).

It is not known if PG activates lymphocytes in G_0 (resting phase) or G_1 (pre-DNA synthetic phase). It is well established, however, that PG drives the cells through all the steps of cell activation, inducing cell proliferation (mitogenic effect), which can be measured by the increase in thymidine incorporation into DNA (43,78,88,89,93,130-135,138-149), and differentiation and expression of effector functions (2,78,89,92,136,137,142-149). With most activators (antigens or mitogens), activated cells go through one mitotic cycle (G_1, S, G_2 and M phases), return to G_0 or early G_1 phase, and do not enter another mitotic cycle unless they are again stimulated (150-155). This also seems to be the case in PG-induced polyclonal activation, because continuous presence of PG during most of the culture period is required for maximal polyclonal activating effect (143).

In mice, PG is a T cell independent B cell mitogen and activator of polyclonal antibodies (78,130-132,136,137,141). In cultures of human peripheral blood lymphocytes (PBL), PG is a T cell dependent mitogen for both B and T cells, and relatively T cell independent activator of polyclonal antibodies (88,89,117). Similarly, in guinea pigs, both B and T cells are activated by PG (133).

The B cell subpopulation(s) activated by PG have not yet been characterized. From the work on other polyclonal B cell activators it is known that different subpopulations of B cells (B cells that are presumably at different stages of differentiation) may be activated by different activators (157-160). Several studies, however, indicate that these subpopulations are often largely overlapping (161,162). Moreover, the results from these experiments are frequently controversial and difficult to interpret, because direct reliable methods to study this problem are not available, and the situation is further complicated by the induction of numerous lymphokines and monokines and stimulation of various regulatory cells by polyclonal activators (153-155).

All B cells (99.9%) activated by PG in cultures of human PBL require continued DNA synthesis for the generation of Ig-secreting cells (149). Similar requirement is seen in the activation of human PBL by other polyclonal B cell activators (156). In the activation of mouse splenocytes by PG, the requirement for DNA synthesis is not so stringent, and approximately 10% of Ig-secreting cells still develop when DNA synthesis is inhibited (149). This finding indicates that in the mouse spleen there are partially differentiated and less differentiated B cells, and that cells in both of these populations can be activated by PG.

Experiments on the ontogenic development of polyclonal activation responses in mice indicate that the B cell subpopulations activated by PG and other polyclonal activators may be at least partially different (148). For example, the polyclonal antibody responses to lipopolysaccharide (LPS) are high at birth and mature completely during the first two weeks of life; whereas the responses to PG are low at birth and continue to develop throughout the first 8 weeks of life. Responsiveness to other polyclonal B cell activators is still lower at birth and follows different patterns of development (148). These data, therefore, indicate that PG activates more mature B cells than LPS, which appear later in ontogeny.

The minimal structure of PG responsible for polyclonal activation is not known and seems to depend on the species of the host, the type of cells studied and culture conditions. Based on the available data (2,43,78,88,89,92-95,117,118, 130-149,163-167), the following generalizations can be made: (i) large insoluble polymers are most effective polyclonal activators; (ii) the smaller the size, the lower the activity; (iii) when the size of PG preparation decreases, the T cell-activating properties are lost first (in humans and guinea pigs); (iv) then, the B cell-activating properties in humans are lost; (v) MDP and low MW glycopeptides are not active in humans, weakly or not active in mice, and weakly active in guinea pigs; (vi) polymerization increases the activity of MDP and makes inactive glycopeptides active.

It should be noted that whereas PG from most bacterial species do show lymphocyte-activating properties, PG from some bacterial species are inactive (2,43,130,137,141), and the differences in the composition of the peptides crosslinking their glycan strands (168) most likely account for this lack of activity. Similarly, other immunological properties of PG and low MW glycopeptides also depend on the structure of their peptide parts (127). However, there is no full correlation between polyclonal activating and other immunomodulating properties of these compounds (95,165).

The Significance of Polyclonal Activation

B cell activation results primarily in the generation of cells secreting polyclonal antibodies (2,78,89,92,136,137,143-155). There are also some indications that activated B cells secrete lymphokines (75,88,169-172), but these lymphokines have not yet been well characterized and purified. Activation of T cells leads to the secretion of lymphokines and/or generation of cytotoxic T cells (88,112,155).

Polyclonal activators are very useful in *in vitro* studies of the immune system. They serve as: (i) models for the antigen-induced responses; (ii) tools for analyzing lymphocyte activation and differentiation; (iii) probes of lymphocyte maturation and differentiation; and (iv) indicators of lymphocyte function in diseases affecting the immune system (155,173-178).

The role of polyclonal activation *in vivo* is not clear, and at least three hypotheses can be proposed to explain this phenomenon: (i) polyclonal activation is a primitive host defense mechanism, which can lead to quick induction of lymphokines and polyclonal antibodies cross-reacting with the antigens of the invading microorganisms during the early stages of infections (179-181); (ii) polyclonal activation could serve as a microbial virulence factor causing an aversion of the specific immune responses, i.e. polyclonally activated lymphocytes would not be able to mount specific immune responses against microbial antigens (there is no experimental evidence to support this hypothesis); (iii) polyclonal activation may play a role in the pathogenesis of some autoimmune diseases by inducing autoantibodies of various specificities and causing excessive polyclonal B cell activation; these two phenomena are the most prominent immunologic features of systemic lupus erythematosus (118,144-148,182-185).

The role of polyclonal activation in autoimmune diseases has been extensively studied, and it appears that PG and other polyclonal activators (e.g. LPS) can (under certain conditions) preferentially induce secretion of certain autoantibodies, such as anti-DNA and anti-autologous red cell antibodies (118, 144-148,183-185). Since this subject has been recently reviewed (185), here I will only point out some differences between PG and other polyclonal activators in the studies on autoimmunity-related polyclonal activation. One such difference was discovered when the ability of PG and other polyclonal activators to stimulate lymphocytes from various normal and genetically autoimmune strains of mice was studied (118). A selective defect in *in vivo* polyclonal antibody and autoantibody responses to PG of genetically autoimmune mice (both young and old) was discovered, whereas the *in vitro* responses of these mice to PG and other polyclonal activators as well as their *in vivo* responses to LPS were normal. The reason for this defective response of autoimmune mice to PG is not known. Another defect in polyclonal antibody responses to PG was observed in patients with rheumatoid arthritis (117), whose PBL have depressed responsiveness to PG, while their responses to another polyclonal activator, pokeweed mitogen, are normal. Suppressor T cells seem to be responsible for this abrogation of response to PG (117).

Immunomodulating Activities of PG

The adjuvant activity of PG has been extensively studied both *in vitro* and *in vivo* in a variety of systems. Most preparations of PG are adjuvant-active, including particulate PG and high and low MW glycopeptides, antigen-glycopeptide conjugates, water solutions and suspensions and oil emulsions (78,98,127,134,186-200). Covalent binding of PG to the antigen increases the adjuvant effect of PG (201). The adjuvant effect of PG is closely related to its structure, as type A PG are adjuvant-active, whereas type B PG are not (202). Similar relationship between structure and function is seen in adjuvant activity of low MW glycopeptides (127). Most types of immune responses are enhanced by PG, including specific primary and secondary systemic and secretory humoral and cell mediated responses, and nonspecific mitogenic responses (78, 98,127,134,186-203). *In vivo*, both parenteral and oral routes are effective (98,196).

Despite extensive research efforts, the mechanism(s) of adjuvant activity of PG and low MW PG fragments and its synthetic analogs is (are) still not known. Most of these studies were performed using MDP and other synthetic glycopeptides, and they show that multiple target cells are involved in the immunoenhancing effects of PG. The mechanism(s) of interaction(s) of PG with these cells at the molecular level is (are) still not known.

One target cell for the adjuvant activity of PG is the macrophage (20). The mechanism of this adjuvant effect seems to be mediated by IL-1 (106), which probably acts on T-helper cells (20).

Another target cell is the B cell, since MDP is also adjuvant-active in macrophage- and T cell-depleted cultures (103,107,108). In this system, MDP acts directly on the B cells in late stages of their differentiation and seems to have a lymphokine-like effect or act synergistically with lymphokines.

T-helper cells can also serve as targets for adjuvant activity of PG, since MDP was shown to enhance the activity of carrier-specific T-helper cells (99,102).

Most adjuvants can also have immunosuppressive effects under appropriate conditions of timing and dosage in relation to the antigenic stimulation (204). This is also true for PG, which, under appropriate conditions, can suppress specific antibody responses, generation of cytotoxic T cells, mitogenic responses and lymphokine production (21,78,134,204-215). The adjuvant, immunosuppressive, mitogenic and pyrogenic activities of PG can be dissociated by using different preparations, dosages, timings and routes of administration (125,127,134,217,218). This indicates that different mechanisms are involved in these activities.

The mechanisms of immunosuppressive activity of PG are not well understood. As is the case with its adjuvant effect, multiple target cells and multiple mechanisms seem to be involved. One of the target cells is the macrophage (21,216), in which MDP can induce synthesis of prostaglandin E_2, which in turn can suppress IL-2 production through suppressor T cells (216). MDP can also act directly on suppressor T cells (208,212-214), which results in the suppression of IL-2 production and generation of antigen-specific suppressor T cells (208,214). PG-induced suppression of mitogenic responses is in turn mediated by B cells (209).

The immunomodulating effects of PG and its synthetic analogs are important because of the potential application of these compounds in human and veterinary vaccines or as nonspecific immunostimulants in cancer immunotherapy, in patients with depressed immune responses, or even in patients with sleep disorders (125-127). It also seems likely that PG, as an ubiquitous component of bacterial cell walls, plays a natural immunoregulatory role as an enhancing agent in immune responses to microbial antigens and as a natural regulator of the development of the immune and hemopoietic system (127). On the other hand PG may also serve as a microbial virulence factor by suppressing the host immune responses.

Future Perspectives

Substantial amount of information has accumulated on the immunologic, biologic and pathologic effects of PG. Relatively little, however, is known about the mechanisms involved in these phenomena. As pointed out in this article, it is very likely that several diverse biologic effects of PG are mediated by just a few lymphokines (e.g. IL-1). Careful reevaluation of our knowledge of the biologic effects of PG and a synthetic approach to this knowledge are, therefore, needed.

Virtually nothing is known about the molecular mechanisms of interactions of PG with lymphocytes, macrophages and other mammalian cells. One of the possibilities is that PG (127), similarly to IL-1 (73), may act as a calcium ionophore. Elucidation of these mechanisms is necessary for our understanding of the diverse biologic effects of PG.

References

1. Oppenheim, J.J., A. Togawa, L. Chedid, S. Mizel. 1980. Cell. Immunol. 50, 71.

2. Mishell, R.I., Y.U. Chen, G.C. Clark, M.R. Gold, J.L. Hill, E. Kwan, D.A. Lee. 1981. In: Chemistry and Biological Activities of Bacterial Surface Amphiphiles. Academic Press, p. 327.

3. Damais, C., G. Riveau, M. Parant, J. Gerota, L. Chedid. 1982. Int. J. Immunopharmac. 4, 451.

4. Vacheron, F., M. Guenounou, C. Nauciel. 1983. Infect. Immun. 42, 1049.

5. Abehsira-Amar, O., C. Damais, M. Parant, L. Chedid. 1985. J. Immunol. 134, 365.

6. Oken, M.M., P.K. Peterson, B.J. Wilkinson. 1981. Infect. Immun. 31, 208.

7. Dinarello, C.A., R.J. Elin, L. Chedid, S.M. Wolff. 1978. J. Infect. Dis. 138, 760.

8. Krueger, J.M., J. Walter, C.A. Dinarello, S.M. Wolff, L. Chedid. 1984. Am. J. Physiol. 246, R994.

9. Krueger, J.M., J. Walter, C. Levin. 1985. In: Brain Mechanisms of Sleep (D.J. McGinty, ed.). Raven Press, p. 253.

10. Staber, F.G., L. Tarcsay, P. Duktor. 1978. Infect. Immun. 20, 40.

11. Wuest, B., E.D. Wachsmuth. 1982. Infect. Immun. 37, 452.

12. Galelli, A., L. Chedid. 1983. Infect. Immun. 42, 1081.

13. Galelli, A., P. Lefrancier, L. Chedid. 1984. Infect. Immun. 46, 495.

14. Hamilton, J.A., J.B. Zabriskie, L.B. Lachman, Y.-S., Chen. 1982. J. Exp. Med. 155, 1702.

15. Learn, D.B., I.S. Snyder. 1983. Infect. Immun. 39, 948.

16. Wahl, S.M., L.M. Wahl, J.B. McCarthy, L. Chedid, S.E. Mergenhagen. 1979. J. Immunol. 122, 2226.

17. Wahl, L.M., S.M. Wahl, J.B. McCarthy. 1980. In: Macrophage Regulation of Immunity. Academic Press, p. 491.

18. Iribe, H., T. Koga, K. Onoue. 1982. J. Immunol. 129, 1029.

19. Iribe, H., T. Koga, S. Kotani, S. Kusumoto, T. Shiba. 1983. J. Exp. Med. 157, 2190.

20. Fevrier, M., J.L. Birrien, C. Leclerc, L. Chedid, P. Liacopoulos. 1978. Eur. J. Immunol. 8, 558.

21. Dziarski, R. 1979. J. Reticuloendothel. Soc. 26, 239.

22. Davies, P., R.C. Page, A.C. Allison. 1974. J. Exp. Med. 139, 1262.

23. Tanaka, A., S. Nagao, K. Imai, R. Mori. 1980. Microbiol. Immunol. 24, 547.

24. Cummings, N.P., M.J. Pabst, R.B. Johnston. 1980. J. Exp. Med. 152, 1659.

25. Takada, H., M. Tsujimoto, K. Kato, S. Kotani, S. Kusumoto, M. Inage, T. Shiba, I. Yano, S. Kawata, K. Yokogawa. 1979. Infect. Immun. 25, 48.

26. Imai, K., M. Tomioka, S. Nagao, K. Kushima, A. Tanaka. 1980. Biomed. Res. 1, 300.

27. Pabst, M.J., R.B. Johnston. 1980. J. Exp. Med. 151, 101.

28. Pabst, M.J., N.P. Cummings, T. Shiba, S. Kusumoto, S. Kotani. 1980. Infect. Immun. 29, 617.

29. Wilson, C.B., J. Bohnsack, W.M. Weaver. 1982. Clin. Exp. Immunol. 49, 371.

30. Kaku, M., K. Yagawa, S. Nagao, A. Tanaka. 1983. Infect. Immun. 39, 559.

31. Lopez-Bernstein, G., K. Mehta, R. Mehta, R.L. Juliano, E.M. Hersh. 1983. J. Immunol. 130, 1500.

32. Schultz, R.M., S. Nanda, M.G. Altom. 1984. J. Immunopharmacol. 6, 257.

33. Yagawa, K., M. Kaku, Y. Ichinose, S. Nagao, A. Tanaka, A. Tomoda. 1984. Infect. Immun. 45, 82.

34. Nauciel, C., A.F. Goguel. 1977. J. Natl. Canc. Inst. 59, 1723.

35. Smialowicz, R.J., J.H. Schwab. 1977. Infect. Immun. 17, 599.

36. Taniyama, T., H.T. Holden. 1979. Cell. Immunol. 48, 369.

37. McLaughlin, C.A., S.M. Schwartzman, B.L. Horner, G.H. Jones, J.G. Moffatt, J.J. Nestor, D. Tegg. 1980. Science 208, 415.

38. Sone, S., I.J. Fidler. 1980. J. Immunol. 125, 2454.

39. Fidler, I.J., S. Sone, W.E. Fogler, Z.L. Barnes. 1981. Proc. Natl. Acad. Sci. USA 78, 1680.

40. Sone, S., I.J. Fidler. 1981. Cell. Immunol. 57, 42.

41. Fidler, I.J. 1981. J. Immunol. 127, 1719.

42. Bruley-Rosset, M., I. Florentin, G. Mathe. 1981. Molec. Cell. Biochem. 41, 113.

43. Goguel, A.F., G. Lespinats, C. Nauciel. 1982. J. Natl. Canc. Inst. 68, 657.

44. Fogler, W.E., J.E. Talmadge, I.J. Fidler. 1983. J. Reticuloendothel. Soc. 33, 165.

45. Sone, S., S. Mutsuura, M. Ogawara, E. Tsubura. 1984. J. Immunol. 132, 2105.

46. Fidler, I.J., A.J. Schroit. 1984. J. Immunol. 133, 515.

47. Sone, S., K. Tachibana, M. Shono, F. Ogushi, E. Tsubura. 1984. J. Biol. Resp. Modifiers 3, 185.

48. Koff, W.C., I.J. Fidler, S.D. Showalter, M.K. Chakrabarty, B. Hampar, L.M. Ceccorulli, E.S. Kleinerman. 1984. Science 224, 1007.

49. Leclerc, C., G.M. Bahr, L. Chedid. 1984. Cell. Immunol. 86, 269.

50. Jacobs, R.F., C.W. Wilson, A.L. Smith, J.E. Haas. 1983. Am. Rev. Respir. Dis. 128, 862.

51. Hotta, H., S. Hotta, H. Takada, S. Kotani, S. Tanaka, M. Ohki. 1983. Infect. Immun. 41, 462.

52. Dziarski, R. 1977. Zbl. Bakt. Hyg. I Abt. A238, 320.

53. Tanaka, A., S. Nagao, R. Nagao, S. Kotani, T. Shiba, S. Kusumoto. 1979. Infect. Immun. 24, 302.

54. Waters, R.V., R.W. Farraresi. 1980. J. Reticuloendothel. Soc. 28, 457.

55. Caravano, R., J. Oberti. 1981. Ann. Immunol. (Inst. Pasteur) 132C, 257.

56. Cromartie, W.J., J.G. Craddock. 1960. Science 154, 285.

57. Ohanian, S.H., J.H. Schwab, W.J. Cromartie. 1969. J. Exp. Med. 129, 27.

58. Heymer, B., B. Bultmann, O. Haferkamp. 1971. J. Immunol. 106, 858.

59. Heymer, B., H. Schafer, W. Schachenmayr, O. Haferkamp, W.C. Schmidt. 1971. Arch. Pathol. 92, 84.

60. Emori, K., A. Tanaka. 1978. Infect. Immun. 19, 613.

61. Tanaka, A., K. Emori. 1980. Am. J. Pathol. 98, 733.

62. Ogawa, T., S. Kotani, K. Fukuda, Y. Tsukamoto, M. Mori, S. Kusumoto, T. Shiba. 1982. Infect. Immun. 38, 817.

63. Ogawa, T., S. Kotani, S. Kusumoto, T. Shiba. 1983. Infect. Immun. 39, 449.

64. Smialowicz, R.J., J.H. Schwab. 1978. Infect. Immun. 20, 258.

65. Heymer, B., B. Bultmann, W. Schachenmayr, R. Spanel, O. Haferkamp, W.C. Schmidt. 1973. J. Immunol. 111, 1743.

66. Heymer, B. 1975. Z. Immunitaetsforsch. Exp. Klin. Immunol. 149, 245.

67. Bultmann, B., B. Heymer, K.H. Schleifer, H.P. Seidl, O. Haferkamp. 1975. Z. Immunitaetsforsch. Exp. Klin. Immunol. 20, 289.

68. Grov, A., S. Helgeland, C. Endersen. 1976. Zbl. Bakt. Hyg. I Abt. Suppl. 5, 389.

69. Yamamoto, Y., S. Nagao, A. Tanaka, T. Koga, K. Onoue. 1978. Biochem. Biophys. Res. Commun. 80, 923.

70. Adam, A., V. Souvannavong, E. Lederer. 1978. Biochem. Biophys. Res. Commun. 85, 684.

71. Nagao, S., A. Tanaka, Y. Yamamoto, T. Koga, K. Onoue, T. Schiba, K. Kusumoto, S. Kotani. 1979. Infect. Immun. 24, 308.

72. Vermeulen, M.W., G.R. Gray. 1984. Infect. Immun. 46, 476.

73. Dinarello, C.A. 1984. Rev. Infect. Dis. 6, 51.

74. Gery, I., J.L. Lepe-Zuniga. 1984. Lymphokines 9, 109.

75. Durum, S.K., J.A. Schmidt, J.J. Oppenheim. 1985. Ann. Rev. Immunol. 3, 263.

76. Iribe, H., Y. Koga. 1984. Cell. Immunol. 88, 9.

77. Kampschmidt, R.F. 1984. J. Leukocyte Biol. 36, 341.

78. Dziarski, R. 1981. In: Immunomodulation by Bacteria and Their Products (H. Friedman, T.W. Klein, A. Szentivanyi, eds.). Plenum Press, p. 95.

79. Krueger, J.M., J.R. Pappenheimer, M.L. Karnovsky. 1982. J. Biol. Chem. 257, 1664.

80. Krueger, J.M., J. Walter, M.L. Karnovsky, L. Chedid, J.P. Choay, P. Lefrancier, E. Lederer. 1984. J. Exp. Med. 159, 68.

81. McAdam, K.P.W.J., N.T. Foss, C. Garcia, R. DeLellis, L. Chedid, R.J.W. Rees, S.M. Wolff. 1983. Infect. Immun. 39, 1147.

82. Leong, P.A., J.H. Schwab, M.S. Cohen. 1984. Infect. Immun. 45, 160.

83. Miossec, P., C.-L. Yu, M. Ziff. 1984. J. Immunol. 133, 2007.

84. Lensgraf, E.J., J.J. Greenblatt, J.W. Bawden. 1979. Archs Oral Biol. 24, 495.

85. Gowen, M., D.D. Wood, E.J. Ihrie, M.K.B. McGuire, G.G. Russell. 1983. Nature 306, 387.

86. Dewhirst, F.E. 1982. Infect. Immun. 35, 133.

87. Raisz, L.G., C. Alander, G. Eilon, S.P. Whitehead, K. Nuki. 1982. Calcif. Tissue Int. 34, 365.

88. Rasanen, L., H. Arvilommi. 1981. Infect. Immun. 34, 712.

89. Levinson, A.I., A. Dziarski, B. Zweiman, R. Dziarski. 1983. Infect. Immun. 39, 290.

90. Howard, M., S.B. Mizel, L. Lachman, J. Ansel, B. Johnson, W.E. Paul. 1983. J. Exp. Med. 157, 1529.

91. Wood, D.D., E.K. Bayne, M.B. Goldring, M. Gowen, D. Hamerman, J.L. Humes, E.J. Ihrie, P.E. Lipsky, M.J. Staruch. 1985. J. Immunol. 134, 895.

92. Babu, U.M., A.R. Zeiger. 1983. Infect. Immun. 42, 1013.

93. Damais, C., M. Parant, L. Chedid. 1977. Cell. Immunol. 34, 49.

94. Specter, S., R. Cimprich, H. Friedman, L. Chedid. 1978. J. Immunol. 120, 487.

95. Wood, D.D., M.J. Staruch. 1981. Int. J. Immunopharmac. 3, 31.

96. Staruch, M.J., D.D. Wood. 1983. J. Immunol. 130, 2191.

97. Curtis, J.L., A.A. Nordin. 1984. Immunology 52, 711.

98. Chedid, L., F. Audibert, P. Lefrancier, J. Choay, E. Lederer. 1976. Proc. Nat. Acad. Sci. USA 73, 2472.

99. Lowy, I., C. Bona, L. Chedid. 1977. Cell. Immunol. 29, 195.

100. Floc'h, F., D. Migliore-Samour, R. Maral, G.H. Werner, P.J. Jolles. 1978. Immunol. Commun. 7, 41.

101. Leclerc, C., I. Lowy, L. Chedid. 1978. Cell. Immunol. 38, 286.

102. Sugimoto, M., R.N. Germain, L. Chedid, B. Benacerraf. 1978. J. Immunol. 120, 980.

103. Watson, J., C. Whitlock. 1978. J. Immunol. 121, 383.

104. Leclerc, C., E. Bourgeois, L. Chedid. 1979. Immunol. Commun. 8, 55.

105. Staruch, M.J., D.D. Wood. 1982. J. Immunol. 128, 155.

106. Wood, D.D., M.J. Staruch, P.L. Durette, W.V. Melvin, B.K. Graham. 1983. In: Interleukins, Lymphokines, and Cytokines. Academic Press, p. 691.

107. Souvannavong, V., L. Rimsky, A. Adam. 1983. Biochem. Biophys. Res. Commun. 114, 721.

108. Souvannavong, V., A. Adam. 1984. Biochem. Biophys. Res. Commun. 125, 431.

109. Farrar, J.J., M.L. Hilfiker. 1982. Fed. Proc. 41, 263.

110. Masihi, K.N., I. Azuma, W. Brehmer, W. Lange. 1983. Infect. Immun. 40, 16.

111. Masihi, K.N., W. Brehmer, I. Azuma, W. Lange, S. Muller. 1984. Infect. Immun. 43, 233.

112. Goguel, A.F., B. Payelle, G. Lespinats, C. Nauciel. 1983. J. Natl. Canc. Inst. 71, 325.

113. Goguel, A.F., B. Payelle, P.C. Quan, G. Lespinats. 1984. J. Natl. Canc. Inst. 73, 697.

114. Sharma, S.D., V. Tsai, J.L. Krahenbuhl, J.S. Remington. 1981. Cell. Immunol. 62, 101.

115. Chedid, L., G.M. Bahr, G. Riveau, J.M. Krueger. 1984. Proc. Natl. Acad. Sci. USA 81, 5888.

116. Kimball, E.S., S.F. Pickeral, J.J. Oppenheim, J.L. Rossio. 1984. J. Immunol. 133, 256.

117. Pardo, I., C. Carafa, R. Dziarski, A.I. Levinson. 1984. Clin. Exp. Immunol. 56, 253.

118. Dziarski, R. 1985. J. Clin. Lab. Immunol. 16, 93.

119. Alcocer-Varela, J., A. Laffon, D. Alarcon-Segovia. 1984. Clin. Exp. Immunol. 55, 125.

120. Falkoff, R.J.M., A. Muraguchi, J.X. Hong, J.L. Butler, C.A. Dinarello, A.S. Fauci. 1983. J. Immunol. 131, 801.

121. Ben-Zvi, A., S.B. Mizel, J.J. Oppenheim. 1981. Clin. Immunol. Immunopathol. 19, 330.

122. Farram, E., C.L. Geczy, D.K. Moon, K. Hopper. 1983. J. Immunol. 130, 2750.

123. Bevilacqua, M.P., J.S. Pober, G.R. Majeau, R.S. Cotran, M.A. Gimbrone. 1984. J. Exp. Med. 160, 618.

124. Matsushima, K., M. Bano, W.R. Kidwell, J.J. Oppenheim. 1985. J. Immunol. 134, 904.

125. Chedid, L. 1983. In: Progress in Immunology V. Academic Press, p. 1349.

126. Leclerc, C., L. Chedid. 1984. Int. J. Tiss. Reac. 6, 213.

127. Adam, A., E. Lederer. 1984. Med. Res. Rev. 4, 111.

128. Cromartie, W.J., J.G. Craddock, J.H. Schwab, S.K. Anderle, C.H. Yang. 1977. J. Exp. Med. 146, 1585.

129. Fox, A., R.R. Brown, S.K. Anderle, C. Chetty, W.J. Cromartie, H. Gooder, J.H. Schwab. 1982. Infect. Immun. 35, 1003.

130. Damais, C., C. Bona, J. Fleck, C. Nauciel, J.P. Martin. 1975. J. Immunol. 115, 268.

131. Ciorbaru, R., J.F. Petit, E. Lederer, C. Bona, L. Chedid. 1977. Ann. Immunol. (Inst. Pasteur) 128C, 41.

132. Dziarski, R., A. Dziarski. 1979. Infect. Immun. 23, 706.

133. Takeda, H., M. Tsujimoto, S. Kotani, S. Kusumoto, M. Inage, T. Shiba, S. Nagao, I. Yano, S. Kawata, K. Yokogawa. 1979. Infect. Immun. 25, 645.

134. Dziarski, R. 1979. Infect. Immun. 26, 508.

135. Dziarski, R., A. Dziarski, A.I. Levinson. 1980. Int. Archs Allergy Appl. Immunol. 63, 383.

136. Dziarski, R. 1980. J. Immunol. 125, 2478.

137. Saito-Taki, T., M.J. Tanabe, H. Mochizuki, T. Matsumoto, M. Nakano, H. Takada, M. Tsujimoto, S. Kotani, S. Kusumoto, T. Shiba, K. Yokogawa, S. Kawata. 1980. Microbiol. Immunol. 24, 209.

138. Saito-Taki, T., M.J. Tanabe, H. Mochizuki, M. Nakano, M. Tsujimoto, S. Kotani, K. Yokogawa, S. Kawata. 1980. Microbiol. Immunol. 24, 249.

139. Takada, H., S. Nagao, S. Kotani, S. Kawata, K. Yokogawa, S. Kusumoto, T. Shiba, I. Yano. 1980. Biken J. 23, 61.

140. Vendrell, J.P., A. Escande, A. Cannat, M.F. Huguet, A. Serre. 1980. Immunology 40, 41.

141. Guenounou, M., A.F. Goguel, C. Nauciel. 1982. Ann. Immunol. (Inst. Pasteur) 133D, 3.

142. Rasanen, L., U.P. Mustikkamaki, H. Arvilommi. 1982. Immunology 46, 481.

143. Dziarski, R. 1982. Infect. Immun. 35, 507.

144. Dziarski, R. 1982. J. Immunol. 128, 1018.

145. Dziarski, R. 1982. J. Immunol. 128, 1026.

146. Dziarski, R. 1984. J. Immunol. 133, 2537.

147. Dziarski, R. 1984. Immunology 53, 563.

148. Dziarski, R. 1985. Develop. Compar. Immunol. 9, 119.

149 Dziarski, R. 1985. Immunol. Letters 9, 161.

150. DeFranco, A.L., J.T. Kung, W.E. Paul. 1982. Immunol. Rev. 64, 161.

151. DeFranco, A.L., J.D. Ashwell, R.H. Schwartz, W.E. Paul. 1984. J. Exp. Med. 159, 861.

152. Wetzel, G.D., S.L. Swain, R.W. Dutton, J.R. Kettman. 1984. J. Immunol. 133, 2327.

153. Paul, W.E. 1984. In: Progress in Immunology V. Academic Press, p. 727.

154. Kehrl, J.H., A. Muraguchi, J.L. Butler, R.J.M. Falkoff, A.S. Fauci. 1984. Immunol. Rev. 78, 75.

155. Ashman, R.F. 1984. In: Fundamental Immunology (W.E. Paul, ed.). Raven Press, p. 267.

156. Jelinek, D.F., P.E. Lipsky. 1983. J. Immunol. 130, 2597.

157. Gronowicz, E., A. Coutinho. 1974. Transplant. Rev. 24, 3.

158. Gronowicz, E., A. Coutinho. 1974. Eur. J. Immunol. 4, 771.

159. Diamanstein, T., E. Blitstein-Willinger, G. Schulz. 1974. Nature 250, 596.

160. Bona, C., A. Yano, A. Dimitru, R.G. Miller. 1978. J. Exp. Med. 148, 136.

161. Melchers, F., J. Andersson. 1974. Eur. J. Immunol. 4,687.

162. Andersson, J., A. Coutinho, F. Melchers. 1979. J. Exp. Med. 149, 553.

163. Takada, H., S. Kotani, S. Kusumoto, Y. Tarumi, K. Ikenaka, T. Shiba. 1977. Biken J. 20, 81.

164. Damais, C., M. Parant, L. Chedid, P. Lefrancier, J. Choay. 1978. Cell. Immunol. 35, 173.

165. Damais, C., M. Parant, L. Chedid. 1978. In: Application of Cancer Chemotherapy, Antibiotics Chemotherapy. S. Karger, p. 19.

166. Parant, M., C. Damais, F. Audibert, F. Parant, L. Chedid, E. Sache, P. Lefrancier, J. Choay, E. Lederer. 1978. J. Infect. Dis. 138, 378.

167. Bahr, G.M., F.Z. Modabber, M. Morin, M. Terrier, A. Eyquem, L. Chedid. 1984. Clin. Exp. Immunol. 57, 178.

168. Schleifer, K.H., O. Kandler. 1972. Bacteriol. Rev. 36, 407.

169. Rasanen, L., E. Karhumaki, H. Arvilommi. 1978. J. Immunol. 121, 418.

170. Gordon, J., S.C. Ley, M.D. Melamed, L.S. English, N.C. Hughes-Jones. 1984. Nature 310, 145.

171. Brooks, K.H., J.W. Uhr, E.S. Vitetta. 1984. J. Immunol. 133, 3133.

172. del Guercio, P., M.F. del Guercio, D.H. Katz. 1985. J. Immunol. 134, 996.

173. Moller, G. 1975. Transplant. Rev. 23, 126.

174. Gronowicz, E., A. Coutinho. 1975. Transplant. Rev. 24, 3.

175. Oppenheim, J.J., D.L. Rosenstreich (eds.). 1976. Mitogens in Immunobiology. Academic Press.

176. Melchers, F. 1977. In: Development of Host Defenses (M.D. Cooper, D.M. Dayton, eds.). Raven Press, p. 11.

177. Fauci, A.C., R. Ballieux (eds.). 1979. Antibody Production in Man, In Vitro Synthesis and Clinical Implications. Academic Press.

178. Vitetta, E.S., K. Brooks, P. Isakson, J. Layton, E. Pure, D. Yuan. 1984. In: Fundamental Immunology (W.E. Paul, ed.). Raven Press, p. 221.

179. Clagett, J.A., D. Engel. 1978. Develop. Compar. Immunol. 2, 235.

180. Rynnel-Dagoo, B. 1978. Clin. Exp. Immunol. 34, 402.

181. Rasanen, L., U.P. Mustikkamaki, P. Leinikki. 1981. Acta Path. Microbiol. Scand. C89, 373.

182. Izui, S., P.J. McConahey, F.J. Dixon. 1978. J. Immunol. 121, 2213.

182. Teodorescu, M. 1981. Immunol. Rev. 55, 155.

183. Hang, L.M., J.H. Slack, C. Amundson, S. Izui, A.N. Theofilopoulos, F.J. Dixon. 1983. J. Exp. Med. 157, 874.

184. Theofilopoulos, A.N., G.J. Prud'homme, T.M. Fieser, F.J. Dixon. 1983. Immunol. Today 4, 287.

185. Dziarski, R. 1985. Clin. Immunol. Newsletter 6, 89.

186. Holton, J.B., J.H. Schwab. 1966. J. Immunol. 96, 134.

187. Adam, A., R. Ciorbaru, F. Elluoz, J.F. Petit, E. Lederer. 1974. Biochem. Biophys. Res. Commun. 56, 561.

188. Nauciel, C., J. Fleck, J.P. Martin, M. Mock, H. Nguyen-Huy. 1974. Eur. J. Immunol. 4, 352.

189. Schleifer, K.H., B. Heymer (eds.). 1975. Z. Immunitaetsforsch. Exp. Klin. Immunol. 149, 103.

190. Kotani, S., T. Narita, D.E.S. Stewart-Tull, T. Shimono, Y. Watanabe, K. Kato, S. Iwata. 1975. Biken J. 18, 77.

191. Kotani, S., Y. Watanabe, T. Shimono, F. Kinoshita, T. Narita, K. Kato, D.E.S. Stewart-Tull, I. Morisaki, K. Yokogawa, S. Kawata. 1975. Biken. J. 18, 93.

192. Kotani, S., Y. Watanabe, F. Kinoshita, T. Shimono, I. Morisaki, T. Shiba, S. Kusumoto, Y. Tarumi, K. Ikenaka. 1975. Biken J. 18, 105.

193. Pryjma, J., K. Pryjma, A. Grov, P.B. Heczko. 1976. Zbl. Bakt. Hyg. Abt. I Suppl. 5, 873.

194. Kotani, S., Y. Watanabe, F. Kinoshita, K. Kato, H. Perkins. 1977. Biken J. 20, 87.

195. Igarashi, T., M. Okada, I. Azuma, Y. Yamamura. 1977. Cell. Immunol. 34, 270.

196. Chedid, L., F. Audibert. 1977. J. Infect. Dis. 136, S246.

197. Ohkuni, H., Y. Norose, M. Ohta, M. Hayama, Y. Kimura, M. Tsujimoto, S. Kotani, T. Shiba, S. Kusumoto, K. Yokogawa, S. Kawata. 1979. Infect. Immun. 24, 313.

198. Hrsak, I., J. Tomasic, K. Pavelic, Z. Valinger. Z. Immunitaetsforsch. Exp. Klin. Immunol. 155, 312.

199. Saiki, I., K.I. Kamisango, Y. Tanio, H. Okumura, Y. Yamamura, I. Azuma. 1982. Infect. Immun. 38, 58.

200. Kamisango, K.I., I. Saiki, Y. Tanio, H. Okumura, Y. Araki, I. Sekikawa, I. Azuma, Y. Yamamura. 1982. J. Biochem. 92, 23.

201. Langbeheim, H., R. Arnon, M. Sela. 1978. Immunology 35, 573.

202. Kotani, S., Y. Watanabe, F. Kinoshita, K. Kato, K.H. Schleifer, H.R. Perkins. 1977. Biken J. 20, 1977.

203. Taubman, M.A., J.L. Ebersole, D.J. Smith, W. Stack. 1983. Ann. N. Y. Acad. Sci. 409, 637.

204. Schwab, J.H. 1975. Bacteriol. Rev. 39, 121.

205. Dziarski, R. 1978. Int. Archs Allergy Appl. Immunol. 57, 304

206. Leclerc, C., D. Juy, L. Chedid. 1979. Cell. Immunol. 42, 336.

207. Leclerc, C., D. Juy, E. Bourgeois, L. Chedid. 1979. Cell. Immunol. 45, 199.

208. Kishimoto, T., Y. Hirai, K. Nakanishi, I. Azuma, A. Nagamatsu, Y. Yamamura. 1979. J. Immunol. 123, 2709.

209. Dziarski, R., 1980. Infect. Immun. 30, 431.

210. Souvannavong, V., A. Adam. 1980. Eur. J. Immunol. 10, 654.

211. Lowy, I., C. Leclerc, E. Bourgeois, L. Chedid. 1980. J. Immunol. 124, 320.

212. Leclerc, C., E. Bourgeois, L. Chedid. 1982. Eur. J. Immunol. 12, 249.

213. Leclerc, C., L. Chedid. 1983. In: Interleukins, Lymphokines, and Cytokines. Academic Press, p. 699.

214. Ferguson, T.A., N.J. Krieger, A. Pesce, J.G. Michael. 1983. Infect. Immun. 39, 800.

215. Gouby, A., R. Caravano, M. Ramuz. 1983. Immunobiol. 164, 343.

216. Leclerc, C., A. Morin, E. Deriaud, L. Chedid. 1984. J. Immunol. 133, 1996.

217. Specter, S., H. Friedman, L. Chedid. 1977. Proc. Soc. Exp. Biol. Med. 155, 349.

218. Parant, M. 1983. In: Regulation of the Immune Response (P.L. Ogra, D.M. Jacobs, eds.). S. Karger, p. 195.

SUBOPTIMAL CONCENTRATIONS OF LPS ARE NECESSARY FOR IN VITRO ACTIVATION OF RAT ALVEOLAR MACROPHAGES BY MURAMYLPEPTIDES.

G. Barratt, J.-P. Tenu, J.-F. Petit
Institut de Biochimie, Bât. 432, Université de Paris-Sud,
91405 Orsay, France

D. Nolibe
Laboratoire de Toxicologie des Transuraniens, I.P.S.N.,
C.E.A., 91680 Bruyères-le-Châtel, France.

Introduction

Among the many interesting properties of the bacterial immunomodulator N-acetyl-muramyl-L-alanyl-D-isoglutamine (MDP, 1) is its capacity for activating macrophages for cytotoxicity against tumour cells (2). However, the results *in vivo* are disappointing, since the free compound is rapidly excreted (3), and the uptake of MDP by macrophages is a relatively slow process. Tenu *et al.* (4) were unable to find evidence of surface receptors for a tritiated MDP derivative on mouse peritoneal macrophages. Recently, Karnovsky *et al.* detected specific binding of an iodinated derivative to these cells (Silverman and Karnovsky, submitted). These two results are compatible when one considers the higher specific activity of the latter product.

Fidler *et al.* (5, 6) have encapsulated muramylpeptides in liposomes, and this preparation, injected intravenously, can activate alveolar macrophages and protect mice against lung metastases of experimental tumours. We were therefore interested by the interaction of MDP with alveolar macrophages, and began by investigating these effects *in vitro*, in a rat model. We payed particular attention to the effects of bacterial lipopolysaccharide (LPS) in our system, since this ubiquitous substance has also been reported to activate macrophages (7).

Biological Properties of Peptidoglycan

Methods

Alveolar macrophages were lavaged from the lungs of female WAG rats and allowed to adhere in 96 well plates, in culture medium containing 5 % serum. The LPS content of the serum was measured by the LAL assay using a fluorogenic substrate which allowed picogram amounts to be detected (8). We used one batch of serum containing 300 ng LPS/ml and another which was free of detectable LPS.

The macrophages were then treated for 24 hours with medium containing immunomodulators, then washed and cocultured with syngeneic tumour cells from the P77 radio-induced fibrohistiocytoma. At the same time [^{3}H]-thymidine, 1.2 μM, specific radioactivity 1 Ci/mmole was added and the incorporation over 20 hours was measured. The results are expressed as percentage growth inhibition (% G.I.).

$$\% \text{ G.I.} = 100 \left(I - \frac{X}{R}\right)$$

when X is the incorporation into the DNA of tumour cells cultured with treated macrophages and R the incorporation into tumour cells cocultured with macrophages exposed to medium only. The latter value was always 80 % or more of the incorporation into tumour cells cultured alone, thus the macrophages were not intrinsically cytotoxic.

Results

Figure 1 shows a typical dose response curve obtained when MDP was presented to macrophages in medium containing about 15 ng/ml of LAL-positive material derived from the supplementing serum. At concentrations above 1 μg/ml MDP always produced maximal activation of the macrophages, leading to 80-90 % growth inhibition at an effector : target ratio of 8. Similar results were obtained at an E : T ratio of 4. The concentration of MDP necessary to produce half this maximum effect was between 0.1 μg/ml and 0.004 μg/ml depending on the particular macrophage preparation ; in the experiment shown in figure 1, this value was about 0.05 μg/ml.

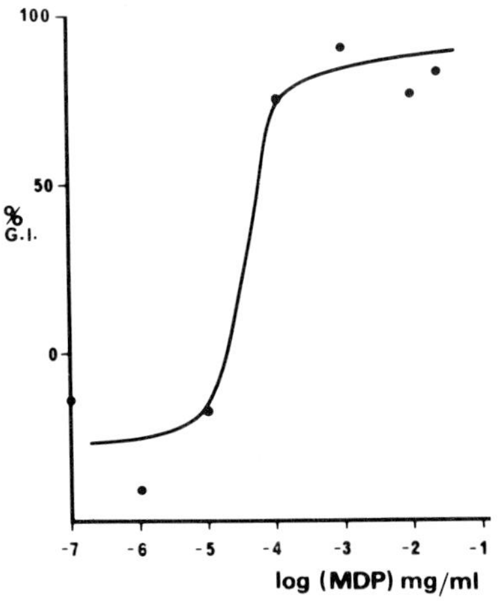

Figure 1. Activation of rat alveolar macrophages by MDP in endotoxin-containing serum.

Rat alveolar macrophages were incubated at 8 x 10^5/ml in 96-well plates with MDP at various concentrations in medium supplemented with 5 % serum (final LPS concentration approximately 15 ng/ml). After 24 hours, the macrophages were washed and 10^5 target cells/ml were added in the presence of [^{3}H]-thymidine for 20 hours. Incorporation of label into tumour cell DNA was measured and percentage G.I. was calculated as described in Methods.

Figure 2 shows that when macrophages are treated with 25 µg/ml of MDP in medium supplemented with endotoxin-free serum, they do not become cytotoxic. However, in the presence of 25 µg/ml MDP and 20 ng/ml of bacterial endotoxin the cytotoxicity was similar to that of MDP in endotoxin-containing medium (Figure 1). 20 ng/ml of LPS alone was without effect. In fact LPS was not active at doses up to 1 µg/ml. (data not shown).

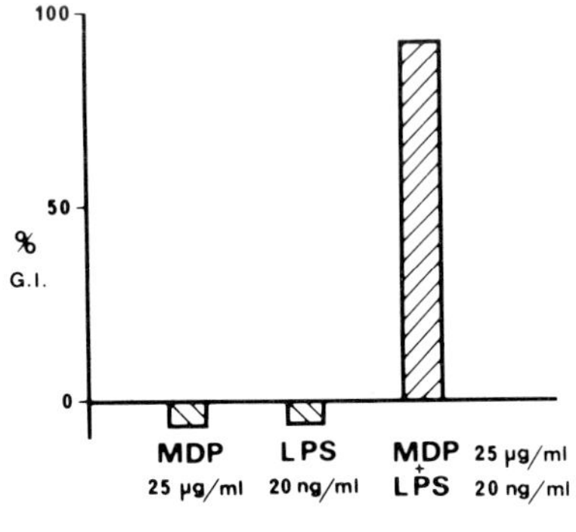

Figure 2. Effects of MDP and LPS on rat alveolar macrophages in endotoxin-free medium.

Rat alveolar macrophages were incubated at 8 x 10^5/ml in 96-well plates with either 25 µg/ml MDP, or 20 ng/ml LPS, or 25 µg/ml MDP mixed with 20 ng/ml LPS in medium supplemented with endotoxin-free serum. After 24 h, the macrophages were washed and 10^5 target cells/ml were added in the presence of [^{3}H]-thymidine for 20 h. Incorporation into tumour cell DNA was measured and percentage G.I. was calculated as described in Methods.

We have performed preliminary experiments to determine whether macrophages can respond to MDP and LPS presented in sequence. For maximal activation, it is necessary for MDP and LPS to be present together during the period immediately after adherence, since macrophage sensitivity diminished after several hours.

Discussion

The data presented above demonstrate clearly that in order to activate WAG rat alveolar macrophages for cytotoxicity against a syngeneic tumour target both muramyl dipeptide and lipopolysaccharide are necessary. It is evident from Figure 1 that 1 μg/ml of MDP associated with about 15 ng/ml of LPS invokes a maximal response in these cells, while 25 μg/ml of MDP, or 1 μg/ml of LPS, are ineffective alone (Fig. 2). Thus, there is a strong synergism between MDP and LPS in this experimental model.

There have been several reports of complementary actions of MDP and other biological response modifiers for the induction of macrophage tumour-directed cytotoxicity. For example, Grand-Perret et al. (9) have shown that MDP can bring mouse peritoneal macrophages primed by TDM in vivo to the fully activated state. Fidler et al. (10) have demonstrated a synergism between MDP and lymphokines encapsulated in liposomes for the activation of murine alveolar macrophages in vivo.

In our experiments, we found a variable threshold for the MDP effects. This may be due to the presence of muramylpeptides and similar compounds within the macrophages, as a result of bacterial degradation, since it is known that macrophages play an important role in the elimination of bacteria after infection (11). Sen and Karnovsky (12) have made a qualitative demonstration of endogenous muramic-acid-containing low molecular weight compounds in rat tissues by extraction and thin layer chromatography.

Such products might sensitise the macrophages to exogenous MDP, if a threshold concentration is necessary for activation ; alternatively they might desensitise the cells, for example by the down regulation of a receptor. Thus, the response to MDP would depend on the animals' previous exposure to micro-organisms.

The precise role of LPS in this system could be elucidated by the use of other bacterial lipidic products such as lipoteichoic acids, and by the use of the Lipid A and core portions of LPS ; in particular the synthetic Lipid A which has recently become available (13). The role of liposomes, as used by Fidler et al. (5-7, 10), is interesting in this regard. It may be that, as well as promoting the entry of MDP into the cells, liposomes might also provide an essential "second signal"which bypasses the requirement for LPS for macrophage activation.

Such a detailed study of the MDP and LPS effects in vitro should aid the design of immunomodulator therapy of metastatic tumours in vivo.

Acknowledgements

This work was supported by C.N.R.S (U.A. 1116) and by a grant from the Ligue Nationale Française contre le Cancer. Dr. Barratt is the recipient of European Science Exchange Programme Fellowship from the Royal Society, U.K.

References

1. Adam, A., Lederer, E. 1984. Med. Res. Rev. 4, 111

2. Juy, D., Chedid, L. 1975. Proc. Natl. Acad. Sci. U.S.A. 72, 4105.

3. Parant, M., Parant, F., Chedid, L., Yapo, A., Petit, J.-F., Lederer, E. 1979. Int. J. Immunopharmacol. 1, 35.

4. Tenu, J.-P., Roche, A.-C., Yapo, A., Kieda, C., Monsigny, M., Petit, J.-F. 1982. Biol. Cell 44, 157.

5. Fidler, I.J., Sone, S., Fogler, W.E., Barnes Z.L. 1981. Proc. Natl. Acad. Sci. U.S.A. 78, 1680.

6. Fidler, I.J., Barnes, Z.L., Fogler, W.E., Kirsh, R., Bugelski, P., Poste, G. 1982. Cancer Res. 42, 496.

7. Sone, S., Moriguchi, S., Shimizu, E., Ogushi, F., Tsubura, E. 1982. Cancer Res. 42, 2227.

8. Monsigny, M., Kieda, C., Maillet, T. 1982. EMBO J. 3, 303.

9. Grand-Perret, T., Lepoivre, M., Petit, J.-F., Lemaire, G.L. 1985. Eur. J. Immunol. in press.

10. Fidler, I.J., Schroit, A.J. 1984. J. Immunol. 133, 515.

11. Ginsburg, I., Lahav, M. 1983. Clin. Immunol. News, 4, 147.

12. Sen, Z., Karnovsky, M.L. 1984. Infect. Immun. 43, 937.

13. Kotani, S., Takada, H. Tsujimoto, M., Ogawo, T., Takahashi, I., Ikeda, T., Otsuka, K., Shimauchi, H., Kasai, N., Mashimo, J., Nagao, S., Tanaka, A., Tanaka, S., Harada, K., Nagaki, K., Kitamura, H., Shida, T., Kusumoto, S., Imoto, M., Yoshimura, H. 1985. Infect. Immun. 49, 225.

ANTIVIRAL AND ANTITUMOR EFFECTS OF LIPOSOME-ENTRAPPED MTP-PE, A LIPOPHILIC MURAMYLPEPTIDE

G. Schumann
Research Department, Pharmaceuticals Division,
CIBA-GEIGY Limited, Basel, Switzerland

Introduction

Stimulation of the immune system by immunomodulators such as muramylpeptides for supporting the body's own defence against cancer and against viral diseases has been a long-term goal of clinical therapy. The lipophilic muramylpeptide MTP-PE [N-acetyl-muramyl-L-alanyl-D-isoglutaminyl-L-alanine-(1',2'-dipalmitoyl-sn-glycero-3'-hydroxyphosphoryloxy)-ethylamine], solubilized in phosphate buffered saline, has pronounced immunomodulating activities *in vivo* and *in vitro* (1) particularly characterized by beneficial effects in various experimental viral and tumor systems (2-9). The most important antiviral and antitumor activities of solubilized MTP-PE given via the peroral, intranasal or intravenous route are:

- Prophylactic and therapeutic effectiveness after single administration in a broad spectrum of experimental viral infections (Influenza A and B, Encephalomyocarditis, Herpes simplex types 1 and 2, Vaccinia, Parainfluenza) in mice and guinea pigs (2,3,9).
- Therapeutic effectiveness in experimental tumor systems (4-8).
- Effectiveness at low dose levels amounting in several instances to 1 µg/kg or less (3,7).

Fidler et al. were the first investigators showing that MTP-PE entrapped in liposomes was effective in a tumor system (10). These results may be of particular interest for future applications in man, since i.v. injection of MTP-PE dissolved in saline shows toxicity at low doses in sensitive species (3) while encapsulated in liposomes the compound appears to be less toxic (11).

Biological Properties of Peptidoglycan

In the following chapters the biological activities of MTP-PE encapsulated in liposomes, so-called multilamellar vesicles (MLV) consisting of phosphatidylcholine and phosphatidylserine (PC/PS molar ratio 7:3), will be reviewed.

Induction of virucidal monocytes and antiviral effects with MTP-PE encapsulated in liposomes (MTP-PE/MLV)

Human peripheral blood monocytes could be activated *in vitro* by MTP-PE/MLV (0.4 µg drug in 100 nmol lipid/0.2 ml culture). Its ability to activate monocytes to a cytotoxic state capable of discriminating between virus-infected and uninfected cells was shown by activated monocytes recognizing and destroying Herpes simplex virus type 2-infected cells while leaving uninfected cells unharmed (12).

Intravenous injection of MTP-PE/MLV on days -2, 0 and +2 of infection with Herpes simplex virus type 2 protected mice significantly (9). Mice infected with Rift Valley Fever Virus were treated with 10 µg MTP-PE encapsulated in MLV on days -6, -4, 0, 3 and 7 of infection. The survival rates for these various treatment groups were 0, 10, 20, 50 and 60%, respectively, indicating that the drug induced a highly efficacious antiviral state when the therapy was initiated as late as 3-7 days after infection (13).

Induction of tumoricidal macrophages and anti-metastatic effects with MTP-PE/MLV

The activation to the tumoricidal state of rat or mouse alveolar macrophages (AM) by MTP-PE/MLV could be demonstrated in vitro at a dose of 0.05 µg drug in 100 nmol lipid per 10^5 AM in 0.2 ml medium. These activated AM remained tumoricidal for up to 5 days in culture (14).

I.v. injections of the liposome entrapped drug led to high levels of AM-mediated cytotoxicity as measured by *in vitro* assays. Such macrophages maintained their tumoricidal activity for 4 days in doses up to 5 µg/kg in mice (0.1 µg drug in 2.5 µmol lipid per mouse) (10,14,15).

For both the in vitro and in vivo studies a variety of tumor target cells have been used, e.g. B16/BL6 melanoma of C57 Bl/6 mice, sarcoma 1 of A/J mice, MADB-100 and MADB-200 mammary adenocarcinomas of F344 rats (10,14,16). Normal cells (e.g. fibroblasts) were not killed by tumoricidal macrophages.

The ability of peritoneal exudate macrophages from drug-liposome treated mice to inhibit the growth of syngeneic T 241 sarcoma was examined in a Winn-type assay (16). The primary tumor as well as the median number of lung metastases were significantly reduced 50 days after onset of the experiment. This effect was achieved by a single i.p. injection of 12.5 µg drug in 2.5 µmol PC/PS liposomes per mouse.

Multiple systemic injections of MTP-PE in liposomes eradicated spontaneous lung and lymph node metastases in the B16/BL6 melanoma system in mice. Systemic treatment of mice began 3 days after surgical resection of the primary tumor by biweekly i.v. injections of 10 µg MTP-PE in 2.5 µmol PC/PS liposomes per mouse for 4 weeks. By day 90 of the experiment, 90% of the animals treated with empty liposomes had died. In contrast, 70% of mice injected with drug-liposomes survived at least 120 days after the last injection. These mice were shown to be tumor free. In further experiments, a significant reduction of the median number of lung metastases could be demonstrated 3 weeks after the last injection of the drug (5,10,15,17).

UV-induced skin tumors which include squamous cell carcinomas, fibrosarcomas, hemangiosarcomas, rabdomyosarcomas and other less differentiated sarcomas, are induced 22 to 50 weeks following the onset of UV-irradiation. The therapeutic protocol of twice a week i.v. injections for 3 weeks with PC/PS liposomes incorporating MTP-PE (10 µg drug in 2.5 µmol lipid per mouse) delayed the growth of UV-induced tumors and prolonged host survival as compared to mice treated with salt solution (18).

Experiments with human peripheral blood monocytes have demonstrated that they can be activated *in vitro* by MTP-PE/MLV to become tumoricidal (19,20,21). These monocytes were isolated from normal human donors and cultured *in vitro* with PC/PS liposomes that contained the drug. Monocytes that have phagocytosed liposomes containing MTP-PE were rendered tumoricidal. They lysed cells of three different tumorigenic lines (A 375 melanoma, HT-29 colon, Natusch) but not cells of three non-tumorigenic lines (Flow 2000 embryonic lung, Flow 4000 embryonic kidney, IMR-90 lung fibroblasts) up to a dose of 0.06 μg CGP 19835 A in 25 nmol lipid per 0.2 ml culture.

Table: Antiviral and Antitumor Effects of MTP-PE/MLV

Species	Tumor cells/Virus	Treatment/Cells	References
1. Antiviral effects:			
- man	HSV-2	in vitro, PBM[1]	12
- mouse	HSV-2	in vivo, i.v.	9
- mouse	RVF	in vivo, i.v.	13
2. Antitumor effects:			
- man	A 375 HT-29 Natusch	in vitro, PBM	19,20,21
- rat, mouse	B16/BL 6 MADB-100 and -200	in vitro, AM[2]	14
- rat, mouse	B16/BL 6 MADB-100 and -200 Sarcoma-1	ex vivo, AM	10,14,15
- mouse	T241 sarcoma	Winn-type assay	16
- mouse	B16/BL 6	in vivo, i.v.	5,10,15,17
- mouse	UV-induced	in vivo, i.v.	18

[1]peripheral blood monocytes

[2]alveolar macrophages

Conclusions

MTP-PE/MLV injected intravenously is effective in curing tumor and virus diseases in animals. Since intravenous injections of MTP-PE solubilized in saline are less effective and more toxic when compared with MTP-PE/MLV, the latter formulation may be of interest for applications in man. Moreover, experiments with human peripheral blood monocytes have demonstrated that they can be activated by MTP-PE/MLV *in vitro* to become tumoricidal and virucidal. This indicates that MTP-PE/MLV may also be effective in combating human diseases such as metastatic cancer and virus infections.

References

1. Gisler, R.H., G. Schumann, W. Sackmann, C. Pericin, L. Tarcsay, F.M. Dietrich. 1982. In: Immunomodulation by Microbial Products and Related Synthetic Compounds (Y. Yamamura and S. Kotani, eds.). Excerpta Medica, 167.

2. Dietrich, F.M., B. Lukas, K.H. Schmidtruppin. 1983. In: Proc. of the 13th Int. Congress of Chemotherapy 91, Abstract No. PS 3.1/1-15, 50.

3. Braun, D.G., P. Dukor, B. Lukas, G. Schumann, L. Tarcsay, M. Court, J.C. Schaffner, T. Skripsky, M. Fischer, P. Graepel. In: Intern. Seminar on the Immunological System as a Target for Toxic Damage (in press).

4. Schumann, G., A.F. Brownbill. 1983. Int. Symp. on Immunomodulation by Chemically-Defined Adjuvants, Sapporo, Japan, Abstract No. W-IV-7

5. Schumann, G., A.F. Brownbill, P. Dukor, L.Z. Tarcsay, D.G. Braun. In: Proc. of the 14th Int. Congress of Chemotherapy (in press).

6. Brownbill, A.F., D.G. Braun, P. Dukor, G. Schumann. 1985. Cancer Immunol. Immunother. 20, 11.

7. Brownbill, A.F., D.G. Braun, P. Dukor, G. Schumann. 1985 Int. J. Immunopharmacol. 7, 630.

8. Xu, Z., I.J. Fidler. 1984. Cancer Immunol. Immunother. 18, 118.

9. Koff, W.C., S.D. Showalter, B. Hampar, I.J. Fidler. 1985. Science 228, 495.

10. Fidler, I.J., W.E. Fogler, L. Tarcsay, G. Schumann, D.G. Braun, A.J. Schroit. 1983. In: Advances in Immunopharmacology 2 (J.W. Hadden et al., eds.), Pergamon Press, 235.

11. Fidler, I.J., N.O. Brown, I.R. Hart. J. Biol. Resp. Modif. 4, 298-309

12. Koff, W.C., I.J. Fidler, S.D. Showalter, M.K. Chatzrabarty, B. Hampar, L.M. Ceccorulli, E.S. Kleinerman. 1984. Science 224, 1007.

13. Kende, M., A. Schroit, W. Rill, P.G. Canonico. 1983. 23rd Interscience Conference on Antimicrobial Agents and Chemotherapy, American Society of Microbiology, Las Vegas, Abstract 153.

14. Fidler, I.J., S. Sone, W.E. Fogler, D. Smith, D.G. Braun, L. Tarcsay, R.H. Gisler, A.J. Schroit. 1982. J. Biol. Resp. Modif. 1, 43.

15. Schroit, A., E. Galligioni, I.J. Fidler. 1983. Biol. Cell 47, 87.

16. Barna, B.P., S.D. Deodhar, S. Gantam, B. Yen-Lieberman, D. Roberts. 1984. Cancer Research 44, 305.

17. Key, M.E., Y.E. Talmadge, W.E. Fogler, C. Bucana, I.J. Fidler. 1982. J. Nat. Cancer Inst. 69, 1189.

18. Talmadge, E., B.F. Lenz, M.S. Collins, K.A. Uithoven, M.A. Schneider, J.S. Adams, J.W. Pearson, W.J. Agee, R.E. Fox, R.K. Oldham. 1984. Behring Inst. Mitt. 74, 219.

19. Bucana, C.D., L.C. Hoyer, A.J. Schroit, E. Kleinerman, I.J. Fidler. 1983. Am. J. Pathol. 112, 101.

20. Kleinerman, E.S., K.L. Erickson, A.J. Schroit, W.E. Fogler, I.J. Fidler. 1983. Cancer Res. 43, 2010.

21. Sone, S., S. Mutsuura, M. Ogawara, E. Tsubura. 1984. J. Immunol. 132, 2105.

EFFECTS OF STAPHYLOCOCCAL PEPTIDOGLYCAN ON PHAGOCYTIC CELLS AND HOST MEDIATION SYSTEMS.

A. Fleer

University Children's Hospital "Het Wilhelmina Kinderziekenhuis", and Laboratory of Microbiology, State University Utrecht, The Netherlands.

F.C.A. Jaspers

St. Doetinchemse Ziekenhuizen, Kruisgersestraat 14, Doetinchem, The Netherlands.

J. Verhoef

Laboratory of Microbiology and The University Hospital of Utrecht, 3511 GV Utrecht, The Netherlands.

Introduction

Serious staphylococcal infections may be accompanied by severe disturbances in hematologic parameters, signs and symptoms of shock. In its full-blown state the clinical presentation of staphylococcal sepsis resembles the syndrome of "endotoxic shock" of serious gram-negative infections (1). The hematologic manifestations of "endotoxic shock" have been shown to be due to either direct triggering of host mediation systems by endotoxin (complement, coagulation, and kallikrein-kinin systems), or to indirect effects of endotoxin mediated through activation of complement (leukopenia and thrombocytopenia) (2).

Whether cell wall constituents from gram-positive bacteria, particularly Staphylococcus aureus have similar capabilities, has been one of the subjects of investigation in our group in recent years. In this report a brief review of these studies, as summarized in table 1, will be presented.

Biological Properties of Peptidoglycan

Table 1:
Effect of Staphylococcal Peptidoglycan on Phagocytic Cells and Host Mediation Systems: Overview of studies.

Studies *in vitro*	: PMN phagocytosis
	chemotaxis
	chemiluminescence
	Complement activation
	Contact activation system
Studies in *animal models*	: Leukocyte counts
	Platelet counts
	Complement activation

Materials & Methods

For details of the materials and methods used and the experimental design, the reader is referred to the original articles (3-8).

Leukocyte function studies. The effect of *S. aureus* peptidoglycan, teichoic acid and protein A on the function of polymorphonuclear leukocytes (PMN) was studied by Musher et al. (3). For studies of the effect on PMN phagocytosis and chemotaxis, the various cell wall components were preincubated with PMN for 30 min at 37°C; the components were added directly to PMN suspensions at the start of the assay for PMN chemiluminescence.

Complement activation. The activation of complement by *S. aureus* whole cells, cell walls and cell wall components was studied by Verbrugh et al. (4,5) using various *S. aureus* strains, normal and MgEGTA-chelated human serum in addition to serum from a patient with an inherited C2 deficiency and serum from a patient with a primary X-linked (Bruton type) agammaglobulinemia.

Activation of the contact system. This was studied by Kalter et al. (6) by determining activation of human plasma kallikrein by *S. aureus* cell wall components (peptidoglycan, teichoic acid, and protein A).

Animal studies. These studies were conducted both in rats (7) and in guinea pigs (8). For hemodynamic monitoring and to facilitate blood sampling a catheter was inserted into one of the carotid arteries under general anesthesia.

Results & Discussion

Effect of peptidoglycan (PG) on PMN function.
Table 2 shows that PG caused a dose-dependent inhibition of PMN phagocytosis and chemotaxis. On the other hand, PG induced a chemiluminescence response in PMN, also in a dose-dependent fashion (table 2). Since PMN chemiluminescence is associated with the generation of toxic oxygen species ("respiratory burst") by these cells, a process not only essential for bactericidal activity of PMN but also potentially autotoxic (9), it is conceivable
that this PG-induced process leads to damage of the PMN membrane. As a consequence, PMN functions dependent on membrane integrity, such as phagocytosis and chemotaxis, may suffer in the process.

Table 2:
Effect of staphylococcal peptidoglycan on human PMN function.

PMN function	PG concentration (µg/ml)		
(% of control)	2.5	25	100
Phagocytosis	88	44	32
Chemotaxis	61	19	-
Chemiluminescence	-	700	1600

Similar findings were reported by Henricks et al. (10), who studied the effects of E. coli lipopolysaccharides (LPS) on human PMN function. These authors presented convincing evidence in favour of the hypothesis that induction of PMN chemiluminescence by LPS led to membrane damage resulting in a diminished phagocytic and chemotactic capacity.

Effect of PG on complement activation.

Table 3 shows that purified PG activated complement both in normal and MgEGTA-chelated human serum. The latter indicates activation of the alternative pathway of complement. Since PG also activated complement in serum from a patient with agammaglobulinemia, it was concluded that activation of the alternative pathway of complement in human serum by PG occurred in an antibody-independent fashion.

Table 3:
Activation of complement induced by staphylococcal peptidoglycan.

Activating agent	Amount of activating agent (µg/ml) required to consume 50% of complement hemolytic activity in: Normal human serum	MgEGTA serum	Agamma serum
Peptidoglycan	45	125	200
Teichoic acid	1	2000	N.D.*
Protein A	110	2000	N.D.

* N.D.: Not determined

Table 3 also shows that staphylococcal teichoic acid and protein A did activate complement in normal but not in MgEGTA-chelated human serum. These results are at variance with those obtained by Loos et al. (11) who showed that S. aureus PG prepared by a different procedure did not activate complement through the alternative pathway. Further studies are needed to resolve the question whether the discrepancy is due to the PG purification procedure or to the complement assay used.

Effect of PG on the contact activation system.

Both S. aureus PG and teichoic acid induced 70 to 100% of maximal activation of prekallikrein at a concentration of 5 mg/ml, whereas protein A lacked such activity (data not shown).

Effect of PG on hematologic parameters in vivo and on blood pressure.

Table 4 shows that PG induced a drop in blood pressure, leukopenia, thrombocytopenia, and the activation of complement. Further experiments in C4-deficient guinea pigs and cobra venom treated animals revealed that early thrombocytopenia and leukopenia required an intact alternative pathway of complement, but were independent from the classic pathway. In contrast, complement was not essential for the induction of late leukopenia by PG.

Table 4:
Effect of Staphylococcal Peptidoglycan on Leukocyte and Thrombocyte Counts, Complement Activity and Blood Pressure In Vivo.

Parameters	PG-induced changes after 5 minutes
Blood pressure	- 29 mmHg
Leukocyte counts	- 47 %
Platelet counts	- 64 %
Complement activity	- 33 %.

Together these findings indicate that PG from S. aureus affects PMN function and host mediation systems in a way similar to endotoxins (LPS) from gram-negative bacteria. Thus, like endotoxin, PG suppressed PMN phagocytosis and chemotaxis, stimulated PMN chemiluminescence and activated complement and prekallikrein. Whether these effects of PG in vitro pertain to derangements in hematologic and hemodynamic parameters as observed during severe staphylococcal infections remains to be established, but studies in rats and guinea pigs revealed that injection of PG induced changes in these parameters closely resembling those occurring in severe cases of S. aureus bacteremia in man. Moreover, the similarity of the phenomena induced by PG and endotoxin both in vitro and in vivo may explain the resemblance of the shock-like state occurring in severe staphylococcal infections to "endotoxic shock".

References

1. Sheagren, J. N.. 1984. Staphylococcus aureus. The persistent pathogen. N.Engl.J.Med. 310, 1368-1373.

2. Morrison, D. C. and R. J. Ulevitch. 1978. The effects of bacterial endotoxins on host mediation systems. Am.J.Pathol. 93, 527-618.

3. Musher, D. M., H. A. Verbrugh and J. Verhoef. 1981. Suppression of phagocytosis and chemotaxis by cell wall components of Staphylococcus aureus. J.Immunol. 127, 84-88.

4. Verbrugh, H. A., W. C. van Dijk, R. Peters, M. E. van der Tol and J. Verhoef. 1979. The role of Staphylococcus aureus cell wall peptidoglycan, teichoic acid and protein A in the processes of complement activation and opsonization. Immunol. 37, 615-621.

5. Verbrugh, H. A., W. C. van Dijk, R. Peters, M. E. van Erne, M. R. Daha, P. K. Peterson and J Verhoef. 1980. Opsonic recognition of staphylococci mediated by cell wall peptidoglycan: antibody-independent activation of human complement and opsonic activity of peptidoglycan antibodies. J.Immunol. 124, 1167-1173.

6. Kalter, E. S., W. C. van Dijk, A. Timmerman, J. Verhoef and B. N. Bouma. 1983. Activation of purified human plasma prekallikrein triggered by cell wall fractions of Escherichia coli and Staphylococcus aureus. J.Infect.Dis. 148, 682-691.

7. Verhoef, J., D. M. Musher, J. S. Spika, H. A. Verbrugh and F. C. A. Jaspers. 1983. The effect of staphylococcal peptidoglycan on polymorphonuclear leukocytes in vitro and in vivo. Scand.J.Infect.Dis. Suppl. 41, 79-85.

8. Spika, J. S., P. K. Peterson, B. J. Wilkinson, D. E. Hammerschmidt, H. A. Verbrugh, J. Verhoef and P. G. Quie. 1982. Role of peptidoglycan from Staphylococcus aureus in leukopenia, thrombocytopenia, and complement activation associated with bacteremia. J.Infect.Dis. 146, 227-234.

9. Tsan, M-F. 1980. Phorbol myristate acetate induced neutrophil autotoxicity. J.Cell.Physiol. 105, 327-334.

10. Henricks, P. A. J., M. E. van der Tol, R. M. W. M. Thijssen, B. S. van Asbeck and J. Verhoef. 1983. Escherichia coli lipopolysaccharides diminish and enhance cell function of human polymorphonuclear leukocytes. Infect.Immun. 41, 294-301.

11. Loos, M. et al. 1985. Teichoic acid-free peptidoglycan of Staphylococcus aureus does not activate complement. (This conference).

CELLULAR RESPONSE TO STREPTOCOCCAL PEPTIDOGLYCAN AND TETANUS TOXOID IN GLOMERULONEPHRITIS-PATIENTS: CORRELATION TO COMPLEMENT GENOTYPE

H.E. Feucht[1,2], G.J. O'Neill[2], G. Riethmüller[2], A. Brase[2] and P.H. Seidl[3]

[1]Medizinische Klinik Innenstadt, Ludwig-Maximilians-Universität München, FRG;

[2]Institut für Immunologie, Ludwig-Maximilians-Universität München, FRG;

[3]Lehrstuhl für Mikrobiologie, Technische Universität München, FRG.

Introduction

Primary Glomerulonephritis (GN) appears as an immunologically mediated disease of remarkable clinical and histopathological heterogeneity. In most cases the underlying pathogenesis remains unknown. However, apart from the well recognized postinfectious acute forms, streptococcal infections seem to promote also some of the chronic forms, such as membranoproliferative GN (1). The 'quality' of a patient's immune response to various stimuli obviously is important for the development and progression of the disease. Since the immune response to streptococcal cell wall and tetanus toxoid is under genetic control (2), it was of interest to study immunogenetic conditions in GN-patients.

The HLA-region is assumed to control immune responsiveness in man, and several studies have examined possible associations between HLA-class I and class II antigens and primary GN (3). In addition, alleles of the HLA-linked complement genes (class III products), C4, C2 and factor B, are also used as genetic markers of disease. The fourth component of human complement, C4, which is controlled by two highly polymorphic genes, C4A and C4B, has been particular useful in this respect. We have recently shown that a rare variant of the C4B locus, termed C4B2.9, is a significant immunogenetic marker in patients with primary GN (4). About 25% of all GN-patients carry this complement variant, furthermore, it is highly

Biological Properties of Peptidoglycan

associated with membranoproliferative and membranous GN. Complement C4 displays multiple biological functions and also seems to be involved in antigen presentation (5). We were therefore interested to determine the cellular immune response to bacterial antigens in correlation to the C4B2.9 allotypic marker. Due to the widespread immunity against tetanus toxoid, we used this compound as a potent T-cell dependent antigen having no apparent relation to primary GN. In addition, streptococcal peptidoglycan was used as an antigen because of its possible nephritogenic property.

Patients and Methods

Patients with primary forms of glomerulonephritis were investigated in the nephrological department of the Medical Clinic Innenstadt. Diagnosis was in the majority confirmed by kidney biopsy. In patients without kidney biopsy, the diagnosis of primary GN was based on the clinical symptoms (and their combination) of nephrotic syndrome, proteinuria, haematuria, hypertension and impaired renal function.

C4 electrophoretic variants were determined in neuraminidase-treated EDTA-plasma by immunofixation electrophoresis in agarose gels. The C4A and C4B locus products were distinguished by a specific in-gel overlay containing sheep erythrocytes sensitised with an IgG antibody and C4-deficient guinea pig serum (6).

Peptidoglycan from Streptococcus pyogenes A-variant A486 (strain obtained from R.C. Lancefield, The Rockefeller University, New York, U.S.A.) was isolated and sonicated according to published procedures (7). Tetanus toxoid was a gift from Behringwerke AG, Marburg, FRG.

Peripheral mononuclear cells (PMN) from patients and healthy volunteers were isolated by density gradient centrifugation as described previously (8). Cellular immune response was determined as the proliferative capacity of PMN after *in vitro* stimulation with tetanus toxoid and peptidoglycan for 6 d. Antigen stimulation was performed in microtitreplates as triplicates, one well containing 10^5 PMN in 100 µl RPMI 1640 medium (Gibco, Karlsruhe, FRG) supplemented with 10% pooled human serum. Antigen-concentrations required for maximal stimulation were 20 µg/ml for tetanus toxoid and 100 µg/ml for peptidoglycan. Microcultures were pulsed with 1µCi ^{3}H-thymidine and triplicates were harvested after 8 hrs onto nitrocellulose filters using an automatic harvesting device. Incorporated ^{3}H-thymidine was measured in a ß-scintillation counter.

Stimulation indices (SI) were calculated from triplicates as:

$$SI = \frac{\text{cpm PMN + antigen}}{\text{cpm PMN + medium}}$$

Results

The antigen-driven proliferative response of peripheral mononuclear cells from glomerulonephritis-patients was compared with cells from age and sex matched healthy volunteers. GN-patients were further subdivided into those having the C4B2.9 immunogenetic marker and those without this marker.

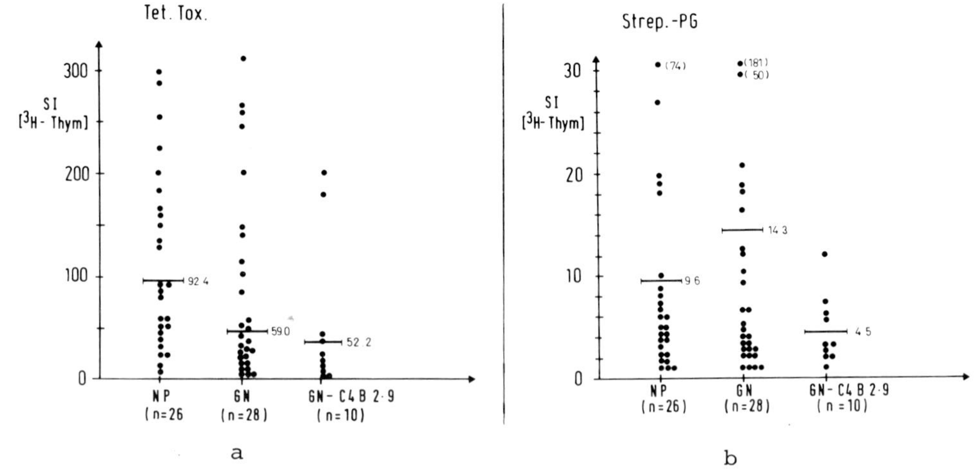

Fig. 1 a + b: ^{3}H-thymidine uptake by peripheral mononuclear cells (10^5) following 6d culture with (a) tetanus toxoid (20µg/ml) and with (b) streptococcal peptidoglycan (100µg/ml)

When the T-cell dependent antigen tetanus toxoid was used, normal persons as well as GN-patients showed a wide variation in their cellular response. Both GN-groups had a markedly lower response than normals. Mean values of stimulation indices were 92.4 in the normal group, 59 and 52 in the GN-group and the C4B2.9-group, respectively (Fig. 1a).

When streptococcal peptidoglycan was used as an antigen, the proliferative response was significantly lower in all groups. Normal persons showed a mean stimulation index of 9.6, the GN-group was definitely higher with a mean value of 14.3. In contrast, patients with the C4B2.9 immunogenetic marker exhibited a significantly lower response, reaching a mean value of only 4.5 (Fig. 1b).

Discussion

Recent experimental work suggests the involvement of peptidoglycans in the development of several chronic autoimmune phenomena such as rheumatoid arthritis (9). With respect to the development of glomerulonephritis, a pathogenic role of peptidoglycans has yet to be shown. There is, however, evidence that chronic forms of GN, especially membranoproliferative GN, which carries a poor clinical prognosis, are initiated and maintained by streptococcal infections (1). Thus, immune responses against the biologically pluripotent peptidoglycans certainly deserve attention in those patients. The allotypic complement marker C4B2.9 is also highly associated with membranoproliferative GN (4). Therefore, the impaired cellular response against peptidoglycan in this patient group is of interest. It is possible that the diminished response against peptidoglycan might be antigen-specific, since no such differences were found when using tetanus toxoid as an antigen. Stimulation with tetanus toxoid revealed an equally diminished T-cell resposne in both GN-groups. Whether this impaired T-cell reactivity is pre-existent or acquired during chronic renal disease remains to be determined.

Depressed immune responses, both cellular and humoral, have already been recognized in other forms of GN, e.g. focal sclerosis and membranous GN (10,11). However, these studies used mitogen-dependent assays, therefore no conclusions can be drawn with respect to the pathogenic role of infections. It appears, nevertheless, that in certain diseases a suppressed immune response, eventually resulting in delayed antigen-clearance, may be more harmful to the organism than an augmented response. Whether the impaired immune response against peptidoglycan is due to the action of suppressor cells, as has been reported for tetanus toxoid and streptococcal cell walls remains to be shown.

It can be assumed from our study that the cellular response to peptidoglycan is under genetic control. It is not clear, however, whether the complement variant C4B2.9 is merely linked to a relevant immune response gene, or whether it is directly involved in this particular immune response.

References

1. Heptinstall, R.H. 1983. Pathology of the kidney, Vol. 1. Boston: Little, Brown, pp. 479-518.

2. Nishimura, Y. and T. Sasazuki. 1983. Suppressor T-cells control the HLA-linked low responsiveness to streptococcal antigen in man. Nature 302, 67-69.

3. Rashid, H.U., S.S. Papiha, B. Agroyannis, A.R. Morley, M.K. Ward, D.F. Roberts and D.N.S. Kerr. 1983. The associations of HLA and other genetic markers with glomerulonephritis. Hum. Genet. 63, 38-44.

4. Wank, R., D.J. Schendel, G.J. O'Neill, G. Riethmüller, E. Held and H.E. Feucht. 1984. Rare variant of complement C4 is seen in high frequency in patients with primary glomerulonephritis. Lancet I, 872-874.

5. Egwang, T.G. and A.D. Befus. 1984. The role of complement in the induction and regulation of immune responses. Immunology 51, 207-224.

6. Awdeh, Z.L. and C.A. Alper. 1980. Inherited structural polymorphism of the fourth component of human complement. Proc. Natl. Acad. Sci. USA 77, 3576-3580.

7. Schleifer, K.H. and R.M. Krause. 1971. The immunochemistry of peptidoglycan. I. The immunodominant site of the peptide subunit and the contribution of each of the amino acids to the binding properties of the peptides. J. Biol. Chem. 246, 986-993.

8. Feucht, H.E., M.R. Hadam, F. Frank and G. Riethmüller. 1980. Efficient separation of human T lymphocytes from venous blood using PVP-coated colloidal silica particles (Percoll). J. Immunol. Meth. 38, 43-51.

9. Hunter, N., S.A. Anderle, R.R. Brown, F.G. Dalldorf, R.L. Clark, W.J. Cromartie and J.H. Schwab. 1980. Cell-mediated immune response during experimental arthritis induced in rats with streptococcal cell walls. Clin. exp. Immunol. 42, 441-449.

10. Matsumoto, K., K. Osakabe, H. Katayama and M. Hatano. 1982. Impaired cell-mediated immunity in focal glomerular sclerosis mediated by suppressor cells. Nephron 32, 263-264.

11. Ooi, B.S., Y.M. Ooi, A. Hsu and P.E. Hurtubise. 1980. Diminished synthesis of immunoglobulin by peripheral lymphocytes of patients with idiopathic membranous glomerulopathy. J. Clin. Invest. 65, 789-797.

ACUTE AND CHRONIC INFLAMMATION INDUCED BY PEPTIDOGLYCAN STRUCTURES AND POLYSACCHARIDE COMPLEXES

S.A. Stimpson, J.H. Schwab, M.J. Janusz, S.K. Anderle, R.R. Brown, and W.J. Cromartie

Department of Microbiology and Immunology, University of North Carolina, Chapel Hill, N.C. 27514

Introduction

Peptidoglycan (PG) is a major toxic structural component common to most bacteria (1). In experimental adjuvant-induced arthritis in rats (2), the minimum effective structure of PG has been reported to be a disaccharide-dipeptide or muramyl dipeptide (MDP). Studies using derived or synthetic peptide subunits, and oil suspensions, have provided much useful information on structure-function relationships. However, the phlogistic, as well as adjuvant and other activities of PG or muramyl peptides, are greatly modulated by their molecular configuration and associated polymers. Furthermore, these structural requirements are altered in lipid emulsions or liposomes. Therefore, to understand the natural history and mechanisms of bacterial cell wall-induced inflammatory diseases, we believe it is necessary to employ polymers which the host is most likely to encounter, and which are introduced in aqueous suspensions.

In addition to polymer structure and the suspending vehicle, other variables which determine the phlogistic potential of a cell wall polymer, and the quantitative and qualitative nature of the host inflammatory response, are listed in Table 1.

Biological Properties of Peptidoglycan

Table 1. Variables Which Affect the Qualitative and Quantitative Nature of the Inflammatory Response to Peptidoglycan.

Polymer	Host genetics
Fine structure	Species
Associated polymers	Susceptible/resistant strains
Vehicle	Age and/or previous antigenic experience of host
Oil	Host immune response
Water	Antibody
Route, dose, and number of injections	Cell mediated immunity
Intraperitoneal, intravenous, intraarticular	Tissue examined

It is also essential to carefully define the qualitative nature of the inflammation induced by a given polymer under prescribed conditions. For example, consistent induction of chronic, recurrent, progressive, irreversible destructive arthritis has been well documented following intraperitoneal injection of certain rat strains with an aqueous suspension of peptidoglycan-polysaccharide (PG-PS) polymers of a mol. wt. greater than 5×10^6, derived from group A streptococci (3,4,5,6). Alteration of any of the variables mentioned above can affect the course of inflammation and prevent the evolution into the irreversible destructive stage (7).

Comparison of Joint Inflammation Induced in the Rat by Intraperitoneal (IP) Injection of Aqueous Suspensions of Cell Wall Polymers Derived from Several Species of Bacteria

The phlogistic activities of various cell wall preparations are summarized in Table 2. Four categories of joint inflammation are defined by gross evaluation, histology, and radiology. Edema is measured by extravasation of radiolabelled albumin.

Table 2. Phases of Joint Inflammation Induced by Intraperitoneal Injection of Aqueous Suspensions of Cell Wall Structures

		Arthritis		
Structure Injected	Edema	Acute	Chronic	Recurrent destructive (>90 days)
Peptidoglycan-polysaccharide (PG-PS)				
Streptococcus pyogenes (group A)	±	+	+	+
Streptococcus agalactiae (group B)	-	+	±	-
Streptococcus faecium (group D)	-	+	+	-
Lactobacillus casei	-	+	+	?
Peptostreptococcus productus	-	+	±	-
Propionibacterium acnes	-	±	-	-
Group A streptococcal PG-PS				
Mutanolysin-derived fragments	+	-	-	-
N-acetylated	-	+	±	-
De-O-acetylated	-	+	±	-
Group A streptococcal PG	-	+	-	-
Group A streptococcal PS	+	-	-	-
Muramyl dipeptide	-	-	-	-
Pseudomurein-polysaccharide				
Methanobacterium formicicum	-	-	-	-

A single IP injection of covalently bound peptidoglycan-polysaccharide polymers isolated from the cell walls of group A streptococci induces an acute exudative inflammation of joints which develops over the first 5 days and then recedes. This acute phase is followed by a chronic phase consisting of a series of recurrences and remissions which progresses over a period of 2 to 6 months into an irreversible destructive process with extensive erosion and bone resorption. PG-PS isolated from group D streptococci, and which has been stabilized by heating and extraction with detergent to eliminate autolytic degradation, can induce acute and chronic phases of inflammation which can last for as long as 60 days (8). However, this is not sustained and the arthritis eventually resolves with relatively little radiographic or histologic evidence of joint destruction. If autolytic enzymes have not been removed from the PG-PS of group D streptococci, only acute inflammation is induced. PG-PS from group B streptococci (9) and Lactobacillus casei (10) can also induce acute and chronic phases of arthritis, but whether a relapsing, erosive synovitis persisting beyond 90 days can occur, as in the case of group A streptococcal PG-PS-induced disease, has not been reported. Peptostreptococcus productus induces acute joint inflammation of 10 days duration followed by a brief chronic phase (8). PG-PS from Propionibacterium acnes produces only a minimal acute inflammation and the pseudomurein-polysaccharide complex from Methanobacterium formicicum induces no inflammation after systemic injection (8). N-acetylation or de-O-acetylation of group A streptococcus PG-PS also greatly reduces the capacity to induce chronic arthritis, which correlates with the increased susceptibility to lysozyme after these chemical modifications (11).

An important variable which affects the kind of inflammatory response is the size of the PG-PS fragments. All of the cell wall preparations described above were sonicated to produce arthropathic fragments of PG-PS. Sonication produces a wide spectrum of fragment sizes which have different arthropathic potentials (12). Therefore, to make a meaningful comparison of PG-PS from different bacteria a fraction of the sonicated cell walls which sediments at 100,000 x g in 60 min., but not 10,000 x g, was injected. This fraction has an average mol. wt. of 5×10^7 and is referred to as the 100p60 fraction (12).

PG-PS from group A streptococci are resistant to degradation by lysozyme, but can be degraded in vitro by a muramidase of the same specificity called mutanolysin M-1, derived from Streptomyces globisporus (13). The products of this degradation have mol. wts. ranging from 5×10^5 to 3×10^3 and neither the high nor the low mol. wt. fraction is arthropathic (13). However, these mutanolsyin-derived fragments of PG-PS express a different phlogistic property, namely, the induction of edema which reaches a peak within 30 min and is resolved by 2 hr (14). The edematous response is quantitated by extravasation of radiolabelled albumin and is largely confined to the limbs. The active moiety is the group-specific A polysaccharide; peptidoglycan completely lacks this property (14).

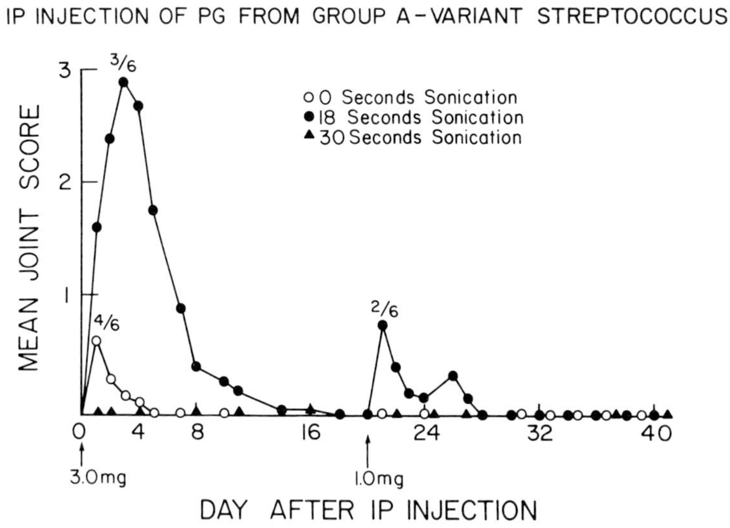

Figure 1. Mean joint scores of three groups of 6 Sprague-Dawley rats injected IP with 3.0 mg PG from group A-variant streptococcal cell wall. PG was sonicated for 0 sec, ○, 4/6 positive; 18 sec, ●, 3/6 positive; or 30 sec, ▲, 0/6 positive. Also tested was PG sonicated for 150 sec (0/6 positive, data not shown). All rats were given a second IP injection of 1.0 mg PG on day 20. Only 2/6 of the group injected with PG sonicated for 18 sec developed grossly evident arthritis. Mean joint scores include the non-responding rats in each group.

IP injection of the isolated PG induces only the acute phase of arthritis and even after a second injection the inflammation is not sustained (Figure 1). This phlogistic activity is increased by sonication for 18 sec, but greater fragmentation by sonication for 30 sec or 150 sec destroys activity. The loss of phlogistic activity parallels the loss of activation of the alternate complement pathway (15).

Intravenous (IV) injection of PG or PG-PS (5 µg/g body wt) induces a unique hemorrhagic reaction confined to the Peyer's patches, mesenteric lymph nodes and lamina propria of the intestinal tract (16).

Comparison of Inflammation Induced by Intraarticular (IA) Injection of Bacterial Cell Wall Polymers.

IA injection allows analysis of cell wall-induced inflammation without complications such as influx from depots in the liver or spleen, transportation in the blood, and entry into synovial tissue. These variables contribute to the pathogenesis of arthritis following systemic injection.

A detailed description of the histopathology is presented elsewhere (17). PG-PS, in doses of 0.2 to 4.0 µg per joint, induces an acute, exudative response which reaches a maximum in 24 hrs. This is replaced over the next 2 weeks with chronic inflammation including some pannus formation and marginal erosions. The important difference from IP injection is that the reaction does not progress to joint destruction and episodes of remission and recurrence are not observed. This may be due to the absence of a reservoir in the liver and spleen, which exists after IP injection, and which would provide a source of PG-PS for continued deposition in the joint tissue.

The relative inflammatory response of susceptible Lewis and Sprague-Dawley rats and the resistant Buffalo rat to IA injection of 0.5 or 2.5 µg of PG-PS, paralleled their responses to IP injection (17,18). This means that the difference between these inbred strains reflects a genetic control of regulation of inflammation rather than the amount of PG-PS which is distributed to the joint. This conclusion is consistent with quantitative measurements showing no difference in amount of PG-PS in the joints of these rat strains after IP injection (19), and is also consistent with studies of the air pouch model of inflammation in these rat strains (33).

IA injection of 50 µg of PG from group A-variant streptococci induced a reaction which resembled that induced by PG-APS, but resolved much more rapidly. This is consistent with the greater susceptibility of PG to tissue enzymes (20).

Group A streptococcal polysaccharide induced an intense edema in the synovial stroma over the first 30 min. A feature of this response is the absence of infiltration by neutrophils, which may reflect the inability of this material to activate the alternative complement pathway, in contrast to PG (15).

IA injection of MDP in doses up to 1000 µg induced only a mild transient inflammation with some focal proliferation of synovial lining cells, which completely subsided by 3 days. Injection of 0.1 to 1.0 µg of LPS induced an acute inflammation which reached a peak within 24 hr and rapidly subsided to become undetectable by 5 days (17).

Reactivation of Cell Wall-Induced Arthritis

The inflammation induced by IA injection of 5 µg of PG-APS largely resolves by 3 weeks. Inflammation can be reactivated by IV injection of normally sub-arthropathic doses (300 µg per rat) of: homologous PG-APS fragments or PG from group A streptococci; PG-PS from group D streptococci; or MDP, in a much higher dose (30 mg per rat). Group A streptococcal polysaccharide, PG-PS or PG from P. acnes, and dextran sulfate are inactive. Inflammation only recurs in the joint previously inflamed by PG-PS from group A streptococci. Saline-injected control joints or joints inflamed initially by other cell wall polymers cannot be reactivated (21).

Of considerable interest is the fact that IV injection of lipopolysaccharide (LPS) 3 weeks after an IA injection of group A streptococcal PG-APS, also can reactivate inflammation (22). The lowest effective dose which can induce recurrence in 100% of the test joints is 10 µg/rat. At this dose inflammation is not induced in contralateral control joints. Reactivation does not occur in joints initially injected with LPS. LPS from Salmonella typhimurium wild type, Escherichia coli 0111:B4 and glycolipid from S. typhimurium Re mutant are all active. These results suggest that LPS derived from a variety of indigenous or pathogenic bacteria could be involved in chronic arthritis by perpetuating joint injury initiated by peptidoglycan polymers.

Reversal of the Progression of Inflammation by Treatment with Muramidase

IV injection of mutanolysin (a muramidase) immediately after an IP injection of PG-APS has a dramatic effect on the course of arthritis (23). Severe, transient edema develops in the limbs during the first 2 hr, demonstrating the in vivo degradation of cell wall to fragments which have this phlogistic activity. The severity of acute arthritis over the next few days is significantly reduced, and no chronic phase of disease occurs (Figure 2). Even if the injection of enzyme is delayed for two weeks, the chronic phase of inflammation is greatly reduced (23). The loss of arthropathic activity is not due to a more rapid elimination of cell wall from the joints. Rather, the fragments persist in all tissues in a polymer size which is not arthropathic (24). This degradation to low mol. wt. polymers is accompanied by the reduced binding of C3b to the joint tissue (25).

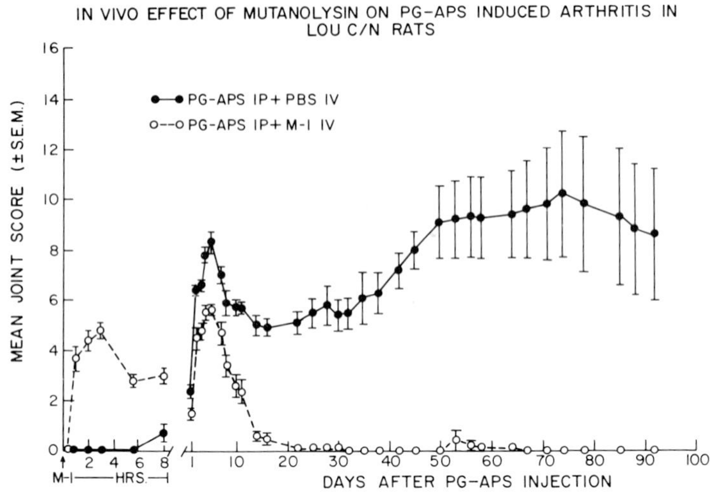

Figure 2. Mean joint scores ± SEM of two groups of Lou C/N rats injected intraperitoneally on day 0 with group A streptococcal peptidoglycan-polysaccharide (PG-APS, 20 µg rhamnose/g body wt). Immediately after PG-APS injection, rats were injected intravenously with either buffered saline (PBS, ●) or 400 µg of mutanolysin (M-1, ○).

Anti-Inflammatory Effect of Low Molecular Weight Fragments of Cell Wall

Zidek et al (26) have reported that MDP incorporated with the phlogistic agent can suppress acute inflammation in rats and mice. They demonstrated that carrageenan-induced pleurisy and dextran or Bayol-Aracel-induced paw edema were suppressed 30 to 40% with 0.1 to 0.2 mg of MDP per animal. We have shown that rats injected IP with 3.0 mg of MDP on each of 3 days before injection of arthropathic PG-PS developed much more severe acute and chronic arthritis (Figure 3). Injection of the same dose of MDP for 3 days after PG-APS reduced the severity of the chronic phase of arthritis (Figure 3). Neither effect of MDP was correlated with alteration of serum antibody responses (Table 3).

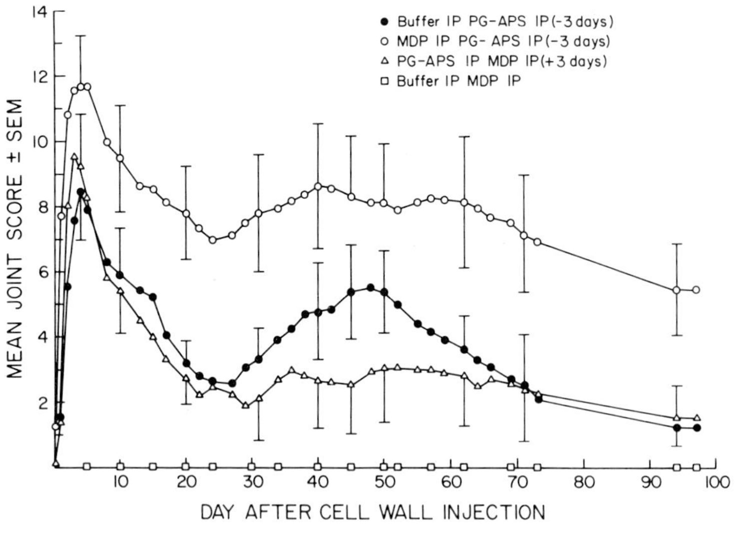

Figure 3. Mean joint scores ± SEM of four groups of Sprague-Dawley rats, showing the effect of the systemic injection of MDP on the arthropathic activity of group A streptococcal peptidoglycan-polysaccharide (PG-APS). All injections were intraperitoneal. Dosages: PG-APS, 20 µg rhamnose per g body weight; MDP, 3 mg per rat each day. ●, 9 rats, buffered saline on days -3, -2, and -1, then PG-APS on day 0; ○, 9 rats, MDP on days -3, -2, and -1, then PG-APS on day 0; △, 9 rats, PG-APS on day 0, then MDP on days 0, +1, and +2; □, 5 rats, MDP on day 0, +1, and +2 (no PG-APS).

Table 3. Effect of MDP Injected Before or After Group A Streptococcal Peptidoglycan-Polysaccharide (PG-APS) on the Serum Concentration of IgM Anti-PG Antibodies[a]

Day after PG-APS injection	PBS -3 days + PG-APS	MDP -3 days + PG-APS	PG-APS + MDP +3 days	MDP only
37	65.2 ± 20.8[b]	60.4 ± 40.2	66.8 ± 17.4	30.4 ± 15.5
64	47.9 ± 24.5	51.2 ± 37.4	63.5 ± 11.3	24.2 ± 10.1
100	40.2 ± 24.5	40.5 ± 22.0	65.5 ± 29.4	24.3 ± 11.1

[a]Serum obtained from the animals in the experiment described in Figure 3.

[b]Mean (±SD) µg antibody specific for PG.

Based upon these observations, a partial explanation of the capacity of muramidase to reverse PG-PS induced arthritis could be the inhibitory effects of the

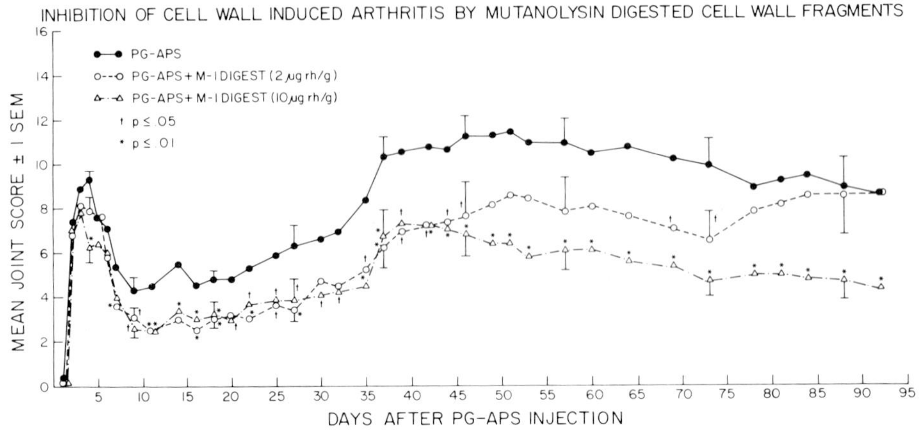

Figure 4. Mean joint scores ± SEM of 3 groups of 8 Lewis rats injected intraperitoneally on day 0 with group A streptococcal peptidoglycan-polysaccharide (PG-APS) alone (20 µg rhamnose/g body wt, ●) or a mixture of PG-APS (20 µg rhamnose/g body wt) and mutanolysin M-1 digested PG-APS fragments [2 (○) or 10 (△) µg rhamnose/g body wt]. Days on which the mean joint score for rats injected with the PG-APS-M-1 digest mixture was significantly lower than the score for rats injected with PG-APS alone are indicated by an asterisk ($P \leq 0.01$) or a dagger ($P \leq 0.05$).

in vivo generated degradation products of PG. In support of this hypothesis, we have found that injection of rats IP with a mixture of arthropathic PG-PS and PG-PS degraded _in vitro_ by mutanolysin results in significantly reduced arthritis (Figure 4). These results are consistent with the hypothesis of Chedid (27) that the products of macrophage degradation of bacterial cell wall can function as immunopharmacological mediators.

Molecular and Cellular Mediators of Cell Wall-Induced Inflammation.

Soluble mediators whose production is induced by peptidoglycan polymers or muramyl peptides, and inflammatory cells which are affected by these cell wall structures, are listed in Table 4.

Table 4. Mediators of Inflammation Induced by Peptidoglycan and Peptidoglycan-Polysaccharide Polymers.

Soluble mediators	Cellular mediators
Interleukin-1	Macrophages
Arachidonic acid metabolites	Effector
PGE_2, LTB_4	Suppressor
Histamine	Polymorphonuclear leukocytes
Complement	Platelets
Oxygen radicals	Mast cells
	T-cells

Interleukin 1 (IL-1) is a product of mononuclear cells to which numerous activities relevant to regulation of inflammation have been ascribed (28,29,30). PG from Staphylococcus aureus and Bacillus megaterium stimulate the production of IL-1 by mouse peritoneal macrophages and human mononuclear cells (31). In our laboratory, Bristol has demonstrated the secretion of IL-1 by rat macrophages stimulated in vivo or in vitro with PG-PS fragments (unpublished results).

Wahl, et al (32) have shown that MDP can stimulate PGE_2 production in vitro by rabbit or guinea pig macrophages. In our laboratory, Yoshino has measured the in vivo kinetics of PGE_2 and LTB_4 production in a rat air pouch stimulated by PG-PS fragments (33). He observed a correlation between quantity of both PGE_2 and LTB_4 and severity of acute inflammation, as measured by several parameters in genetically susceptible or resistant rats. However, the sustained production of LTB_4 relative to PGE_2 suggested that LTB_4 may be a factor in the progression of inflammation into a chronic stage.

Histamine can be released from rat peritoneal mast cells stimulated by PG from group A streptococci (34). The group A streptococcal cell wall polysaccharide, on the other hand, does not affect peritoneal mast cells but does induce transient edema which is largely confined to the limbs (14). We have recently shown that this early edema is related to the degranulation of mast cells in the joints and periarticular tissues following IV injection (Dalldorf, Anderle, Brown, and Schwab, in preparation). Mast cells have also been proposed to play an important role in late phase reactions and the mononuclear inflammatory response (35). PG from Staphylococcus aureus will also induce release of histamine from human basophils (36).

Several studies have demonstrated that both the classical and alternative complement pathways are activated by PG (15). The activities of the C3a and C5a products in vascular permeability and chemotaxis are well known (37). PG-PS and PG can activate both pathways in vivo (38), and this is related to the acute phase of arthritis induced by PG-PS in rats. We have recently shown that C3b forms stable complexes with arthropathic PG-PS fragments in joint tissue, but not in liver or spleen (39).

Oxygen radicals are secreted by human monocytes stimulated in vitro with MDP (40). PMN are also stimulated by PG-PS in the presence of complement to have increased oxygen consumption, hexose monophosphate shunt activity, and secretion of oxygen radicals (41). It is generally presumed that oxygen metabolites can contribute to inflammation (42).

The macrophage is activated by phagocytosis of cell wall structures (43,44) and it has many roles in the chronic inflammation induced by PG-PS. In addition to most of the secreted mediators described above, macrophages stimulated with PG-PS or PG can produce a large number of phlogistic agents such as lysosomal enzymes (45), plasminogen activator (46), and collagenase (47). They are directly involved in the resorption of bone (48), and we have shown that peritoneal and joint macrophages from PG-PS injected rats can solubilize ^{45}Ca-labeled devitalized bone chips in vitro (Bristol and Schwab, in preparation). Suppressor macrophages which are responsible for the anergy and depressed T cell function in rats injected with arthropathic PG-PS have also been described (49).

PMN are essential cells in acute inflammation and they probably accomplish this role in part by secretion of O_2 metabolites and lysosomal enzymes (50). Periodic appearance of focal accumulation of PMN is a consistent feature throughout the active chronic phase of arthritis induced by PG-PS, and it has been proposed that they are associated with the renewed episodes of joint inflammation (3,41).

Ryc and Rotta (51,52) reported the capacity of PG to clump and lyse platelets. We have observed a severe thrombocytopenia within a few minutes after injection of rats with arthropathic doses of PG-PS, but this is due to a sequestration since there is a restoration to normal levels within one hour. In guinea pigs the early thrombocytopenia induced by PG from S. aureus requires complement (53). Bacterial cell wall structures also cause the release of serotonin from rabbit platelets (54). Although platelets can secrete vasoactive amines and chemotactic factors, a normal edematous response to carrageenin and passive cutaneous anaphylaxis occurs in thrombocytopenic rats (55). Therefore, the importance of PG-induced platelet aggregation or lysis in the inflammatory response to cell wall polymers remains uncertain.

The role of T lymphocytes in the inflammatory responses to PG or PG-PS polymers is complex. Neonatal thymectomy has little effect on the development of cell wall-induced arthritis (56). Indeed, rats given arthropathic doses of PG-PS develop an anergy and loss of T cell function measured by responses to mitogens (49,56). This anergy persists for at least 3 months and is related to the development of adherent suppressor cells (49). Recently, however, it has been observed that athymic nu/nu rats develop the acute phase of arthritis, but less severe chronic disease than euthymic littermates (57). Leclerc, et al (58) reported that MDP and some analogs can suppress synthesis of IL-2 by human peripheral blood cells, and this was due to activation of a radioresistant suppressor cell and stimulation of an adherent monocyte to release PGE_2.

Conclusions

1. The nature of the inflammatory response induced by PG structures is determined by both the fine structure of the PG and the associated polymers. PG-PS from group A streptococci is one of the few structures which can induce an irreversible, destructive arthritis.

2. Persistence in tissue, in a physical state capable of binding complement, is essential for expression of phlogistic activities of PG.

3. Inflammation initiated by PG-PS from group A streptococci can be reactivated by systemic injection of homologous and heterologous cell wall polymers, including LPS. This may reflect the mechanism of recurrent, chronic arthritis.

4. The course of PG-PS induced arthritis can be reversed by injection of a muramidase, even after the chronic phase of inflammation has begun. This may be partially ascribed to in vivo degradation to polymers which cannot bind complement, and to the anti-inflammatory effect of degradation products.

5. Analysis of the regulation of acute and chronic inflammation induced by PG polymers is being pursued by measurement of inflammatory cell activities and secreted mediators.

Acknowledgement

This research was supported by grant number AM25733 from the National Institute of Arthritis, Diabetes, Digestive and Kidney Diseases.

References

1. Chetty, C. and J.H. Schwab. 1984. In: Handbook of Endotoxin, Vol. 1, Chemistry of Endotoxin. (E.T. Rietschel, ed.) Elsevier, Amsterdam, pp. 376-410.

2. Kohashi, O., A. Tanaka, S. Kotani, T. Shiba, S. Kusumoto, K. Yokogawa, S. Kawato, A. and A. Ozawa. 1980. Arthritis-inducing ability of a synthetic adjuvant, N-acetylmuramyl peptides, and bacterial disaccharide peptides related to different oil vehicles and their composition. Infect. Immun., 29:70-75.

3. Cromartie, W.J., J.G. Craddock, J.H. Schwab, S.K. Anderle, and C.H. Yang. 1977. Arthritis in rats after systemic injection of streptococcal cells or cell walls. J. Exp. Med. 146:1585-1602.

4. Clark, R.L., J.T. Cuttino, S.K. Anderle, W.J. Cromartie, and J.H. Schwab. 1979. Radiologic analysis of arthritis in rats after systemic injection of streptococcal cell walls. Arthritis Rheum. 22:25-35.

5. Dalldorf, F.G., W.J. Cromartie, S.K. Anderle, R.L. Clark, and J.H. Schwab. 1980. The relation of experimental arthritis to the distribution of streptococcal cell wall fragments. Am. J. Pathol. 100:383-402.

6. Wilder, R.L., G.B. Calandra, A.J. Garvin, K.D. Wright, and C.T. Hansen. 1982. Strain and sex variation in the susceptibility to streptococcal cell wall-induced polyarthritis in the rat. Arthritis Rheum. 25:1064-1072.

7. Schwab, J.H. 1979. In: Microbiology-1979. (D. Schlessinger, ed), American Society for Microbiology, Washington, D.C. pp. 209-214.

8. Stimpson, S.A., R.R. Brown, S.K. Anderle, D.G. Klapper, R.L. Clark, W.J. Cromartie, and J.H. Schwab. Arthropathic properties of cell wall polymers from normal flora bacteria. Submitted.

9. Spitznagel, J.K., K.J. Goodrum, and D.J. Warejcka. 1983. Rat arthritis due to whole group B streptococci. Clinical and histopathologic features compared with groups A and D. Am. J. Pathol. 112:37-47.

10. Lehman, T.J.A., J.B. Allen, P.H. Plotz, and R.L. Wilder. 1983. Polyarthritis in rats following the systemic injection of Lactobacillus casei cell walls in aqueous suspension. Arthritis Rheum. 26:1259-1265.

11. Stimpson, S.A., R.A. Lerch, D.R. Cleland, W.J. Cromartie, and J.H. Schwab. Effect of acetylation on the arthropathic properties of group A streptococcal peptidoglycan-polysaccharide polymers. Submitted.

12. Fox, A., R.R. Brown, S.K. Anderle, C. Chetty, W.J. Cromartie, H. Gooder and J.H. Schwab. 1982. Arthropathic properties related to the molecular weight of peptidoglycan-polysaccharide polymers of streptococcal cell walls. Infect. Immun. 35:1003-1010.

13. Chetty, C., D.G. Klapper, and J.H. Schwab. 1982. Soluble peptidoglycan-polysaccharide fragments of the bacterial cell wall induce acute inflammation. Infect. Immun. 38:1010-1019.

14. Chetty, C., R.R. Brown, and J.H. Schwab. 1983. Edema-producing activity of group A streptococcal polysaccharide and its possible role in the pathogenesis of cell wall-induced polyarthritis. J. Exp. Med. 157:1089-1100.

15. Greenblatt, J.J., R.J. Boackle, and J.H. Schwab. 1978. Activation of the alternate complement pathway by peptidoglycan from streptococcal cell wall. Infect. Immun. 19: 296-303.

16. Sartor, R.B., S.K. Anderle, S.A. Stimpson, W.J. Cromartie, and J.H. Schwab. 1985. Bacterial cell wall induced GALT hemorrhage: Effect of indigenous intestinal bacterial cell walls and quantitation of tissue specificity. Gastroenterology 86:1232.

17. Esser, R.E., S.K. Anderle, C. Chetty, S.A. Stimpson, W.J. Cromartie, and J.H. Schwab. Comparison of inflammatory reactions induced by intra-articular injection of bacterial cell wall polymers. Am. J. Pathol., in press.

18. Anderle, S.K., J.J. Greenblatt, W.J. Cromartie, R. Clark, and J.H. Schwab. 1979. Modulation of the susceptibility of inbred and outbred rats to arthritis induced by cell walls of group A streptococci. Infect. Immun. 25:484-490.

19. Anderle, S.K., J.B. Allen, R.L. Wilder, R. Eisenberg, W.J. Cromartie, and J.H. Schwab. 1985. Measurement of streptococcal cell wall in tissues of rats resistant or susceptible to cell-wall induced arthritis. Infect. Immun. 49:836-837.

20. Schwab, J.H. and S.H. Ohanian. 1967. Degradation of streptococcal cell wall antigens in vivo. J. Bacteriol 94:1346-1352.

21. Esser, R.E., S.A. Stimpson, W.J. Cromartie, and J.H. Schwab. Reactivation of streptococcal cell wall-induced arthritis by homologous and heterologous cell wall polymers. Arthritis Rheum. (in press).

22. Stimpson, S.A., R.E. Esser, W.J. Cromartie, and J.H. Schwab. Arthropathic properties of lipopolysaccharide in rats. Submitted.

23. Janusz, M.J., C. Chetty, R.A. Eisenberg, W.J. Cromartie, and J.H. Schwab. 1984. Treatment of experimental erosive arthritis in rats by injection of the muralytic enzyme mutanolysin. J. Exp. Med. 160:1360-1374.

24. Janusz, M.J., R.E. Esser, and J.H. Schwab. In vivo degradation of bacterial cell wall by the muralytic enzyme mutanolysin. Submitted.

25. Janusz, M.J., R.A. Eisenberg, and J.H. Schwab. Treatment of bacterial cell wall-induced arthritis in rats with a muralytic enzyme: Effect on complement activation by cell wall. Submitted.

26. Zidek, Z., K. Masek, and F. Sedivy. 1984. Antiinflammatory effects of muramyl dipeptide in experimental models of inflammation. Agents Actions 14:72-75.

27. Chedid, L. 1983. Muramyl peptides as possible endogenous immunopharmacological mediators. Microbiol. Immunol. 27:723-732.

28. Dinarello, C.A. 1984. Interleukin-1 and the pathogenesis of the acute-phase response. N. Engl. J. Med. 311:1413-1418.

29. Mizel, S.B., J.M. Dayer, S.M. Krane, and S.E. Mergenhagen. 1981. Stimulation of rheumatoid synovial cell collagenase and prostaglandin production by partially purified lymphocyte-activating factor (interleukin 1). Proc. Natl. Acad. Sci. 78:2474-2477.

30. Schmidt, J.A., S.B. Mizel, D. Cohen, and I. Green. 1982. Interleukin 1, a potential regulator of fibroblast proliferation. J. Immunol. 128:2177-2182.

31. Vacheron, F., M. Guenounou, and C. Nauciel. 1983. Induction of interleukin 1 secretion by adjuvant-active peptidoglycans. Infect. Immun. 42: 1049-1054.

32. Wahl, S.M., L.M. Wahl, J.B. McCarthy, L. Chedid, and S.E. Mergenhagen. 1979. Macrophage activation by water soluble components and synthetic muramyl dipeptide. J. Immunol. 122:2226-2231.

33. Yoshino, S., W.J. Cromartie, and J.H. Schwab. 1985. Inflammation induced by bacterial cell wall fragments in the rat air pouch: Comparison of rat strains and measurement of arachidonic acid metabolites. Am. J. Pathol. (In press).

34. Kimural, Y., Y. Norose, T. Nato, M. Furuya, M. Hida, and T. Okabe. 1982. In: Basic Concepts of Streptococci and Streptococcal Diseases. (S.E. Holm and P. Christenssen, eds.) Reedbooks Ltd., Surrey. p. 99-100.

35. Kaliner, M. and R. Lemanske. 1984. Inflammatory response to mast cell granules. Fed. Proc. 43:2846-2851.

36. Espersen, F., J.O. Jarlov, C. Jensen, P.S. Skov and S. Norn. 1984. Staphylococcus aureus peptidoglycan induces histamine release from basophil human leukocytes in vitro. Infect. Immun. 46:710-714.

37. Ward, P.A. 1982. In: Current Topics in Inflammation and Infection (R.S. Cotran and N. Kaufman, eds). Williams and Wilkins, Baltimore. pp. 54-61.

38. Schwab, J.H., J.B. Allen, S.K. Anderle, F.G. Dalldorf, R.A. Eissenberg, and W.J. Cromartie. 1982. Relationships of complement to experimental arthritis induced in rats with streptococcal cell walls. Immunol. 46:83-88.

39. Eisenberg, R.A. and J.H. Schwab. Arthropathic group A streptococcal cell walls require specific antibody for activation of human complement by both the classical and alternative pathways. Submitted.

40. Pabst, M.J. and R.B. Johnston. 1980. Increased production of superoxide anion by macrophages exposed in vitro to muramyl dipeptide or lipopolysaccharide. J. Exp. Med. 151:101-114.

41. Leong, P.A., J.H. Schwab and M.S. Cohen. 1984. Interaction of group A streptococcal peptidoglycan-polysaccharide with human polymorphonuclear leukocytes: Implications for pathogenesis of chronic inflammation. Infect. Immun. 45:160-165.

42. Janoff, A. and H. Carp. 1982. In: Current Topics in Inflammation and Infection (R.S. Cotran and N. Kaufman, eds) Williams and Wilkins, Baltimore. pp. 62-82.

43. Takada, H., M. Tsujimoto, K. Kato, S. Kotani, S. Kusumoto, M. Image, T. Shiba, I. Yano, S. Kawata and K. Yokogawa. 1979. Macrophage activation by bacterial cell walls and related compounds. Infect. Immun. 25:48-53.

44. Smialowicz, R. and J.H. Schwab. 1977. Cytotoxicity of rat macrophages activated by persistent or biodegradable bacterial cell walls. Infect. Immun. 17:599-606.

45. Davies, P., R.C. Page, and A.C. Allison. 1974. Changes in cellular enzyme levels and extracellular release of lysosomal acid hydrolases in macrophages exposed to group A streptoooccal cell wall substances. J. Exp. Med. 139:1262-1282.

46. Drapier, J.C., G. Lamaire, and J-F Petit. 1982. Regulation of plasminogen activator secretion in mouse peritoneal macrophages. II. Inhibition by immunomodulators of bacterial origin. Int. J. Immunopharmacol. 4:21-34.

47. Werb, Z. and S. Gordon. 1975. Secretion of a specific collagenase by stimulated macrophages. J. Exp. Med. 142:346-360.

48. McArthur, W., A.M. Yaari, and I.M. Shapiro. 1980. Bone solubilization by mononuclear cells. Lab. Invest. 42:450-456.

49. Schwab, J.H. 1982. In: Immunomodulation by Microbial Products and Related Synthetic Compounds. (Y. Yamamura, S. Kotani, I. Azuma, A. Koda, and T. Shiba, eds.) Excerpta Medica, Amsterdam. pp. 84-93.

50. Petrone, W.F., D.K. English, K. Wong, and J.M. McCord. 1980. Free radicals and inflammation: Superoxide-dependent activation of a neutrophil chemotactic factor in plasma. Proc. Natl. Acad. Sci. U.S.A., 77: 1159-1163.

51. Ryc, M. and J. Rotta. 1975. The thrombocytolytic activity of bacterial peptidoglycans. Z. Immun.-Forsch. 149:265-272.

52. Ryc, M. and J. Rotta. 1978. Biological characteristics of peptidoglycan of group A streptococcus and some other bacterial species. II. Immunological mechanisms involved in thrombocytolysis. J. Hyg. Epidemiol. Microbiol. Immunol. 22:435-441.

53. Spika, J.S., P.K. Peterson, B.J. Wilkinson, D.E. Hammerschmidt, H.A. Verbrugh, J. Verhoef, and P. Quie. 1982. Role of peptidoglycan from Staphylocococcus aureus in leukopenia, thrombocytopenia, and complement activation associated with bacteremia. J. Inf. Dis. 146:227-234.

54. Harada, K., S. Kotani, H. Takada, M. Tsujimoto, Y. Hirachi, S. Kusumoto, T. Shiba, S. Kawata, K. Yokogawa, H. Nishimura, T. Kitaura, and T. Nakajima. 1982. Liberation of serotonin from rabbit blood platelets by bacterial cell walls and related components. Infect. Immun. 37:1181-1190.

55. Ubatuba, F.B., E.A. Harvey, and S.H. Ferreira. 1975. Are platelets important in inflammation? Agents Action 5:31-34.

56. Hunter, N., S.K. Anderle, R.R. Brown, F.G. Dalldorf, R.L. Clark, W.J. Cromartie, and J.H. Schwab. 1980. Cell-mediated immune response during experimental arthritis induced in rats with streptococcal cell walls. Clin. Exp. Immunol. 42:441-449.

57. Wilder, R.L. and J.B. Allen. 1984. The role of thymus-dependent lymphocytes in the pathogenesis of group A streptococcal cell wall-induced arthritis in rats. Arthritis Rheum (Suppl) 27:536 (Abstract).

58. Leclerc, C., A. Morin, E. Deriand, and L. Chedid. 1984. Inhibition of human IL2 production by MDP and derivatives. J. Immunol. 133:1996-2000.

BIOLOGICAL ACTIVITY OF PEPTIDOGLYCAN IN MAN

B. Heymer
Department of Pathology, University of Ulm, 7900 Ulm/Donau

J. v. Mayenburg
Department of Dermatology, Technical University of Munich, 8000 Munich

Introduction

Since considerable differences in susceptibility to microbial agents exist among mammalian species, findings obtained in experimental animals frequently cannot be said without qualification to hold good for man (5). This also holds true for the vast amount of experimental data that has accumulated on the biological activity of peptidoglycan (PG) both in vitro and in vivo (18, 20, 36). However, exposure of man to PG is - indirectly - shown by the widespread occurrence of PG-antibodies in human sera (10, 11, 20, 34, 43, 45). Furthermore, there is evidence that PG may play a pathogenic role in a number of human diseases (1, 9, 24), the etiology of which is yet unknown. For these reasons, the biological activity of PG was studied directly in man.

Materials and Methods

Four kinds of PG preparations were employed: PG from Staphylococcus aureus (Cowan I) and Staphylococcus epidermidis (CCM 2495), both extracted with trichloroacetic acid (35); PG from Streptococcus pyogenes (B 196), extracted with hot formamide (17, 19), and streptococcal PG additionally treated with dilute NaOH (12) to remove O-formyl groups introduced into PG by formamide (35). Before use, all preparations were solubilized by sonication (16, 17). Amounts of PG varying from 1.3 to 10 µg, which is below the toxic threshold dose (>20 µg; 23, 24), were injected intradermally on the volar side of the forearm in 14 patients with various allergic or infectious diseases and 2 healthy volunteers. The reactions induced by PG were evaluated macroscopically and skin biopsies were taken 20 minutes to 48 hours p.i.. A total of 26 biopsies was investigated by conventional histological techniques using HE,

Biological Properties of Peptidoglycan

Giemsa and Toluidin blue staining of sections (33); 24 biopsies, snap frozen in liquid nitrogen, were studied immunohistologically for deposits of IgG, IgM, IgA, IgE and C_3 (23); 11 biopsies, after appropriate fixation and embedding, were analyzed by electron microscopy (23).

Results

The skin reactions induced by PG, depending on the kind of PG employed and the subject tested, macroscopically showed a surprising variability in regard to onset, intensity and duration. Some of the reaction patterns observed are depicted in Figure 1.

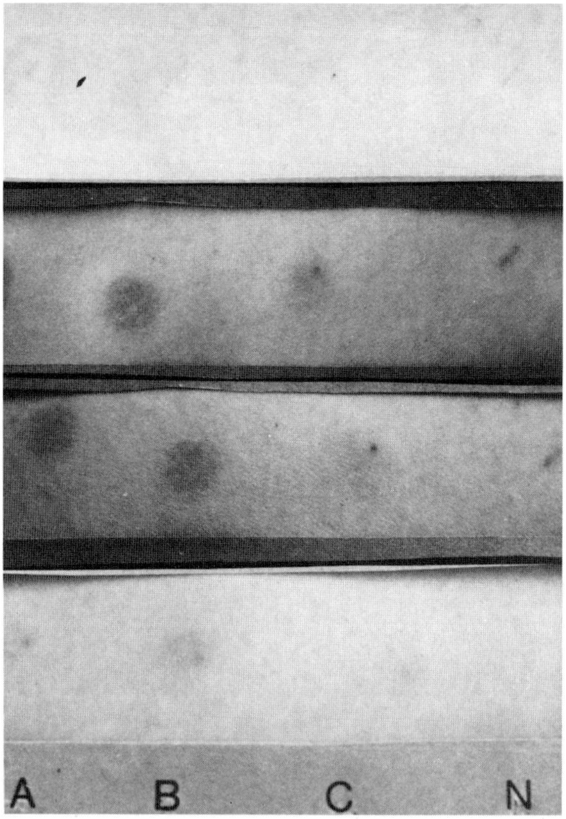

Fig. 1 Skin reactions (from top to bottom) 20 minutes, 4 hours, 24 hours, 48 hours after intradermal injection of 1.3 µg streptococcal (A, B) or 1.4 µg staphylococcal (C) PG in a healthy volunteer. Saline control (N).

Two streptococcal PGs (A and B), one staphylococcal PG (C) and a saline control (N) had been applied in a healthy volunteer. Reactions are shown from 20 minutes to 48 hours after the injection. It is evident that lesions start with edema (top) followed by erythema (middle) and - at least macroscopically - disappear within 48 hours (bottom). Haemorrhage or focal necrosis of skin as in the so called dermonecrotic reaction induced by high dose PG application in experimental animals (16) was never observed in the present studies.

Histomorphologically, the PG reactive sites in the various subjects tested also disclosed considerable differences, both quantitatively and qualitatively. Biopsies taken 20 minutes after the injection showed margination and the onset of vascular emigration of leukocytes. Four to 6 hours p.i. - in most instances - a diffuse infiltration of the dermis by granulocytes was present (Fig. 2).

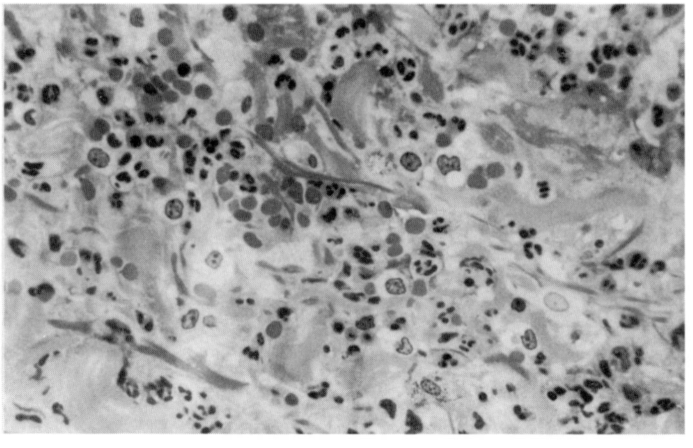

Fig. 2 Skin reaction 4 hours after intradermal injection of 1.4 µg staphylococcal PG in a healthy volunteer. Semithin section, HE, 512 x.

Infiltrates were primarily composed of neutrophils. In contrast, 24 to 48 hours p.i. lesions exhibited a pronounced mononuclear cell infiltration (Fig. 3).

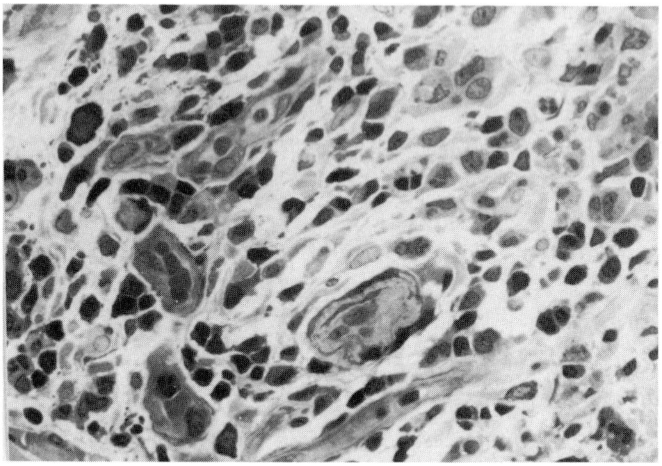

Fig. 3 Skin reaction 24 hours after intradermal injection of 1.3 µg streptococcal PG in a healthy volunteer. Semithin section, HE, 512 x.

Frequently, the PG reactive sites contained - in addition to lymphocytes and monocytes - mast cells and varying numbers of eosinophils. By electron microscopy, degranulation of mast cells could be detected in many of the skin lesions induced by PG (Fig. 4). This indicates release of potent mediators of inflammation such as histamine (38).

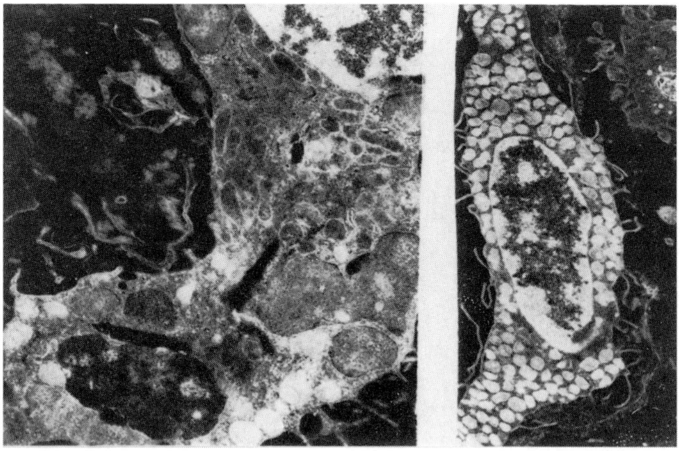

A B

Fig. 4 Degranulation of mast cell (A) 4 hours after intradermal injection of 1.4 μg streptococcal PG in a healthy volunteer. Non-degranulated mast cell for comparison (B). Electron micrograph, A = 14 500 x, B = 5 900 x.

Immunohistologically, only the skin biopsies of 2 patients with allergic vasculitis showed deposition of Ig's (IgG, IgM). However, all 7 biopsies taken 4 to 6 hours after the PG injection - but only 2 out of 17 biopsies taken 24 to 48 hours p.i. - contained C_3-deposits, primarily localized within vessel walls (Fig. 5). In this context it may be interesting to note that the immunohistologic detectability of C_3 in the skin of patients with leukocytoclastic vasculitis also rapidly decreases within 8 to 24 hours after the injection of histamine (13).

A synopsis of part of the studies made is given in Table 1. The subjects tested (healthy volunteers and patients with various infectious diseases), the clinical diagnosis, the type of PG injected, the time of biopsy and the intensity of the reactions observed are listed. From the data presented two facts are obvious: 1. Reactions induced by PG are not uniform; there is a considerable variation in type and intensity of lesions among the subjects tested. 2. When comparing the extent of the skin reactions (diameter of erythema) evident macroscopically and the density of the inflammatory infiltrations (+ to ++++) as detected histologically, there is no strict correlation rather a significant discrepancy (Table 1). Without going into details or starting speculation on this issue, this indicates a variable participation of vascular and cellular pathomechanisms in the skin reactions induced in the individual subjects.

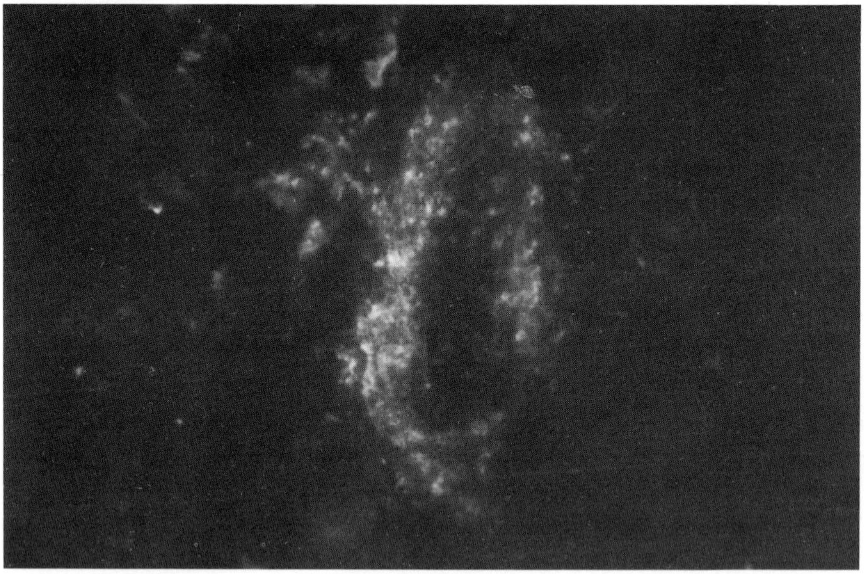

Fig. 5 Immunohistologic demonstration of C_3 in the wall of a postcapillary venule 4 hours after intradermal injection of 1.4 μg staphylococcal PG. There are also some perivascular C_3 deposits. 800 x.

Table 1

Skin Reactions induced by Peptidoglycan (PG) in Healthy Volunteers and in Patients with various Infectious Diseases					
Subject Age (years) Sex (m/f)	Clinical Diagnosis	Pg-Prep. injected (Sp/Sa/Se)*	Time of Skin Biopsy (hrs p.i.)	Intensity of Reaction	
				Macroscop. (mm)**	Histolog. (+)***
B.G. 49 m	Healthy volunteer	Sp	0,4	10	+
		Sp	6	7	+++
		Sp	24	7	++++
		Sp	48	5	+++
M.J. 38 m	Healthy volunteer	Sa	6	14	++
		Sa	24	5	+
H.F. 34 m	Bacterial sinusitis	Sp	6	25	++++
		Sp	48	15	+++
K.P. 32 m	Bacterial rhinitis	Sp	24	30	+
H.E. 28 m	Bacterial bronchitis	Sa	24	30	+++
		Se	24	20	+++
Sch.J. 60 f	Bacterial rhinitis	Sp	48	20	+++
K.Y. 22 m	Pyoderma	Sp	48	6	++++
W.A. 24 m	Immunodeficiency	Sp	24	14	+++
		Sp	48	6	++

*Sp = Streptococcus pyogenes, Sa = Staphylococcus aureus, Se = Staphylococcus epidermidis

** = Diameter of erythema in mm

*** = Grading of inflammatory reactions from + = weak to ++++ = very marked

A summary of the second part of the studies, carried out in patients with various allergic diseases, is given in Table 2. All these patients had strongly elevated total IgE levels (from 210 to 1652 units/ml serum). However, there was no correlation between the IgE titer in serum and the intensity of the reaction induced by PG in skin. As in patients with infectious diseases or in healthy volunteers (Tab. 1) a discrepancy did exist between the inflammatory reaction evident macroscopically (size of erythema) and the degree of inflammation (+ to ++++) present histologically (Tab. 2). Nevertheless, it appears noteworthy that - in contrast to patients with infectious diseases - patients with allergic diseases showed no marked cellular infiltrations in the dermis 24 to 48 hours p.i. (Tab. 2).

Table 2

Skin Reactions induced by Peptidoglycan (PG) in Patients with various Allergic Diseases					
Subject Age (years) Sex (m/f)	Clinical Diagnosis	PG-Prep. injected (Sp/Sa/Se)*	Time of Skin Biopsy (hrs p.i.)	Intensity of Reaction	
				Macroscop. (mm)**	Histolog. (+)***
V.A. 46 m	Allergic bronchitis	Se	6	10	++
		Sa	24	5	++
P.A. 48 f	Allergic vasculitis	Se	4	7	++
		Sa	4	11	+++
H.L. 47 m	Allergic vasculitis	Sa	6	9	+++
D.H. 34 m	Allergic rhinitis	Sp	6	10	++++
K.Ch. 58 f	Allergic bronchitis	Sa	24	20	++
		Sp	48	25	+
G.W. 42 m	Allergic bronchitis	Sp	24	40	++
J.S. 24 m	Allergic rhinitis	Sp	48	8	+
D.G. 39 m	Contact eczema	Sp	48	25	++

*Sp = Streptococcus pyogenes, Sa = Staphylococcus aureus, Se = Staphylococcus epidermidis

** = Diameter of erythema in mm

*** = Grading of inflammatory reactions from + = weak to ++++ = very marked

In addition there was a tendency for staphylococcal PGs to induce "early" (20 minutes to 4 hours p.i.) and for streptococcal PGs to induce "late" or "delayed" (24 hours p.i.) skin reactions.

When trying to define the possible pathomechanism underlying the skin reactions induced by PG in humans - as described here - one should remember that PG has multiple biological activities and different modes of action, immunological as well as non-immunological (18, 20, 36). At least some of these activities have also been studied in man. This holds true e.g. for the activation of complement, B-cells, macrophages and neutrophils (Tab.3).

Table 3

Biological Activity of Peptidoglycan in Man		
Biological Activity	Source of Peptidoglycan	Reference
Complement Activation	Streptococcus pyogenes Staphylococcus epidermidis Staphylococcus aureus Various bacterial species Neisseria gonorrhoeae	Bokisch, 1975 Greenblatt et al., 1978 Semerado et al., 1979 Verbrugh et al., 1980 Wilkinson et al., 1981 Peterson and Rosenthal, 1982
Polyclonal B-cell Activation	Staphylococcus aureus Various bacterial species	Dziarski et al., 1980 Räsänen et al., 1982 Levinson et al., 1983
Macrophage Activation	Staphylococcus aureus Various bacterial species	Oken et al., 1979 Ogawa et al., 1982
Neutrophil Activation (chemiluminescence, capping of C'receptors)	Staphylococcus aureus Streptococcus pyogenes	Musher et al., 1981 Pryzwansky et al., 1983
Induction of inflammatory Skin Reactions	Various bacterial species	v. Mayenburg et al.,1980,1982

In addition it is known that antibodies to PG occur in most - if not all - human sera (11, 20, 34). These "natural" antibodies in conjunction with complement could mediate the inflammatory reactions induced by PG in man.

In this context it might be of interest to note that microbial cell wall teichoic acid (TA), in animal experiments, has recently been found to produce acute transient inflammation whereas PG induces a more delayed but persistent inflammatory process (39, 40). This was attributed to the ability of TA to activate complement through the alternative pathway (21, 39, 40) whereas PG appeares to be relatively inefficient in this respect (3, 18). Since TA within the microbial cell wall is covalently linked to PG (21, 29, 40) extraction procedures for PG which leave TA undegraded might result in the isolation of PG-TA complexes (35, 39, 40). The biological activity of such a preparation could be determined partially or completely by the TA moiety. However, while the presence of small amounts of cell wall TA in the staphylococcal PGs employed in the present study cannot be completely excluded, methods used for the extraction of streptococcal PGs are known to destroy microbial TA (17, 35).

From the data presented here as well as from earlier investigations (18, 20) it appears that the activation of complement and the release of endogenous mediators of inflammation (e.g. of histamine from mast cells) plays an important role in the production of the inflammatory reactions induced by PG. The pathogenic pathways just mentioned can be activated by immunological as well as by non-immunological means (4), resulting in the same or a similar type of inflammation (15, 41). However, it appears important to stress that the mode of action of PG must be quite different from microbial endotoxins as well as exotoxins. In most assay systems PG displays a relatively low activity (20) as compared to the extreme biological activity of many bacterial exotoxins (2) and endotoxins (18, 36). Thus, in rabbits the pyrogenic activity of 50 µg PG/kg body weight is equivalent to 0.008 µg/kg of Lipid A-BSA (18). Whereas the lethal dose of PG in guinea pigs is many mg, an amount of only 10^{-6} mg of botulism exotoxin is sufficient to produce this effect (2). Such profound differences in activity indicate that PGs produce their in vivo effects by indirect mechanisms (e.g. via the immune system) rather than by direct intrinsic toxicity, as originally thought (16). In summarizing the data collected on the biological activity of PG in man one may conclude:

1. PG does possess a distinct biological activity and pathogenicity in man.
2. Lesions induced by intradermal injection of PG exhibit certain common features such as edema, erythema and predominant granulocytic infiltration.
3. Activation of complement and release of endogenous mediators from mast cells appears to play a key role in the inflammatory reactions induced by PG.
4. It is not imperative to decide whether PG activates complement through the alternate and/or the classical pathway. In view of the widespread occurrence of PG antibodies in human sera the prerequisites for the latter mechanism of action are fulfilled.

References

1. Bennett, J.C.. 1978. Arthritis Rheum. 21, 531-538.

2. Bhakdi, S.. 1983. Immun. Infekt. 11, 181-189.

3. Bokisch, V.A.. 1975. Z. Immun.-Forsch. 149, 320-330.

4. Boyle, M.D.P., T. Borsos. 1983. In: The Reticuloendothelial System (N.R. Rose and B.V. Siegel, eds.). Plenum Press, New York-London, pp. 43-76.

5. Cottier, H.. 1980. Pathogenese, Bd. 1 und 2. Springer-Verlag, Berlin-Heidelberg-New York.

6. Dziarski, R.. 1980. Infect. Immun. 30, 431-438.

7. Dziarski, R.. 1980. J. Immunol. 125, 2478-2483.

8. Dziarski, R., A. Dziarski, A.I. Levinson. 1980. Int. Archs. Allergy appl. Immun. 63, 383-395.

9. Ely, P.H.. 1980. J. Am. Acad. Dermatol. 2, 473-487.

10. Esser, R.E., J.H. Schwab, R.A. Eisenberg. 1985. In: Immunology of the Bacterial Cell Envelope (D.E.S. Stewart-Tull and M. Davies, eds.). Wiley, Chichester-New York-Brisbane-Toronto-Singapore, pp. 91-118.

11. Franken, N., P.H. Seidl, T. Kuchenbauer, H.J. Kolb, K.H. Schleifer, L. Weiss, K.-D. Tympner. 1984. Infect. Immun. 44, 182-187.

12. Glick, A.D., J.M. Ranhand, R.M. Cole. 1972. Infect. Immun. 6, 403-408.

13. Gower, R.G., W.M. Sams, E.G. Thorne, P.F. Kohler, H.N. Claman. 1977. J. Invest. Dermatol. 69, 477-484.

14. Greenblatt, J., R.J. Boackle, J.H. Schwab. 1978. Infect. Immun. 19, 296-303.

15. Heymer, B.. 1985. Path. Res. Pract. 180, 143-150.

16. Heymer, B., B. Bültmann, O. Haferkamp. 1971. J. Immunol. 106, 858-861.

17. Heymer, B., B. Bültmann, W. Schachenmayr, R. Spanel, O. Haferkamp, W.C. Schmidt. 1973. J. Immunol. 116, 1743-1754.

18. Heymer, B., E. Th. Rietschel. 1977. In: Microbiology 1977 (D. Schlesinger, ed.). Am. Soc. Microbiol., Washington, pp. 344-349.

19. Heymer, B., W. Schachenmayr, B. Bültmann, R. Spanel, O. Haferkamp, W.C. Schmidt. 1973. J. Immunol. 111, 478-484.

20. Heymer, B., P.H. Seidl, K.H. Schleifer. 1985. In: Immunology of the Bacterial Cell Envelope (D.E.S. Stewart-Tull and M. Davies, eds.). Wiley, Chichester-New York-Brisbane-Toronto-Singapore, pp. 11-46.

21. Hummel, D.S., A.J. Swift, A. Tomasz, J.A. Winkelstein. 1985. Infect. Immun. 47, 384-387.

22. Levinson, A.I., A. Dziarski, B. Zweiman, R. Dziarski. 1983. Infect. Immun. 39, 290-296.

23. von Mayenburg, J., B. Heymer, H. Düngemann, K.H. Schleifer, P.H. Seidl, J. Galle, A. Neiß, S. Borelli. 1980. Z. Hautkr. 55, 710-733.

24. von Mayenburg, J., B. Heymer, H. Düngemann, K.H. Schleifer, P.H. Seidl, A. Neiß, S. Borelli. 1982. Allergy 37, 249-258.

25. Musher, D.M., H.A. Verbrugh, J. Verhoef. 1981. J. Immunol. 127, 84-88.

26. Ogawa, T., S. Kotani, K. Fukuda, Y. Tsukamoto, M. Mori, S. Kusumoto, T. Shiba. 1982. Infect. Immun. 38, 817-824.

27. Oken, M.M., P.K. Peterson, B.J. Wilkinson. 1979. Clin. Res. 27, 352-361.

28. Peterson, B.H., R.S. Rosenthal. 1982. Infect. Immun. 35, 442-448.

29. Peterson, P.K., Wilkinson, B.J., Y. Kim, D. Schmeling. 1978. J. Clin. Invest. 61, 597-609.

30. Pryzwansky, K.B., E.K. MacRae, J.D. Lambris. 1983. J. Immunol. 130, 1674-1677.

31. Räsänen, L., H. Arvilommi. 1982. Infect. Immun. 35, 523-527.

32. Räsänen, L., U.P. Mustikkamäki, H. Arvilommi. 1982. Immunol. 46, 481-486.

33. Romeis, B.. 1968. Mikroskopische Technik. Oldenburg-Verlag, München-Wien.

34. Schachenmayr, W., B. Heymer, O. Haferkamp. 1975. Z. Immun.-Forsch. 149, 179-186.

35. Schleifer, K.H..1975. Z. Immun.-Forsch. 149, 104-117.

36. Schleifer, K.H., B. Heymer. 1975. Z. Immun.-Forsch. 149, 103-356.

37. Semerado, N., M. Colucci, J. Vermylen. 1979. Thrombos. Haemostas. 41, 392-406.

38. Soter, N.A., K.F. Austen. 1976. J. Invest. Dermatol. 67, 313-319.

39. Tuomanen, E., H. Lin, B. Hengstler, O. Zak, A. Tomasz. 1985. J. Infect. Dis. 151, 859-868.

40. Tuomanen, E., A. Tomasz, B. Hengstler, O. Zak. 1985. J. Infect. Dis. 151, 535-540.

41. Turk, J.L.. 1980. Delayed Hypersensitivity. Elsevier-North-Holland Biomedical Press, Amsterdam-New York-Oxford, pp. 45-64.

42. Verbrugh, H.A., W.C. van Dijk, R. Peters, M.E. van Erne, M.R. Daha, P.K. Peterson, J. Verhoef. 1980. J. Immunol. 124, 1167-1173.

43. Wheat, L.J., B.J. Wilkinson, R.B. Kohler, A.C. White. 1983. J. Infect. Dis. 147, 16-22.

44. Wilkinson, B.J., Y. Kim, P.K. Peterson. 1981. Infect. Immun. 32, 216-224.

45. Zeiger, A.R., C.U. Tuazon, J.N. Sheagren. 1981. Infect. Immun. 33, 795-800.

PATHOMECHANISMS OF PEPTIDOGLYCAN IN MAN

J.v.Mayenburg

Department of Dermatology, Technical University of Munich, 8000 Munich

B.Heymer

Department of Pathology, University of Ulm, 7900 Ulm/Donau

Introduction

The pathogenesis of many allergic diseases ascribed to the action of bacterial antigens in man is so far unclarified (6,7,9,10,13, 14,15). On the other hand, bacterial substances are used to examine the immune response (2) in patients with HTLV-III infection (Aids). Since peptidoglycan (PG) is present in all bacteria and therefore occurs almost ubiquitously in the human environment its biological effects in man are of particular interest. Thus the main objective of the present study was to obtain an insight into the pathomechanisms involved in the action of PG in man.

Subjects and Methods

A total of 181 subjects (32 females, 89 males) aged from 12 to 87 years was employed in the studies. These comprised of 18 healthy volunteers and 163 patients from the Department of Dermatology, Technical University of Munich. The patients suffered from various airway and skin diseases: 1) 66 patients with atopic diseases, 2) 34 patients with chronic bacterial infections, 3) 9 patients with atopic diseases and chronic bacterial infections, 4) 54 patients without the diseases specified.

The following microbial preparations were employed in the present studies: 1) PG from Staphylococcus aureus, Cowan I(=PG-Sa), test dose 10 µg, 2) PG from Staphylococcus epidermidis CCM 2495 (= PG-Se) test dose 1.4 µg, 3) PG from Streptococcus pyogenes group A,(= PG-SF) test dose (1,3 µg), 4) PG prepared as in 3, but additionally treated with dilute NaOH (= PG-SNaOH) test dose 1.4 µg, 5) PG teichoic acid complex from Staphylococcus aureus, Cowan I (= PG-TA-Sa), test dose 20 µg. 6) Teichoic acid from Staphylococcus aureus, Cowan I, (= TA - Sa) test dose 10 µg . These microbial preparations were isolated as earlier described (3,12).

In addition to skin testing with the bacterial antigens described above, antibodies to PG and total IgE were determined in the sera of all subjects by RIA (4).

Biological Properties of Peptidoglycan

Results

Due to the small test dose used, no obligatory toxic reactions were observed. Between 15% and 55% of the skintests were negative. The positive reactions showed three different patterns: (Fig. 1) 1) an isolated immediate reaction, 2) an isolated late reaction and 3) a dual reaction with an immediate response followed by a late reaction. Figure 1 shows three curves representative of the three reaction types.

The reactions were macroscopically read off at hourly intervals. Example 1) shows an immediate reaction to PG-Sa not followed by any further inflammatory reaction. Example 2) illustrates a gradually developing late reaction to PG-Sa. Finally 3), an immediate reaction is depicted with subsequent late reaction (dual type) to PG-SF.

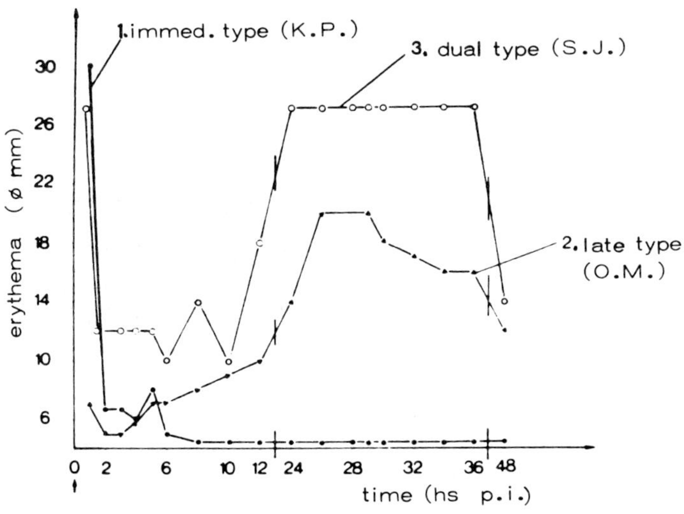

Fig. 1: Schematic representation of the three basic patterns of skin reactions induced by PG:
1. Immediate type (PG-Sa).
2. Late type (PG-Sa).
3. Dual type (PG-SF).

Figure 2 reproduces the profound differences observed in the course of reactions induced by PG (PG - S-NaOH) in eight subjects. It is evident that the reactions vary greatly in intensity, but always correspond to one of the three basic patterns of reaction just described.

Noteworthy is that all reactions induced by intradermal injection of the variouses PG-preparations subsided within two to three days. (Fig. 3, page 308).

Typical tuberculin-like cellular mediated reactions of type IV (COOMBS and GELL) were not observed.

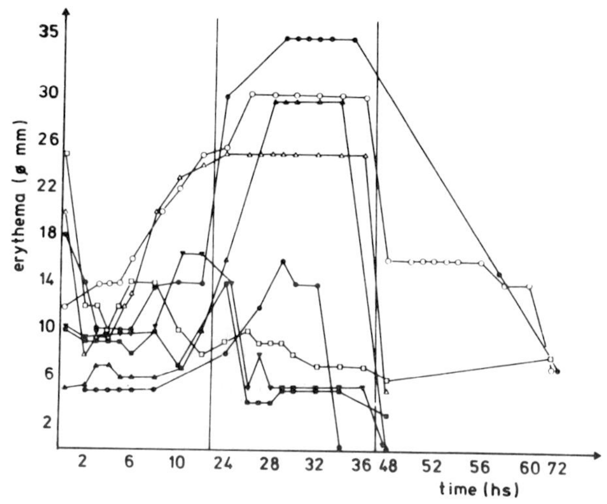

Fig. 2: Time course of skin reactions induced by PG-SNaOH in eight subjects.

The following findings indicate that immunological pathomechanisms are involved in the reactions induced by PG in man:

1. Obligatory toxic reactions did not occur.
2. In two patients with antibody deficiency syndrome, no immediate reactions, but only weak late reactions to streptococcal PG were observed (Fig. 4).

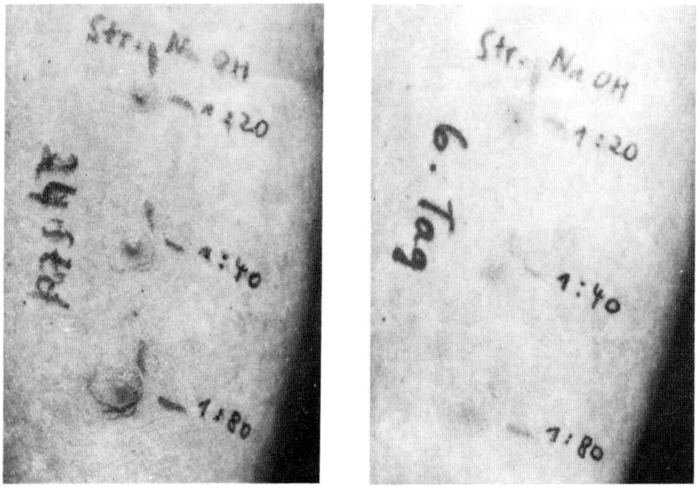

Fig. 3: Skin reactions 24 hours (left) and 6 days (right) after intradermal injection of 1,4 g streptococcal PG (PG-SNaOH).

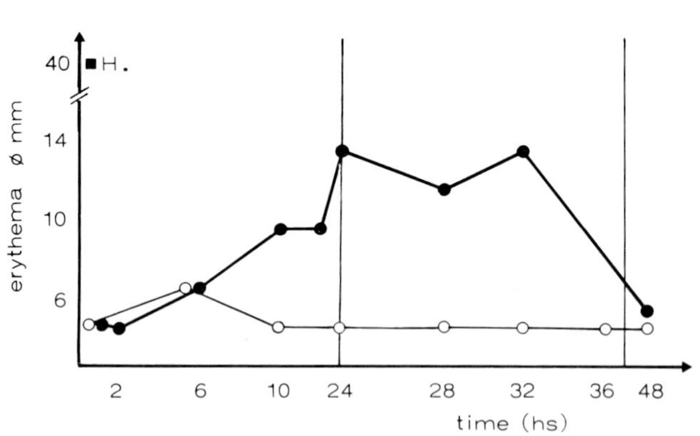

Fig. 4: Skin reactions to different PG-preparations in a patient with antibody deficiency syndrome:
● = reaction to PG-SF.
o = no reaction to all other PG's.

3. In general, the reaction pattern was largely consistent in the individual patients. Fig. 5 discloses the frequencies of the four reaction types (immediate reaction, late reaction, combined immediate and late reaction as well as isolated six-hour reaction) besides the negative reactions as induced by the various bacterial test preparations. The data shown reveal an interesting finding: the incidence of immediate reactions is particulary high when isolated teichoic acid and PG teichoic acid complex (on the left) were employed, whereas late or dual reactions predominate in the response to streptococcal PG (on the right). On the other hand reactions induced by staphylococcal PG, are primarily of the immediate type. Nevertheless it is evident, that there are similarities in the reaction pattern to staphylococcal and streptococcal PG.

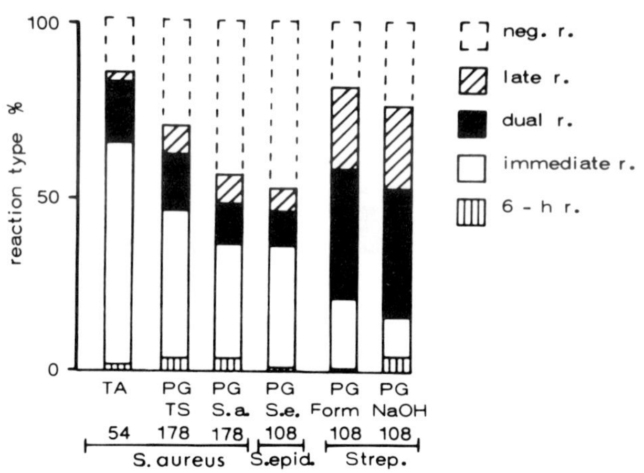

Fig. 5: Frequency of negative, immediate, late, dual and 6-hr skin reactions as induced by different microbial preparations in 181 test subjects.

4. The mean antibody titer of the sera from test subjects with late type or dual type reactions was significantly higher than that of probands with negative skin tests (Fig 6).
   The correlation between late reactions and antibody titers was statistically significant ($P < 0.01$) for the three staphylococcal PGs (PG-Sa, PG-TA-Sa and PG-Se). For the streptococcal PG preparations, no significant correlations could be detected.
5. The total serum IgE concentrations in subjects with immediate reactions to staphylococcal PG were significantly ($P < 0.05$) higher than those in patients with negative skin tests (Fig. 7).

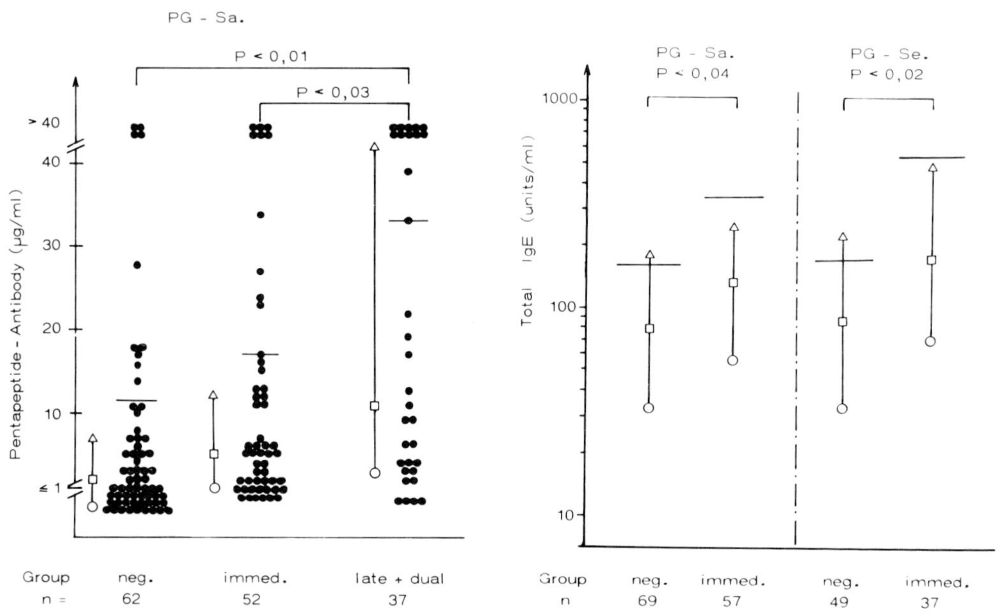

Fig. 6 (left): PG (pentapeptide) antibody titers in test subjects with negative, with immediate and with late or dual skin reactions to PG from Staphylocossus aureas Cowan I(PG-Sa). o = 25%, □ = 50%, △ = 75% (fractiles) of the test subjects.

Fig. 7 (right): Total IgE in test subjects with negative and with immediate reactions to PG-Sa or PG-Se. Representation of the 25% fractiles and mean values. For abbreviations and symbols see legend to Fig. 6.

6. As compared to the two other groups of patients, such patients with chronic bacterial infections had the highest incidence of late reactions. This was true for several kinds of PG (Fig. 8). Some of the differences observed were statistically significant.

When trying to define the immune mechanisms involved in the skin reaction induced by PG, there is good evidence for the participation of an immunpathologic reaction of type III (1,6,9). The positive correlation between late skin reaction and PG serum antibody titers, the rapid regression of the reactions after two to three days and the predominance of dual reactions support this hypothesis. There are interesting parallels between the inflammatory response induced by PG and the type III hypersensitivity produced by molds (8,9).

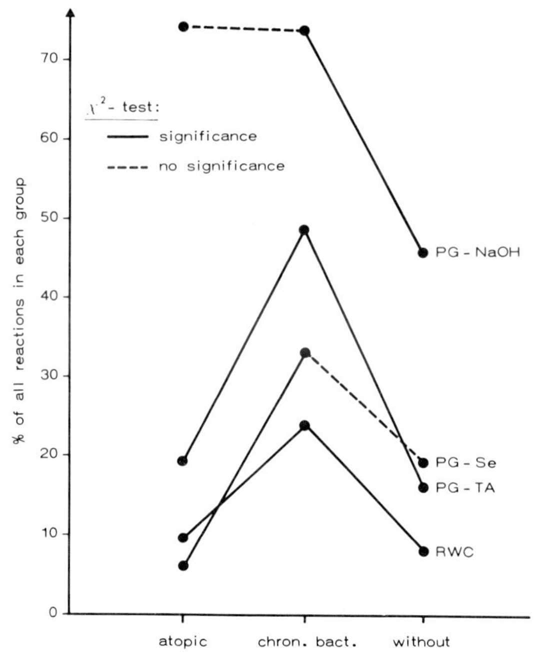

Fig. 8: Incidence of late reactions to PG in three different groups of patients. Significant accumulation of late reactions in patients with chronic bacterial infections.

References

1. Cochrane,C.G.1971 Mechnisms involved in the deposition of immunecomplexes in tissues. J.Exp.Med.134,Part 2,Suppl. 75,75-89.

2. Gottlieb,M.S. et al.1981. Pneumocystis carinii pneumonia and mucosal candidiasis in previously multley homosexual men: evidence of a mer aequired cellular immunodeficiency. N.Engl.J.Med.305,1525-1431.

3. Heymer,B.,Bültmann,B.,Haferkamp,O.1971. Toxicity of streptococcal mucopeptides in vivo and in vivo.J.Immunol.106,858-861.

4. Heymer,B.,Bernstein,D.Schleifer,K.H., Krause,R.M..1975 Measurement of peptidoglycan antibodies by a radioimmunoassay. Z.Immunitaetsforsch.149,168-178.

5. Kraft,B.,Mothersill,M.H.,Nestmann,R.H.1949. A preliminary report on immediate urticarial reactions to intradermal injections of bacterial antigens. Ann.Allergy 7,162-164.

6. Mayenburg v.,J.Heymer,B.,Düngemann,H.,Schleifer,K.H.,Seidl,P.H., Neiß,A., Borelli,S..1982.Studies on Bacterial Hypersensitivity in Man.Allergy,37,249-258.

7. Parish,W.E., Welbourn,E.,Champion,R.H.1976. Hypersensitivity to bacteria in eczema. II. Titre and immunoglobulin class of antibodies to staphylococci and micrococci.Br.J.Dermatol.95, 285-293.

8. Pepys,J.1973. Immunopathology of allergic lung disease. Clin. Allergy 3,1-22,

9. Pepys,J.1979 Allergy. In Dick,G.(ed.): Immunological aspects of infectious diseases,pp215-268. MTP Press,Lancaster.

10. Rajka,E.1963. Epicutane und intracutane Tests mit Mikroben und Mikrobinen. Arch.Klin.Exp.Dermatol.219,939-842,

11. Schleifer,K.H.,Kandler,O.1972. Peptidoglycan types of bacterial cell walls and their taxonomic implications. Bacteriol.Rev. 36,407-477.

12. Schleifer,K.H.1975. Chemical structure of the peptidoglycan, its modifiability and relation to the biological activity. Z.Immunitaetsforsch.149,104-117.

13. Storck,H.1975. Einige Aspekte bakterieller Allergie. Hautarzt 26,507-513

14. Swineford,O.,Holman,J.1949. Studies in bacterial allergy. III. Results of 3860 cutaneous tests with 34 crude polysaccharide and nucleoprotein fractions of 14 different bacteria. J.Allergy 20,420-427

15. Welbourn,E., Champion,R.H.,Parish,W.E.1976. Hypersensitivity to bacteria in exzema. III. Arthus-like responses to bacterial antigens without specific antibody. Br.J.Dermatol.95,379-387.

ROLE OF PEPTIDOGLYCAN IN GONOCOCCAL ARTHRITIS

R. S. Rosenthal, F. R. Montoya, W. Nogami, T. J. Fleming

Department of Microbiology and Immunology
Indiana University School of Medicine
Indianapolis, Indiana USA 46223

Introduction

This laboratory focuses on the role of peptidoglycan (PG) fragments as modulators of immune and inflammatory reactions during natural infections. Several of our previous studies (1-4) have argued that Neisseria gonorrhoeae represents a unique model organism in which PG-host interactions might be particularly direct and extensive in vivo. Accordingly, our central hypothesis is that gonococcal PG fragments, released by bacterial (1-3) and host (4) PG hydrolases, contribute to the intense inflammatory response characteristic of gonococcal disease. Such conditions include gonococcal arthritis (the leading cause of arthritis in young adults), a syndrome in which affected joints are frequently sterile and in which immunological injury likely contributes to tissue damage (5). To test the hypothesis, we have identified and purified numerous gonococcal PG fragments that likely gain access to host tissues in vivo, and have tested their abilities to initiate diverse biological reactions associated with modulation of inflammation. To date, the essential conclusion is that gonococcal PG fragments, collectively, possess numerous such activities, e.g., toxicity for human fallopian tube mucosa (6), activation of human complement (7), and arthritogenicity in a rat model, consistent with their playing a governing role in the host response to gonococcal infection. Current studies continue this approach by addressing the structural requirements and mechanism of gonococcal PG-arthritis in a rat model. We have found that PG-hydrolase resistant, extensively-O-acetylated PG exhibits greatly enhanced arthritogenicity (relative to non-O-acetylated PG) and that the increased arthropathic activity is associated with the relative persistence of PG in certain tissues in vivo.

Biological Properties of Peptidoglycan

Results

Arthritogenicity: enhanced activity of PG hydrolase-resistant O-acetylated PG. In our standard PG-arthritis model, groups of eight male Lewis rats (225 g) were administered a single intradermal injection in the tail of 200 µl of test substances emulsified in a 1:1 mixture of oil (Freund's incomplete adjuvant) and water, or of the oil-water emulsion alone (controls). Unless otherwise noted, the dose of PG fragments was 200 µg per rat. At intervals following injection, hind paw sizes were measured in planes parallel and perpendicular to the plane of the lateral malleoli using a dial thickness gage and expressed, in mm, as the mean score for each group. The net increase in joint score vs time was estimated for each rat using the trapezoidal rule, and the mean areas between groups in a given experiment were compared using a two-sample t test.

Initial studies on the structural requirements of gonococcal PG-arthritis indicated that soluble, heterogeneously sized PG fragments (greater than 10^6 daltons) derived from N. gonorrhoeae strain FA19 had considerably greater arthropathic activity in the rat model than did the corresponding PG fragments derived from strain RD5. This finding was of great interest in light of previous data (4,8) that strain FA19 PG is extensively substituted in the glycan chain with O-acetyl derivatives and is relatively resistant to degradation by diverse PG hydrolases, while strain RD5 PG is non-O-acetylated and is susceptible to PG-degrading enzymes. Current experiments (Fig. 1) provided convincing confirmation that extensive O-acetylation potentiated the arthritogenicity of gonococcal PG. That is, 200 µg of soluble, high-molecular-weight, extensively-O-acetylated PG from strain FA19 (O-PG; Fig. 1) induced significantly greater paw swelling ($p < .01$) than did 200 ug of homologous FA19 PG that had been completely and specifically de-O-acetylated by treatment with mild alkali (de-O-PG).

The previous data suggested that the PG-hydrolase resistance attributable to extensive O-acetylation was necessary for maximal arthropathic activity and, thus, implied that the persistence of macromolecular PG in vivo might be an important determinant of gonococcal PG-arthritis. However, macromolecular PG was not an absolute requirement. Indeed, several classes of low-molecular-weight gonococcal PG fragments, including the novel class of anhydro-muramic acid-containing disaccharide peptide monomers (anhydro-monomer; Fig. 2) retained significant arthropathic activity (p <.05). This was of considerable interest because exponentially growing gonococci turn over and release up to one-third of their existing PG per generation and anhydro-monomers are the principal soluble PG fragments released (1,2). The mixture of fragments comprising anhydro-monomers has now been structurally defined using fast atom bombardment mass spectrometry (see S. A. Martin, K. Biemann, and R. S. Rosenthal this volume). The major components of anhydro-monomers are N-acetylglucosamine-1, 6-anhydro-N-acetylmuramic acid-L-alanine-D-glutamic acid-meso-diaminopimelic acid-D-alanine and the corresponding disaccharide tripeptide lacking the C-terminal D-alanine. Biologically active anhydro-monomers contain no detectable non-PG amino acids. Quantitatively, the low-molecular-weight fragments, e.g., anhydro-monomers, induced paw swelling that was less extensive than that caused by O-acetylated macromolecular soluble PG (S-O-PG) and similar to that caused by commercial muramyl dipeptide (MDP; Fig. 2).

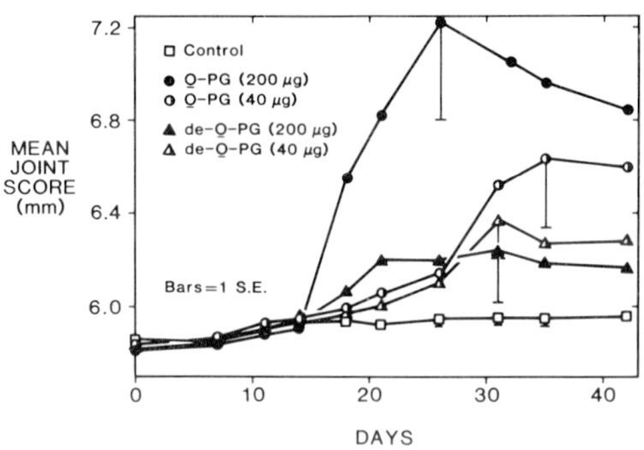

Figure 1. Effect of de-O-acetylation on arthropathic activity of O-PG.

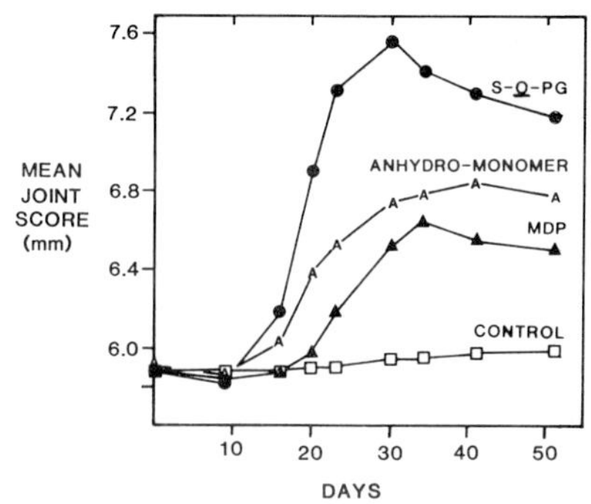

Figure 2. Arthropathic activity of gonococcal PG fragments and of MDP in male Lewis rats.

In other experiments (not shown), PG-induced paw swelling was found to be accompanied by synovitis and arthritis, as detected histologically and radiologically. Further, control experiments indicated that the arthropathic activity of gonococcal PG preparations was not due to trace contamination by gonococcal endotoxin (a known arthritogen in other systems; 5).

Tissue localization of PG in vivo. We hypothesized that the enhanced arthropathic activity of _O_-PG in the rat model (Fig. 1) might be due to the relative persistence of these hydrolase-resistant PG fragments _in vivo_. Accordingly, we examined the distribution of radiolabeled PG fragments in tissues. Preliminary studies compared the initial fate of soluble highly arthropathic S-_O_-PG and of weakly arthropathic non-_O_-acetylated PG (S-non-_O_-PG) following intravenous administration of both PGs simultaneously. To accomplish this, one group of rats received mixed PGs in which the S-_O_-PG was labeled with ^{3}H-glucosamine and the S-non-_O_-PG was labeled with ^{14}C-glucosamine, and the other group of rats received the same two PGs, but with the labels reversed. This dual label method offered the advantage of allowing direct comparison of the metabolic fate of both test PGs simultaneously in individual rats. The pooled results from 23 rats (Table 1) indicated that most of the recoverable radioactivity of both S-_O_-PG and S-non-_O_-PG was excreted within 24 hr after injection of macromolecular PG. However, of the residual PG recovered, there was significant enrichment of _O_-PG (relative to non-_O_-PG) in several organs, most notably the liver, among those tissues collected. Thus, enhanced arthropathic activity of _O_-PG was associated with the relative persistence of PG in certain tissues _in vivo_.

Table 1. Tissue Localization of PG 24 hr after iv Injection of Radiolabeled _O_-PG and Non-_O_-PG (200 µg of each) in Male Lewis Rats (n = 23).

Tissue	% of PG injected (mean) Non-_O_-PG	_O_-PG	p value Non-_O_-PG _vs_ _O_-PG
Cumulative urine	37.40	39.70	NS
Final blood level	0.31	0.53	< .05
Liver	0.96	3.89	<.001
Kidneys	0.26	0.41	<.001
Spleen	0.03	0.09	<.001
Lungs	0.02	0.06	NS
Brain	0.02	0.02	NS
Hind ankles	0.01	0.01	NS

In conclusion, these data indicate that extensive *O*-acetylation potentiates PG-induced arthritis and suggest that the arthropathic potential of gonococcal PG may be governed, in part, by the extent of its intrinsic resistance to degradation and its tendency to persist in tissues. The resistance of macromolecular PG *in vivo* is also thought to be important in the pathogenesis of streptococcal cell wall-induced arthritis (9). However, the macromolecular nature of gonococcal PG is not an absolute requirement for arthritis induction, since even low-molecular-weight PG monomers retain at least some arthropathic activity.

Acknowledgement

This work was supported by PHS grants R01 AI-14826 and P01 AI-20110 from the National Institute of Allergy and Infectious Disease.

References

1. Rosenthal, R. S. 1979. Infect Immun. 24, 869.

2. Sinha, R. K., R. S. Rosenthal. 1980. Infect. Immun. 29, 914.

3. Sinha, R. K., R.S. Rosenthal. 1981. Antimicrob. Agents Chemother. 20, 98.

4. Rosenthal, R.S., W.J. Folkening, D.R. Miller, S.C. Swim. 1983. Infect. Immun. 40, 903.

5. Goldenberg, D. L. 1983. Am. J. Med. 74, 925.

6. Melly, M.A., Z.A. McGee, R.S. Rosenthal. 1984. J. Infect. Dis. 149, 378.

7. Petersen, B. H., R. S. Rosenthal. 1982. Infect. Immun. 35, 442.

8. Swim, S. C., M. A. Gfell, C. E. Wilde, III, R. S. Rosenthal. 1983. Infect. Immun. 42, 446.

9. Eisenberg, R., A. Fox, J. J. Greenblatt, S. K. Anderle, W. J. Cromartie, J. H. Schwab. 1982. Infect. Immun. 38, 127.

TEICHOIC ACID-FREE PEPTIDOGLYCAN OF STAPHYLOCOCCUS AUREUS RM 59 DOES NOT ACTIVATE COMPLEMENT

Michael Loos
Institute of Medical Microbiology, University of Mainz, FRG

Peter H. Seidl, K.H. Schleifer
Institute of Microbiology, Technical University of Munich, FRG

Introduction

Peptidoglycan preparations from various Gram-positive bacteria have been reported to activate the alternate pathway of complement (APC) (1-10). On the other hand, several investigations indicate that cell wall teichoic acids rather than peptidoglycan activate the APC (11-13). Since all procedures hitherto employed for cleaving the covalent linkage between peptidoglycan and teichoic acids and /or polysaccharides (1-13) result in severe chemical modifications of peptidoglycan (14), the activities obtained do not reflect the properties of "natural" peptidoglycan. In the present paper, teichoic acids were selectively removed from peptidoglycan by action of hydrogen fluoride on purified staphylococcal cell walls without causing hydrolysis or chemical modification of the peptidoglycan. The resulting peptidoglycan did not activate the APC and was devoid of any detectable phosphate or ribitol, characteristic constituents of the cell wall teichoic acid of *S. aureus* RM59.

Materials and Methods

Bacterial strains used. *Staphylococcus aureus* RM59 was isolated from raw milk and was devoid of any detectable protein A as demonstrated by failing binding of iodine labelled IgG to the cells. *Staphylococcus aureus* 52A5 ("teichoic acid deficient mutant", poor in protein A) was obtained from D. Mirelman, The Weizman Institute, Rehovot, Israel.
Cell wall preparations. Cell walls were obtained in the usual way (15) by mechanical disruption of cells with glass beads and incubation with trypsin (Prep. 1). This preparation was further treated with pepsin, subtilisin and proteinase K (16) (Prep. 2), followed

Biological Properties of Peptidoglycan

by treatment with DNase and RNase (10) (Prep. 3). The lyophilized Prep. 3 was further incubated either with 40% phenol at 65°C for 30 min. to remove lipoteichoic acid (17) (Prep. 4), or incubated with 0.01M EDTA for 30 min. at room temperature, sedimented at 48.000 g and then extracted with 4% boiling sodium dodecylsulphate for 30 min. (14) (Prep. 5). For selective and complete removal of teichoic acid, cell walls were treated with 70% hydrogen fluoride at 0°C for 4 h (18,19) (Prep. 6). For control, teichoic acids were further removed by the action of hot trichloracetic acid (10% TCA, 60°C, 4 h; (20)) (Prep. 7). All extracted cell wall preparations were sedimented at 48.000 g for 30 min., washed 10 times with twice distilled water, and then thoroughly dialyzed against twice distilled water.
<u>Analytical methods</u>. Gas chromatography, amino acid analyses and quantitative determination of phosphate were carried out as described (21), protein was quantified by a modified Lowry method (22).
<u>Complement assays</u>. The methods for preparing sheep erythrocytes (E), cell intermediates (EA, $EAC\overline{1}\overline{4}\overline{2}$), antisera, and Veronal-buffered saline, (VBS, μ = 0.15, pH 7.4), VBS with sucrose (VBS-S, μ = 0.065, pH 7.4), or VBS with ethylenediaminetetraacetate (VBS-EDTA) have been described by Rapp and Borsos (23). Aliquots of pooled normal and C4 deficient guinea pig serum (NIH strain obtained from Dr. D. Bitter-Suermann, Mainz) were stored at -70°C. CH50-activity was determined using $5x10^8$ EA cells/ml (23). C3-activity was tested using $EAC\overline{1}\overline{4}\overline{2}$ and purified components C5, C6, C7, C8 and C9 purchased from Cordis Laboratories, Inc., Miami, Fl. "C3-9"-activity was measured by diluting sera samples 1:25 in VBS-EDTA and testing the capacity to lyse $EAC\overline{1}\overline{4}\overline{2}$-cells. Hemolysis was measured either at 541 nm ($5x10^8$ cells/ml) or at 412 nm ($1.3x10^8$ cells/ml). Z-values were determined by the equation $Z = -\ln(1-y)$, where y is the extent of hemolysis.

Results and Discussion

The wall preparation isolated form <u>Staphylococcus aureus</u> RM59 in different steps are depicted in Fig. 1. Trypsinized cell walls frequently employed in the investigation of biological activities of peptidoglycan may contain considerable amounts of protein (compare Prep. 1), and additional digestion with proteases was necessary for complete removal of proteinacious material (compare Prep. 2). The analyses further revealed that only the hydrogen fluoride treated preparation Prep. 6 was free of any detectable phosphate and ribitol, characteristic constituents of cell wall teichoic acid of <u>Staphylococcus aureus</u> RM59. In comparison, extraction with trichloroacetic acid at 60°C for 4 hours only reduced phosphate and ribitol to 1/20 of the initial value (Prep. 7).

Cell wall preparations 1 to 5 were found to be potent, temperature-dependent activators of C3 in normal and C4 deficient

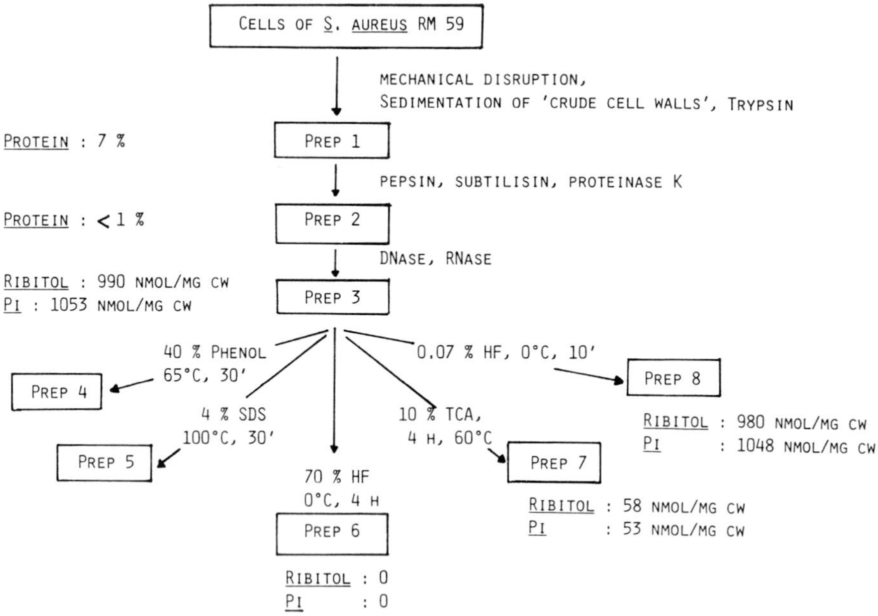

Fig. 1. Wall preparations isolated from Staphylococcus aureus RM59

guinea pig serum via APC (compare Tables 1,2,3). It is evident from the Tables that on the contrary peptidoglycan, completely free of teichoic acid (Prep. 6) did not show any C3 consumption in normal or C4 deficient guinea pig serum.

The activity of cell walls extracted with hot trichloroacetic acid (Prep. 7; still revealing impurifications caused by teichoic acid; see Fig. 1) was clearly diminished in comparison to Prep. 3, but was still moderate as compared with Prep. 6 (Tables 1 and 2). Interference of possible fluoride salt formation of the peptidoglycan molecule in Prep. 6 was excluded comparing the unchanged activity of Prep. 8 (Tables 1, 2 and 3).

The action of 70% hydrogen fluoride at 0°C was used in this paper for selectively splitting the covalent linkages between peptidoglycan and teichoic acid. Due to a pure nucleophilic attack onto phosphodiester linkages (18,19,24), this procedure does not cause chemical modification or hydrolysis of the residual peptidoglycan molecule. In comparison, extraction of polysaccha-

Table 1

CONSUMPTION OF HEMOLYTIC COMPLEMENT ACTIVITY BY DIFFERENT PEPTIDOGLYCAN PREPARATIONS[a]

Conc. of Preparation (number)	2.5 mg/ml		0.5 mg/ml	
	CH50 U/ml	% Inhibition	CH50 U/ml	% Inhibition
1	0	100	0	100
2	0	100	0	100
3	0	100	0	100
4	0	100	0	100
5	0	100	0	100
6	108.4	17.9	123.4	6.5
7	86.0	34.9	98.6	25.4
8	0	100	0	100
9	81.2	38.5	94.6	28.4
control	132	0	132	0

[a] Equal volumes of guinea pig serum and peptidoglycan preparations (suspended in VBS^{++}) were incubated for 30 min at 37 °C. Afterwards the remaining CH-50 activity was determined using EA ($5x10^8$ cells/ml).

Table 2

CONSUMPTION OF C3 IN NORMAL OR C4 DEFICIENT GUINEA PIG SERUM BY DIFFERENT PEPTIDOGLYCAN PREPARATIONS OF STAPHYLOCOCCUS AUREUS

Preparation (Number)	C3 activity in guinea pig serum: normal		C4 deficient	
	2.5 (mg/ml)	0.5 (% inhibition)	2.5 (mg/ml)	0.5
1	94	96	n.t.*	n.t.
2	93	95	n.t.	n.t.
3	92	94	n.t.	n.t.
4	90	93	n.t.	n.t.
5	90	91	n.t.	96
6	0	0	n.t.	0
7	n.t.	n.t.	42	39
8	n.t.	n.t.	97	96
9	n.t.	n.t.	31	4

* n.t. = not tested

Table 3

TEMPERATURE-DEPENDENT CONSUMPTION OF "C3-9" ACTIVITY IN NORMAL GUINEA PIG SERUM BY DIFFERENT PEPTIDOGLYCAN PREPARATIONS OF STAPHYLOCOCCUS AUREUS[a]

Preparation (number) 0.5 mg/ml	Incubation of normal guinea pig serum 30 min 4°C	30 min 37°C
	Z-values (= -ln (1-y)	
1	1.062	0.168
5	0.792	0.205
6	0.645	0.557
8	0.880	0.222
control	0.612	0.607

[a] Equal volumes of gps and peptidoglycans were incubated for 30 min at 4 °C or 37 °C. Then the samples were diluted 1:25 in VBS-EDTA and tested for the capacity to lyse $EAC\overline{1}\overline{4}\overline{2}$-cells.

rides with hot formamide (3-5, 11-13) or treatment with hot trichloroacetic acid for removal of teichoic acids (1,2,4,6-11) results in the introduction of O- and N-formyl groups or in partial acid degradation of the peptidoglycan (14), respectively. Own kinetical studies revealed the increase of reducing sugars and of free amino groups during the action of hot trichloroacetic acid on cell walls thus proving degradation and splitting covalent bindings in the glycan and the peptide moiety of peptidoglycan. Lacking increase of free amino groups or of reducing sugars in the residual peptidoglycan during extraction with hydrogen fluoride again proved, on the contrary, the selective removal of teichoic acid without degrading the peptidoglycan. It is, moreover, important that peptidoglycan, prepared by the action of 70% hydrogen fluoride at 0°C, was immunologically active, i.e. specific antibodies to peptidoglycan were bound by this preparation as demonstrated by appropriate preadsorption studies (25).

The activation of C3 via APC by bacterial cell walls is well established (1-13, 26-30) and was again corroborated in this report. However, we conclude from our results that highly purified peptidoglycan of *Staphylococcus aureus* RM 59 does not activate APC.

The APC activating potency seemed to be strongly associated with the content of phosphate and ribitol, characteristic components of teichoic acid. This presumption was confirmed by the weak reaction of a Prep. 3 from Staphylococcus aureus 52A5 (=Prep. 9; compare Tables 1 and 2), a mutant with a very low content of teichoic acid (31); (phosphate found: 88 nmol/mg cell wall; ribitol found: 72 nmol/mg cell wall).

Conclusion

Differing results have been published concerning the activation of APC by either peptidoglycan or cell wall teichoic acid (1-13, 27, and this paper). However, data obtained for one strain or one bacterial species are not necessarily of general validity. Regarding, furthermore, the well established dose dependency, some reports currently cited as examples for activation of APC by peptidoglycan (1,5) actually failed to observe significant activation of the APC due to the high amounts required. Previous studies indicated that complement activation is dependent upon the particle size of the cell wall material (3,8,27), and in this respect it is again of importance that the action of hydrogen fluoride on cell walls, employed in this report for removal of teichoic acid, does not degrade the peptidoglycan. Certainly, both chemical and physical factors must be considered in studies of complement activation by isolated bacterial cell wall components (27). Further studies with carefully purified and analysed cell wall polymers are necessary to elucidate their contributions to the complement activating potency.

Acknowledgement

This research was supported by the DFG, Lo 188/3-2.

References

1. Bokisch, V.A. 1975. Interaction of peptidoglycans with anti-IgGs and with complement. Z. Immun. Forsch., 149, 320-330.

2. Pryjma, J., K. Pryjma, A. Grov and P.B. Heczko. 1976. Immunological activity of staphylococcal cell wall antigen. In: Zentralbl. Bakteriol. Parasitenkd. Infektionskr. Hyg. Abt. I. Suppl., 5, 873-881.

3. Greenblatt, J., R.J. Boackle and J.H. Schwab. 1978. Activation of the alternate complement pathway by peptidoglycan from streptococcal cell wall. Infect. Immun. 19, 296-303.

4. Semeraro, N., M. Colucci and J. Vermylen. 1979. Complement-dependent and complement-independent interactions of bacterial lipopolysaccharides and mucopeptides with rabbit and human platelets. Thromb. Haemostas. 41, 392-406.

5. Gallis, H.A. 1979. Interactions of streptococci with complement and immunoglobulin, In: Pathogenic Streptococci (M.T. Parker, ed.), pp. 78-80, Reedbooks Ltd., Chertsey, Surrey.

6. Verbrugh, H.A., W.C. Van Dijk, R. Peters, M.E. van Erne, M.R. Daha, P.K. Peterson and J. Verhoef. 1980. Opsonic recognition of staphylococci mediated by cell wall peptidoglycan. Antibody independent activation of human complement and opsonic activity of peptidoglycan antibodies. J. Immunol. 124, 1167-1173.

7. Wilkinson, B., Y. Kim, P.K. Peterson, P.G. Quie and A.F. Michael. 1978. Activation of complement by cell surface components of *Staphylococcus aureus*. Infect. Immun. 20, 388-392.

8. Smith, S.M., R. Burrel and I.S. Snyder. 1978. Complement activation by cell wall fractions of *Micropolyspora faeni*. Infect. Immun. 22,, 569-574.

9. Verbrugh, H.A., W.C. van Dijk, R. Peters, M.E. van der Tol and J. Verhoef. 1979. The role of *Staphylococcus aureus* cell wall peptidoglycan, teichoic acid and protein A in the processes of complement activation and opsonization. Immunology 37, 615-621.

10. Peterson, P.K., B.J. Wilkinson, Y. Kim, D. Schmeling, S.D. Douglas and P.G. Quie. 1978. The key role of peptidoglycan in the opsonization of *Staphylococcus aureus*. J. Clin. Invest. 61, 597-609.

11. Webster, G.F., U.R. Nilsson and W.P. McArthur. 1981. Activation of the alternate pathway of complement in human serum by Proprionibacterium acnes (Corynebacterium parvum) cell fractions. Inflammation 5, 165-176.

12. Winkelstein, J.A. and A. Tomasz. 1978. Activation of the alternative complement pathway by pneumococccal cell wall teichoic acid. J. Immunol. 120, 174-178.

13. Prellner, K. 1981. C1q binding and complement activation by capsular and cell wall components of S. pneumoniae type XIX. Acta path. microbiol. scand. Sect. C 89, 359-364.

14. Schleifer, K.H. 1975. Chemical structure of the peptidoglycan, its modifiability and relation to the biclogical activity. Z. Immun.- Forsch. 149, 104-117.

15. Schleifer, K.H. and O. Kandler. 1967. Zur chemischen Zusammensetzung der Zellwand der Streptokokken. I. Die Aminosäuresequenz des Mureins von Str. thermophilus and Str. faecalis. Arch. Mikrobiol. 57, 335-365.

16. Seidl, P.H., P. Zwerenz, J.R. Golecki, N. Franken and K.H. Schleifer. 1986. Isolation of specific antibodies to the glycan moiety of peptidoglycan and their application in the indirect immunoferritin technique. FEMS Microbiol. Lett., in press.

17. Wicken, A.J., J.W. Gibbens and K.W. Knox. 1973. Comparative studies on the isolation of membrane lipoteichoic acid from Lactobacillus fermenti. J. Bacteriol. 113, 365-372.

18. Fiedler, F., M.J. Schäffler and E. Stackebrandt. 1981. Biochemical and nucleic acid hybridization studies on Brevibacterium linens and related strains. Arch. Microbiol. 129, 85-93.

19. Lipkin, D., B.E. Phillips and J.M. Abrell. 1969. The action of hydrogen fluoride on nucleotides and other esters of phosphorus (V) acids. J. Org. Chem. 34, 1539-1547.

20. Schleifer, K.H. and R.M. Krause. 1971. The immunochemistry of peptidoglycan. I. The immunodominant site of the peptide subunit and the contribution of each of the amino acids in the binding properties of the peptides. J. Biol. Chem. 243, 986-993.

21. Endl, J., Seidl, P.H., Fiedler F. and K.H. Schleifer. 1983. Chemical composition and structure of cell wall teichoic acids of staphylococci. Arch. Microbiol. 135, 215-223.

22. Hartree, E.F. 1972. Determination of protein: a modification of the Lowry method that gives a linear photometric response. Analyt. Biochem. 48, 422-427.

23. Rapp, J.J. and T. Borsos. 1970. Molecular basis of complement action. Appleton-Century-Crofts, New York, N.Y.

24. Mort, A.J. and T.A. Lamport. 1977. Anhydrous hydrogen fluoride deglycosylates glycoproteins. Anal. Biochem. 82, 289

25. Franken, N., P.H. Seidl, T. Kuchenbauer, H.J. Kolb, K.H. Schleifer, L. Weiss and K.-D. Tympner. 1984. Specific immunoglobulin A antibodies to a peptide subunit sequence of bacterial cell wall peptidoglycan. Infect. Immun. 44, 182-187.

26. Tauber, J.W., M.J. Polley and J.B. Zabriskie. 1976. Nonspecific complement activation by streptococcal structures. II. Properdin-independent initiation of the alternate pathway. J. Exp. Med. 143, 1352-1366.

17. Wilkinson, B.J., Y. Kim and P.K. Peterson. 1981. Factors affecting complement activation by Staphylococcus aureus cell walls, their components, and mutants altered in teichoic acid. Infect. Immun. 32, 216-224.

28. Baker, L.A., P.A. Campbell and J.R. Hollister. 1977. Chemotaxigenesis and complement fixation by Listeria monocytogenes cell wall fractions. J. Immunol. 119, 1723-1726.

29. Winkelstein, J.A., A.S. Abramovitz and A. Tomasz. 1980. Activation of C3 via the alternative complement pathway results in fixation of C3b to the pneumococcal cell wall. J. Immunol. 124, 2502-2506.

30. Winkelstein, J.A. and A. Tomasz. 1977. Activation of the alternative pathway by pneumococcal cell walls. J. Immunol. 118, 451-454.

31. Shaw, D.R.D., D. Mirelman, A.N. Chatterjee and J.T. Park. 1970. Ribitol teichoic acid synthesis in bacteriophage-resistant mutants of Staphylococcus aureus H. J. Biol. Chem. 245, 5101-5106.

SOMNOGENIC MURAMYL PEPTIDES

J.M. Krueger

Department of Physiology and Biophysics, University of Tennessee Center for the Health Sciences, Memphis, TN 38163

Introduction

Most people are aware of three general aspects of sleep. They are: 1) one sleeps almost every night; 2) the longer one stays awake the more tired one becomes; and 3) during the course of infectious disease, one often experiences subjective feelings of sleepiness. The first two of these observations have been studied extensively, but the third has not been studied at all. The second observation led to the hypothesis that during wakefulness a sleep-promoting substance(s) accumulates in the brain. The purpose of this paper is to review our work concerning testing this hypothesis and show how our results provide a plausable explanation for the sleepiness associated with infectious disease.

Nearly two decades ago, John Pappenheimer and colleagues (1) described the accumulation of a somnogenic substance (factor S) in cerebrospinal fluid during prolonged wakefulness. More recently, we characterized substances that have physical, chemical, and biological properties similar to those of factor S and found that these naturally occurring somnogens were muramyl peptides (MPs) (2). The urinary material (FSu) was completely characterized and found to be N-acetylglucosaminyl-1, 6-anhydro-N-acetylmuramyl-alanyl-glutamyl-diaminopimelyl-alanine (NAG-1,6 anhydro NAM-ala-glu-dap-ala) (4).

Effects of MPs on Sleep

Sleep is usually divided into two categories, slow-wave sleep (SWS) and rapid-eye movement (REM) sleep. To distinguish these

Biological Properties of Peptidoglycan

states from each other and from wakefulness (W) we record electroencephalograms (EEG), brain temperature (Tbr), and movement of experimental animals. These measurements are used to define states of vigilance (reviewed in ref. 5).

The major effect of MPs on sleep was to enhance duration of SWS (6). For example, administration of 1 pmol of FSu into a cerebral ventricle in rabbits induced excess SWS for several hours. Excess SWS was first observed during the second hour postinfusion, then persisted for 6 or more hours depending upon the dose. Average hourly % SWS typically increased from control values of 40% to 60-80% during this period. Similar results were obtained from cats (7), rats (8) and monkeys (9). During periods of MP-enhanced SWS, average EEG slow-wave (1/2-4 Hz) voltages were also increased (12). These increases were usually 40-50% above control values and resulted primarily from increased amplitudes of individual slow-waves. The significance of these "supranormal" slow-waves is unknown. They were also observed, however, during the deep sleep that follows sleep-deprivation (12) and thus they may be indicative of the intensity of sleep.

MPs also had an effect on REM sleep, although results were dose- and species-dependent. In cats FSu induced an inhibition of REM sleep for 2-4 hours (7). In rats, a similar inhibition of REM sleep was reported after muramyl dipeptide (NAM-L-ala-D-isogln) (MDP) was administered during the day (when rats normally sleep 70-80% of the time) (10). In another study, however, nighttime administration of MDP did not affect REM sleep (11). Further, another laboratory reported that low doses of MDP enhanced REM sleep while higher doses inhibited it (8).

Other aspects of sleep following MP-treatment remained normal. The organization of states of vigilance remained intact, e.g., SWS and REM remained episodic and sleep-cycle length (time between onset of one REM episode to the next) remained close to control values (7). Behavior also remained normal, i.e., typical sleep postures were observed and waking behaviors such as eating and grooming persisted in MP-treated animals. However, high doses of MDP induced abnormal autonomic responses such as excess nasal secretions (6).

MPs have a variety of other biological activities in mammals, e.g., inhibition of angiotensin converting enzyme (13). The most extensively studied are the immunomodulatory and pyrogenic activities of MPs (14) which were separated, in part, from somnogenic activity. For example, NAM-L-ala-D-gln is an immunostimulant but was not somnogenic (15). Co-administration of an antipyretic with MDP attenuated expected febrile responses but failed to affect sleep responses (6). However, another facet of temperature regulation remained normal in MP-treated animals. At the onset of REM sleep Tbr increases relatively rapidly, whereas during the transition between W and SWS there is a regulated decrease in Tbr (16). The magnitude of these Tbr changes is about 0.2-0.4° C, which is substantially less than those associated with febrile responses. The normal Tbr changes associated with state transitions persisted in MP-treated animals even though their base line temperature was increased (unpublished).

MP Structure-Sleep Activity Relationships

Variations in somnogenic activity with MP structure that have been determined to date are summarized as follows.

1) Steric configuration is important for somnogenic activity. The stereoisomers of MDP, MDP-DD, and MDP-LL were inactive as somnogens (6).

2) Modifications of the muramyl moiety of MDP can affect sleep-promoting activity. If NAM was destroyed by periodate oxidation or if the N-acetyl moiety of NAM of MDP was absent, somnogenic activity was lost. However, the disaccharide containing NAG in glycosidic linkage to NAM of MDP was somnogenic (15).

3) The 1,6 anhydro non-reducing NAM moiety of FSu is important for somnogenic potency. One pmol of FSu was somnogenic whereas

10 pmol of a hydrated form of FSu (i.e., containing a free reducing end) was not (unpublished).

4) Conversion of certain carboxyl groups to unsubstituted amides resulted in loss of somnogenic activity. For example, if the γ carboxyl of the isogln moiety of MDP was converted to an unsubstituted amide [i.e., NAM-L-ala-D-glu(NH_2)NH_2], somnogenic activity was lost. This result and others (4,6,15) suggest that a free carboxyl in the vicinity of the isogln moiety is important for somnogenic activity.

Hypotheses

1) There are no known synthetic pathways in mammals for some of the constituents of FSu, e.g., NAM and dap. The role MPs play in normal mammalian physiology remains unknown. However, MPs have multiple actions in mammals and MPs are normal constituents of mammalian tissue including brain (17). For these reasons, it was postulated that MPs may be analogous to vitamins in that they are required by but cannot be synthesized by the host (18,19).

2) MPs can induce the synthesis and release of interleukin-1 (IL1) (20), which is normally present in numerous tissues including the central nervous system (21). Further, IL1 may contain or tightly bind a MP-like moiety (22,23). For these reasons we assayed IL1 and found that it induced sleep effects similar to those described above for MPs (5,24). However, IL1-induced excess SWS more rapidly than MPs, suggesting that MPs may induce sleep through a step involving IL1.

3) As noted above, during the course of infectious disease one often feels tired. This may be related to altered MP and/or IL1 levels. In further pursuit of sleep mechanisms integrated into the immune response, we recently demonstrated that human recombinant interferon-alpha enhances SWS in rabbits (25). Although the function(s) of sleep remain unknown, that these immuno-active endogenous substances can promote sleep suggests

that sleep plays a role in recuperative processes, whether it is recovery from a disease state or a day's activity.

Acknowledgment

This research was supported by the Office of Naval Research (N00014-85-K-0773) and by NIH (GM-33268).

References

1. Pappenheimer, J.R., T.B. Miller, C.A. Goodrich: 1967. Proc. Natl. Acad. Sci. U.S.A. 58, 513.

2. Krueger, J.M., J.R. Pappenheimer, M.L. Karnovsky: 1982. J. Biol. Chem. 257, 1664.

3. Martin, S.A., M.L. Karnovsky, J.M. Krueger, J.R. Pappenheimer, K. Biemann: 1984. J. Biol. Chem. 259, 12652.

4. Krueger, J.M., M.L. Karnovsky, S.A. Martin, J.R. Pappenheimer, J. Walter, K. Biemann: 1984. J. Biol. Chem. 259, 12659.

5. Walter, J., D. Davenne, S. Shoham, C.A. Dinarello, J.M. Krueger. Am. J. Physiol. (in press).

6. Krueger, J.M., J.R. Pappenheimer, M.L. Karnovsky: 1982. Proc. Natl. Acad. Sci. U.S.A. 79, 6102.

7. Garcia-Arraras, J.: 1981. Am. J. Physiol. 241, E269.

8. Masek, K., O. Kadlecova, O. Kadlec, Z. Zidek, H. Farghalli, Z. Machkova. In: Muramyl Dipeptides. Bioscience Ediprint Inc. (in press).

9. Wexler, D.B., M.C. Moore-Ede: 1984. Am. J. Physiol. 247, R672.

10. Fornal, C., R. Markus, M. Radulovacki: 1984. Peptides 5, 91.

11. Inoue, S., K. Honda, Y. Komoda, K. Uchizono, R. Ueno, O. Hayaishi: 1984. Proc. Natl. Acad. Sci. U.S.A. 81, 6240.

12. Pappenheimer, J.R., G. Koski, V. Fencl, M.L. Karnovsky, J.M. Krueger: 1975. J. Neurophysiol. 38, 1299.

13. Bush, K., P.R. Henry, D.S. Slusarchyk: 1984. J. Antibiotics 37, 330.

14. Lederer, E: 1980. J. Med. Chem. 23, 819.

15. Krueger, J.M., J. Walter, M.L. Karnovsky, L. Chedid, J.P. Choay, P. Lefrancier, E. Lederer: 1984. J. Exp. Med. 159, 68.

16. Heller, H.C., S.F. Glotzbach: 1977. In: International Review of Physiology (D. Robertshaw, ed.) University Park, p. 147.

17. Zhai, C., M.L. Karnovsky: 1984. Infect. Immun. 43, 937.

18. Adam, A., E. Lederer: 1984. Med. Res. Rev. 4, 111.

19. Jolles, P.:1976. Biomedicine 25, 275.

20. Dinarello, C.A.: 1984. Rev. Infect. Dis. 6, 51.

21. Nieto-Sampedro, M., K.G. Chandy, M. Berman, C.I. Sandborg, C.W. Cotman, S. Gupta, G.J. Friou: 1985. Soc. Neurosci. Absts. 11, 397.

22. Chedid, L., G.M. Bahr, F.Z. Modabber, C.A. Dinarello, S.M. Wolff: 1983. C.R. Acad. Sci. (Paris) Ser. C. 297, 209.

23. Chedid, L., G.M. Bahr, G. Riveau, J.M. Krueger: 1984. Proc. Natl. Acad. Sci. U.S.A. 81, 5888.

24. Krueger, J.M., J. Walter, C.A. Dinarello, S.M. Wolff, L. Chedid: 1984. Am. J. Physiol. 246, R994.

25. Krueger, J.M., J. Walter, D. Davenne, S. Shoham, C.A. Dinarello: 1985. Soc. Neurosci. Absts. 11, 1283.

SYNTHETIC LIPOPEPTIDE ANALOGUES OF PEPTIDOGLYCAN-ASSOCIATED LIPOPOTEIN ARE POTENT NOVEL B-LYMPHOCYTE MITOGENS

Wolfgang G. Bessler[+*], Bernhard Kleine[+], Marianne Cox[+], Angelika Lex[+], Barbara Suhr[+],Karl-Heinz Wiesmüller[++], Günther Jung[++] and Carlos Martinez-Alonso[+++]

Arbeitsbereich Mikrobiologie und Immunologie[+] and Institut für Organische Chemie[++] der Universität, D 7400 Tübingen, Fed. Rep. of Germany

Institut für Immunbiologie der Universität, D 7800 Freiburg[*]

Dept. of Immunology, Clinica Puerty de Hierro, Madrid, Spain[+++]

Bacterial surface components such as lipoprotein[1,2], lipopolysaccharide (LPS)[3], protein I and protein II[*4] are potent B-lymphocyte activators. The molecular structure of lipoprotein responsible for mitogenicity resides in its N-terminal three fatty acids containing lipopeptide region[5]. In order to precisely determine the minimal structural requirements for biological activity, we prepared oligopeptide analogues to the N-terminal lipid part of lipoprotein by chemical synthesis. Here we demonstrate that the synthetic lipopeptide segment S-(2,3-bis(palmitoyloxy)-(2-RS)-propyl)-N-palmitoyl-(R)-cysteinyl-serine ("Tripalmitoyl-Cys-Ser"; Fig. 1) is fully active in mouse B-lymphocytes to stimulate proliferation and immunoglobulin synthesis. The fragment of this molecule lacking the terminal serine is by far less active.

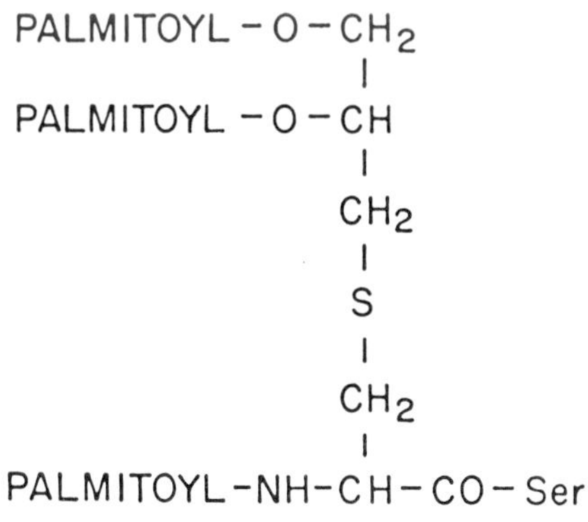

Figure 1: Chemical structure of Tripalmitoyl-Cys-Ser as prepared by chemical synthesis[6].

Biological Properties of Peptidoglycan

Segments carrying additionally parts of the N-terminal amino acid sequence of native lipoprotein, i.e. -Ser, -Ser-Asn, -Ser-Asn-Ala, exhibit stimulatory activities comparable to the dipeptide compound. Thus we could demonstrate that a well defined series of synthetic analogues to a bacterial surface protein exhibits pronounced biological activity.

In previous work we have shown that pronase digestion fragments of bacterial lipoprotein, which carry 3 fatty acids bound to glyceryl-cysteine attached to an oligopeptide chain of 3-5 amino acids, were potent B-lymphocyte mitogens[5]. A corresponding synthetic lipopentapeptide prepared by us synthetically[6] was identical to one of these pronase fragments with respect to its amino acid composition, but differed in its fatty acid composition since it contained solely palmitic acid residues. This product, S-(2,3-bis(palmitoyloxy)-(2-RS)-propyl)-N-palmitoyl-(R)-cysteinyl-seryl-seryl-asparaginyl-alanine (Tripalmitoyl-Cys-Ser-Ser-Asn-Ala) was an active mitogen and polyclonal B-lymphocyte activator in vitro and in vivo for splenocytes from different mouse strains [7]. The product has also been shown to induce the production of interleukin 1 in human mononuclear leukocytes and the release of prostaglandin (PGE_2) from mouse peritoneal macrophages (W. Bessler et al., in preparation).
Our attempts to precisely localize the molecular region responsible for the biological activity of the mitogen are shown in fig. 2. The incorporation of ^{3}H-thymidine into DNA in spleen cells of the LPS nonresponder mouse strain C3H/HeJ[8], induced by a well defined series of our newly synthesized (10 and K.H.Wiesmüller, W.G. Bessler, G. Jung, manuscript in preparation) lipopeptides, is demonstrated. Tripalmitoyl-Cys-Ser was able to activate in vitro splenocyte cultures (cell density 6,6 x 10^5/ml) starting at concentrations around 0,1 µg/ml. The dose optimum for stimulation was around 30 µg/ml. The fragments Tripalmitoyl-Cys-Ser-Ser, -Cys-Ser-Ser-Asn, and -Cys-Ser-Ser-Asn-Ala exhibited a similar mitogenicity. In contrast, a product lacking the serine residue, Tripalmitoyl-Cys, was less active. Control cultures with LPS were not activated (Fig. 2a). Thus, the lipopeptide carrying the two amino acids cysteine and serine constitutes the minimal structure of lipoprotein required for full mitogenic activity. The findings that the synthetic compounds induced activation of C3H/HeJ mouse spleen cells demonstrate that their mitogenicity is not attributable to any contaminating endotoxin.
When spleen cells of athymic nude mice (C3H/Tif/Bom/nu/nu) as a source of B-

lymphocytes were tested for the incorporation of ^{3}H-thymidine in response to the synthetic mitogens, a similar pattern as seen above evolved: Tripalmitoyl-Cys-Ser was a potent mitogen active starting at concentrations around 10^{-3} µg/ml, and maximal stimulatory activity was observed around 30-50 µg/ml (Fig. 2b). The compounds carrying the additional amino acids -Ser, -Ser-Asn, or -Ser-Asn-Ala showed comparable activities, whereas Tripalmitoyl-Cys was less active. The T-lymphocyte mitogen Concanavalin A, which was included to the test as a control, was not stimulatory. In order to determine whether the compounds were able to stimulate B-lymphocytes into immunoglobin secretion, a hemolytic plaque assay was performed using Balb/c mouse spleen cell cultures (Fig. 3). Tripalmitoyl-Cys-Ser, -Cys-Ser-Ser, -Cys-Ser-Ser-Asn, or -Cys-Ser-Ser-Asn-Ala, but not Tripalmitoyl-Cys, induced a marked increase in the number of plaque forming cells against densily trinitrophenylated[9] sheep red blood cells (TNP-SRBC) over a broad concentration range (1-50 µg/ml). A pronounced increase in plaque numbers was also observed in LPS treated control cultures (not shown). Thus, the lipopeptides cause the differentiation of B-lymphocytes into immunoglobulin producing plasma cells, as has been shown before for native lipoprotein[1,2].

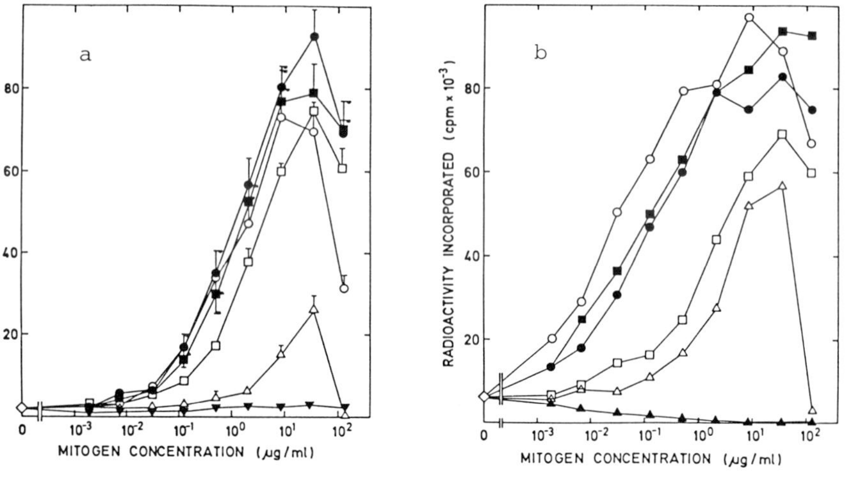

Figure 2: Dose-response curves for the incorporation of ^{3}H-thymidine into DNA in spleen cells from C3H/HeJ (a) or C3H/Tif/Bom/nu/nu mice (b) induced by Tripalmitoyl-Cys-Ser-Ser-Asn-Ala (○), Tripalmitoyl-Cys-Ser-Ser-Asn (●), Tripalmitoyl-Cys-Ser-Ser (□), Tripalmitoyl-Cys-Ser (■), Tripalmitoyl-Cys (△), LPS (▼), and Concanavalin A (▲). Cell cultures were carried out as described[7] for 72 h (a) or 48 h (b) at a cell density of 6,6 x 10^5/ml (a) or 3,3 x 10^6 (b) in 200 µl aliquots. Means of triplicate experiments.

The detailed analysis of the isotype distribution in in vitro stimulated cultures revealed some interesting differences between cultures stimulated with Tripalmitoyl-Cys-Ser-Ser-Asn-Ala and those with lipopolysaccharide. As can be seen in Fig. 4, the number of IgM plaques was two to threefold reduced in lipopeptide activated cultures compared to those stimulated with LPS. There was also an reduced IgG1 production after stimulation with the synthetic product: we observed about five times less IgG1 PFC than in LPS stimulated cultures. The amount of IgA PFC which is also increased after LPS stimulation was not significantly enhanced. The IgG3 and IgG2 PFC numbers were similar in LPS and Tripalmitoyl-pentapeptide activated cultures. These results provide further evidence that both mitogens, native lipoprotein as well as its synthetic analogues, and lipopolysaccharide act on different subpopulations of B lymphocytes: LPS activates lymphocytes in an earlier stage of differentiation and maturation than lipoprotein and its analogues.

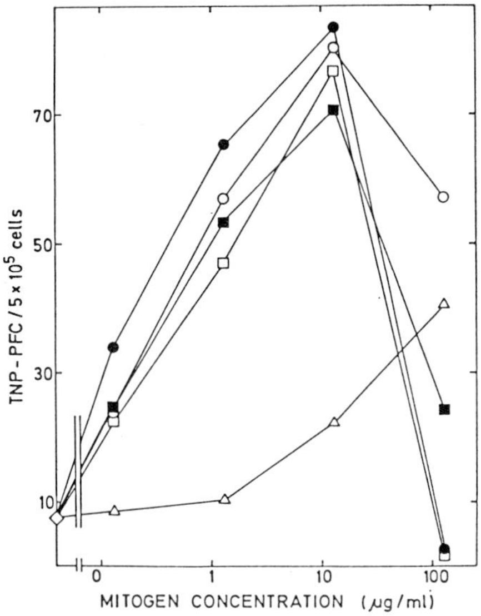

Figure 3: Stimulation of BALB/c splenocytes to the development of plaque forming cells (PFC). Lymphocytes were cultured in the presence of Tripalmitoyl-Cys-Ser-Ser-Asn-Ala, -Cys-Ser-Ser-Asn, -Cys-Ser-Ser, -Cys-Ser and Tripalmitoyl-Cys at a cell density of 5×10^6/ml (1 ml aliquots) in RPMI 1640 supplemented as described [6] Means of triplicate experiments. For symbols compare fig.2.

Summarizing our results, Tripalmitoyl-Cys-Ser, a dipeptide lipid of MW 872, was shown to be a mitogen and polyclonal B-cell activator comparable to native lipoprotein (MW 7,200). Throughout all experiments the segments Tripalmitoyl-Cys-Ser, -Cys-Ser-Ser, -Cys-Ser-Ser-Asn, and -Cys-Ser-Ser-Asn-Ala were by far more active than Tripalmitoyl-Cys, indicating that in order to obtain full biological activity, the presence of a lipophilic part followed by at least the hydrophilic dipeptide Cys-Ser is necessary. Interestingly, these active lipopetides are all diacylglycerol analogues and thus possibly are able to stimulate leukocytes via the activation of protein kinase C. In addition, the compunds exhibit a sequence homology with interleukin-3 (residues 43-45 **Cys-Ser-Ser**)[11], which might indicate a common structure responsible for the induction of cell activation. The three fatty acids in the hydrophobic part of the lipopeptide probably serve to anchor the mitogen in the lymphocyte plasma membrane presenting the lipopeptides to membrane associated proteins. Such binding proteins for lipoprotein, which might be involved in cell activation, were shown by us to include MHC gene products[12]. In very recent experiments we could also show, that the synthetic lipopeptides are able to act directly on single isolated splenic murine B-lymphocytes to induce clonal proliferation (Wetzel and Bessler, unpub-

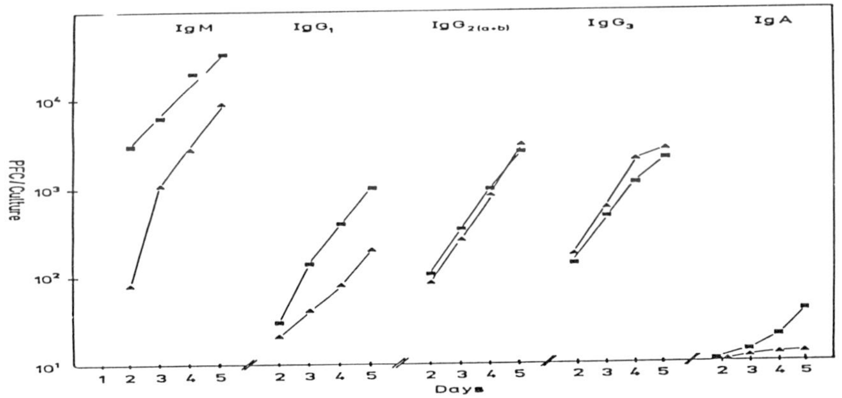

Figure 4: Production of the Ig-isotypes in spleen cells of C57/B16 mice after *in vitro* stimulation with Tripalmitoyl-Cys-Ser-Ser-Asn or LPS. 10^5 cells per 0.2 ml were cultured in the presence of 10 µg/ml Tripalmitoyl-Cys-Ser-Ser-Asn (▲) or 25 µg/ml LPS (■). An isotype specific reverse plaque assay was used as described by E. Gronowizc et al.[13].

lished). Thus, the novel synthetic mitogens will be a valuable tool for the investigation of the molecular mechanism of B-lymphocyte activation and for the study of the effect of bacterial surface components in the host-parasite relationship, particularly prior to the establishment of specific humoral and cellular immune responses, and may be useful in vivo as adjuvants of the immune response.

Acknowledgements: This work was supported by the VW Foundation. We thank L.J. Neumann for excellent editorial help.

References

1. Melchers, F., Braun, V. & Galanos, C. J. Exp. Med. **142**, 473-482 (1975)
2. Bessler, W.G. & Ottenbreit, B.P. Biochem. Biophys. Res. Comm. **76**, 239-246 (1977)
3. Andersson, J.,Sjöberg, O. & Möller, G. Eur. J. Immunol. **2**, 349-353 (1972)
4. Bessler, W.G. & Henning, U. Immunobiol. **155**, 387-398 (1979)
5. Bessler, W., Resch, K., Hancock, E. & Hantke, K. Z.Immun.-Forsch. **153**, 11-22 (1977)
6. Wiesmüller, K.H., Bessler, W., & Jung, G. Hoppe Seyler's Z.Physiol. Chem. **364**, 593-606 (1983)
7. Johnson, R.B., Köhl, S., Wiesmüller, K.H., Jung, G., & Bessler, W.G. Immunobiol. **165**, 27-35 (1984)
8. Sultzer, B.M. & Nilsson, G.S. Nature New Biol. **240**, 198-200 (1972)
9. Rittenberg, M.B. & Pratt, K.L. Proc. Soc. Exp. Biol. Med. **132**, 575-581 (1969)
10. Bessler, W.G., Cox, M., Lex, A., Suhr, B., Wiesmüller, K.-H. & Jung, G. J. Immunol. **135**, 1353-1358 (1985)
11. Fung, M.C., Hapel, A.J., Ymer, S., Cohen, D.R., Johnson, R.M., Campell, H.D., & Young, I.G. Nature **307**, 233-237 (1984)
12. Scheuer, W.V., Müller, H., & Bessler, W.G. Immunobiol. **165**, 349-350 (1983)
13. Gronowicz, E., Coutinho, A., & Melchers, F. Eur. J. Immunol. **6**, 588 (1976)

LIPID A, THE ENDOTOXIC PRINCIPLE OF BACTERIAL LIPOPOLYSACCHARIDES: CHEMICAL STRUCTURE AND BIOLOGICAL ACTIVITY

E. Th. Rietschel, H. Brade, L. Brade, K. Kawahara, Th. Lüderitz, U. Schade, A. Tacken, and U. Zähringer

Forschungsinstitut Borstel, Institut für Experimentelle Biologie und Medizin, 2061 Borstel, FRG

A characteristic feature of Gram-negative bacteria is that in their cell envelope they express a macromolecular glycolipid, termed lipopolysaccharide (LPS). Lipopolysaccharides after being synthesized in the cytoplasmic membrane are transported to the cell surface. Here, together with phospholipids and proteins, they form the outer membrane of Gram-negative bacteria.

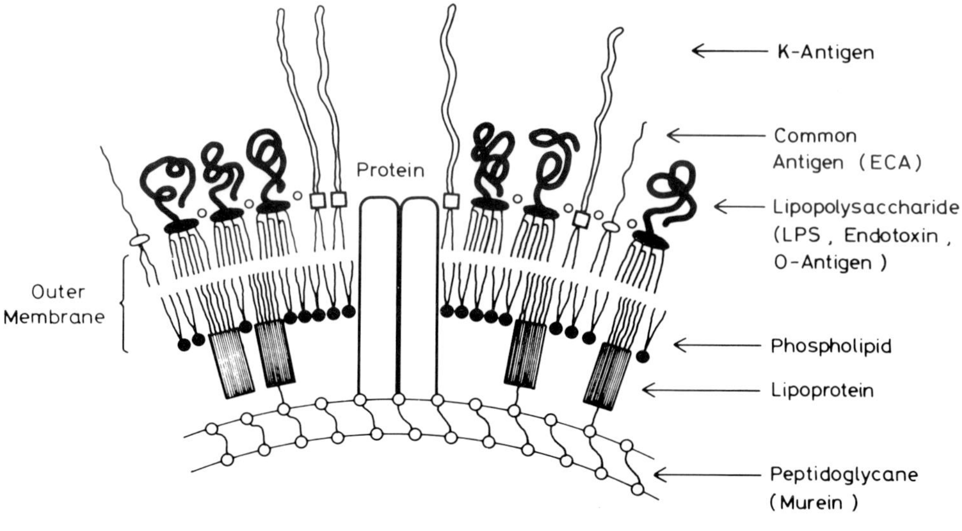

Fig. 1 Schematic Representation of the Gram-negative Bacterial Cell Wall.

Biological Properties of Peptidoglycan

Fig. 1 shows the cell wall of a Gram-negative microbe with the murein layer and the outer membrane. This outer membrane is made up by various amphiphiles, such as the K-antigens, the enterobacterial common antigen, the lipoprotein and, most notably, by lipopolysaccharide. The outer membrane has an asymmetric architecture, i.e. lipopolysaccharides are located exclusively in its outer leaflet through which the bacterial cell senses, interacts and communicates with the environment.

As exposed cell surface components lipopolysaccharides play indeed an important role in the interaction of the bacterial cell with the host, e.g. during infection. Thus, lipopolysaccharides may prevent the engulfment of bacteria by phagocytic cells of the host. On the other hand, contact of Gram-negative bacteria with the immune system of higher organismus leads to the production of antibacterial antibodies which are predominantly directed against determinants embedded in the lipopolysaccharide molecule. Accordingly, lipopolysaccharides belong to the principal cell surface antigens of Gram-negative bacteria, a property to which expression is given by the term O-antigen which is often synonymously used instead of the name lipopolysaccharide.

Injection of Gram-negative bacteria into experimental animals causes a range of nonspecific pathophysiological reactions. These acute effects include fever, hypotension, changes in white blood cell counts, disseminated intravascular coagulation and, if administered in higher doses, irreversible shock. These pathophysiological reactions are equally elicited by isolated and purified lipopolysaccharide preparations derived from different Gram-negative bacteria. Lipopolysaccharides, therefore, represent potent bacterial toxins and this quality is indicated by the term endotoxin which is also used synonymously for the name lipopolysaccharide.

In view of their biological significance it was of great theoretical and practical interest to elucidate the chemical structure of lipopolysaccharides and to characterize those substructures which determine their different biological properties. The investigations of a number of laboratories demonstrated that lipopolysaccharides from different groups of Gram-negative bacteria conform to a common structural principle: they consist of a polysaccharide portion and a covalently bound lipid component, termed lipid A (1).

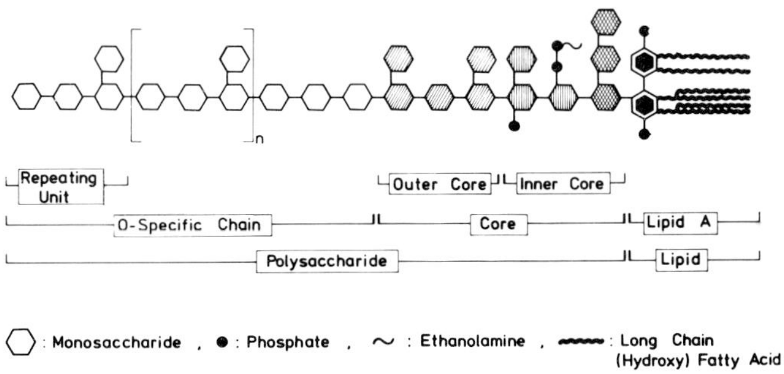

Fig. 2 Schematic Representation of the Structure of a Salmonella Lipopolysaccharide

As an example Fig. 2 shows schematically the architecture of a Salmonella lipopolysaccharide. The polysaccharide component consists of two subregions which are under separate genetical control: the O-specific chain and the core.

The O-specific chain represents a polymer of oligosaccharide molecules, the so called repeating units. The nature and type of linkage of the individual monosaccharide residues within a repeating unit is characteristic and unique for a given lipopolysaccharide and the parental bacterial strain, i.e. the serotype. The core consists of a branched oligosaccharide which carries charged substituents such as carboxyl, phosphoryl and phosphorylethanolaminyl groups. All wild type Salmonella species express a lipopolysaccharide with an identically structured core. However, in other bacterial families such as Escherichia, Proteus and Pseudomonas, structurally distinct core types have been identified which differ structurally in the outer core region (1). The core oligosaccharide is covalently linked to an unusual lipid which has been termed lipid A. Lipid A consists of a bisphosphorylated glucosamine disaccharide backbone to which long chain fatty acids are bound (see below).

For attempts to define those regions responsible for the mediation of the manyfold biological activities of lipopolysaccharides nature was cooperative in that it made, in enterobacterial and many other lipopolysaccharides, the linkage between the polysaccharide portion and lipid A highly susceptible to acidic hydrolysis. Thus, by treatment of lipopolysaccharide with mild acid the polysaccharide portion and the lipid A component can be separated. Serological and biological investigations into the isolated components revealed that the polysaccharide portion, notably the O-specific chain, is the carrier of the O-specificity of lipopolysaccharides and that it determines their quality as O-antigens. On the other hand, the lipid A component was found to represent the endotoxic principle of lipopolysaccharides, i.e. lipid A was thought to determine the quality of lipopolysaccharides to act as endotoxins. During the last 15 years an overwhelming body of data was accumulated confirming this idea (2).

Subsequently, we and also other laboratories have studied the chemical structure of lipid A with the aim to define those structural features which make lipid A such a potent pyrogen and bacterial toxin in general. In the structural elucidation of lipid A lipopolysaccharides from deep-rough mutants were of great help. Very recently the complete structure of the lipopolysaccharide of an Re mutant of Escherichia coli has been determined (3) and its structure is depicted in Fig. 3.

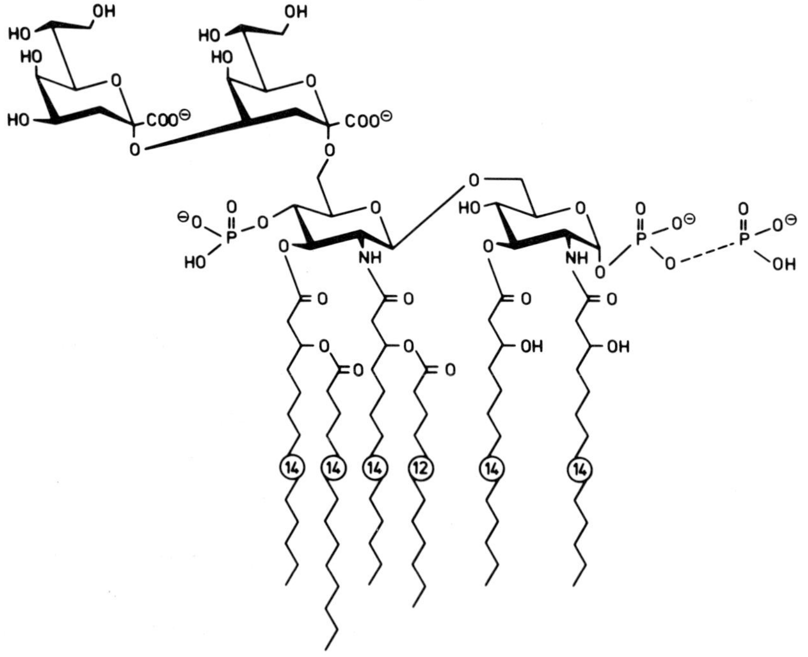

Fig. 3 Structure of an Escherichia coli Re Lipopolysaccharide (3-5). Shown is the structure of lipid A with an attached disaccharide of 2-keto-3-deoxy-octonic acid.

Thus, Fig. 3 shows the chemical structure of E. coli lipid A with an attached α2.4-linked disaccharide of 2-keto-3-deoxy-octonic acid (KDO, 4). E. coli lipid A consists of a ß (1→6) linked-D-glucosamine disaccharide GlcN II-ß1.6-GlcN I), which carries phosphate residues in positions 1 (GlcN I) and 4' (GlcN II). This hydrophilic backbone carries a number of substituents. The phosphate group at position 1 is substituted in a nonstoichiometric manner by a phosphoryl group. The hydroxyl groups in positions 3 and 3' and the two amino functions of the central disaccharide are acylated by (R)-3-hydroxymyristic acid residues. The 3-hydroxyacyl groups of GlcN II are specifically substituted by lauric and myristic acid. The primary hydroxyl group of GlcN II (position 6') represents the attachment site of the KDO-disaccharide and, thus of the polysaccharide portion in intact lipopolysaccharide. In free lipid A therefore, the hydroxyl group at position 6' (as well as that at position 4) are free (for literature compare ref. 5).

The results of a great number of chemical analyses performed in various laboratories (summarized in (5)) show that biologically active lipid A's of different bacterial origin are structurally closely related and that they share common structural elements. Accordingly, endotoxic lipid A is characterized by a central 1,4'-bisphosphorylated, ß(1→6)-interlinked D-glucosamine disaccharide which carries two moles of ester- and two moles of amide-linked (R)-3-hydroxy fatty acids (in positions 2, 3, 2' and 3') of which, in general, two residues are 3-O-acylated. A limited variability of lipid A structures among Gram-negative bacteria is due to the presence of different polar headgroups bound to the phosphate residues of the lipid A backbone. These groups include phosphate, phosphorylethanolamine, D-glucosamine (at GlcN I) and 4-amino-4-deoxy-L-arabinose (at GlcN II). In some cases, like in E. coli and Salmonella, these headgroups are not present in stoichiometric amounts implying a certain microheterogeneity of lipid A preparations. Heterogeneity is also due to the variable presence of acyl groups (6) as well as to variations in the type and chain lengths of fatty acids bound to the 3-hydroxyl groups of the 3-hydroxy fatty acids (5). Nevertheless, these marginal variations do not seem to be relevant for biological activity since various lipid A preparations are known to express similar endotoxic activity. A general structure of the lipid A component of endotoxically active lipopolysaccharides is shown in Fig. 4 (7). It should be mentioned that certain Gram-negative bacteria (notably photosynthetic bacteria) form lipopolysaccharides which are biologically inactive. Their lipid A component has been studied by the group of H. Mayer and shown to differ significantly in structure from those lipid A's described above (8).

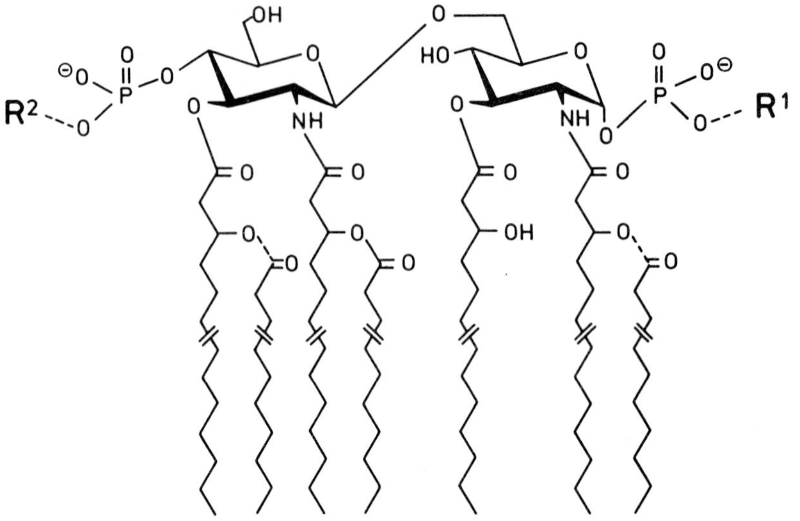

Fig. 4 Proposal of a General Structure of biologically active Lipid A's (7). Symbols R^1 and R^2 represent polar substituents of the lipid A backbone or hydrogen. Backbone-bound 3-hydroxy fatty acids may have 10 to 18 carbon atoms. In general, six moles of acyl residues are present per mol glucosamine disaccharide (for details compare ref. 5).

As Fig. 3 and 4 show, lipid A represents a complex molecule which contains regions of opposing physicochemical nature resulting in the amphiphilic character of lipid A. This chemical and physicochemical complexity and the tendency to form micelles and aggregates of various sizes and orientations have previously prevented the direct successful application of modern noninvaisve physicochemical methods, such as nuclear magnetic resonance (NMR) to the analysis of unfractionated lipid A or lipopolysaccharide. Instead, fragments of lipopolysaccharide or lipid A had to be prepared by chemical and enzymatic degradation procedures (9). Then the chemical make-up of these fragments was elucidated and this information integrated into a structural formula of lipid A. Although this strategy has led to proposals of the lipid A structure it harboured the danger that during isolation and purification of one component another - perhaps important one - may have escaped detection which would be obiously lacking in the final structural proposal for lipid A. It was, therefore, felt since a long time that for confirming the proposed lipid A structure and also for deciding whether this structure contained the endotoxic principle, chemically synthesized lipid A was needed.

Chemically synthesized lipid A remained over many years only a hope. But in 1983 a series of lipid A part structures were prepared by the research team of S. Kusumoto and T. Shiba at the Osaka University (10).

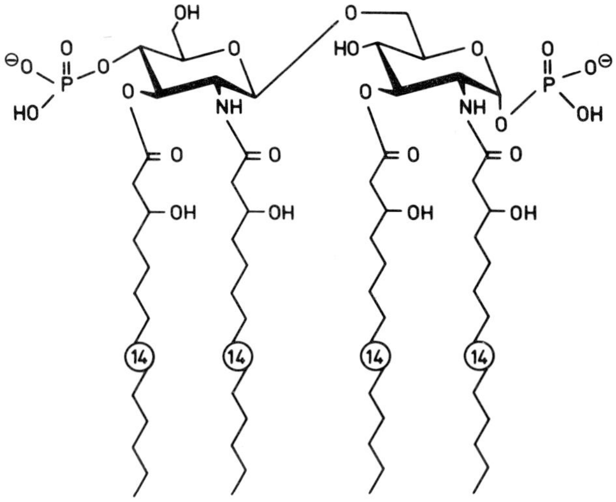

Fig. 5 Chemical Structure of Synthetic Lipid A Precursor Ia (10).

One of the synthetic preparations (preparation 406) structurally corresponds to a precursor of lipid A biosynthesis (precursor Ia) which represents an incomplete lipid A being devoid of polar substituents of phosphate groups as well as of non-hydroxylated fatty acids. Its structure is shown in Fig. 5. The synthetic precursor (compound 406) was then analyzed for biological activity in the Max-Planck-Institut für Immunbiologie (Freiburg) and our group in Borstel, in various endotoxin test systems _in vivo_ and _in vitro_, in comparison with the bacterial precursor and free lipid A and some of the results obtained are summarized in Table 1.

Biological Activities of Synthetic (Prep. 406) and Bacterial Lipid A Precursor Ia and Free E. coli Lipid A (11)

Endotoxin Test System	Dose required to induce biological effect with		
	Synthetic Lipid A Precursor (Prep. 406)	Bacterial Lipid A Precursor	Free Lipid A E.coli
	μg		
B-LYMPHOCYTE MITOGENICITY (Index of 15)	5	5	2
MACROPHAGE ACTIVATION (Release of 10 ng/ml PGE_2)	5	8	1
CLOTTING OF AMOEBOCYTE LYSATE[1]	10^{-5}	10^{-5}	10^{-5}
ENDOTOXIN CROSS-TOLERANCE (100% Survival)	30	30	30
LETHAL TOXICITY[2] (LD_{50})	0.01	0.01	0.005

[1] Kotani et al., 1985 [2] In D-GalN-sensitized mice

It was found that the synthetic preparation 406, the bacterial precursor Ia and free lipid A were of comparable activity in inducing lethal toxicity (in galactosamine-sensitized mice), B-lymphocyte mitogenicity, endotoxin tolerance, prostaglandin formation in macrophages, in gelating the Limulus amoebocyte lysate and in expressing the determinant which is recognized by lipid A specific antibodies (11, 12, 13). These results demonstrated that the structural elements required for the expression of all the above biological activities are present in a molecule consisting of a ß(1→6)-linked D-glucosamine disaccharide carrying 4 moles of (R)-3-hydroxymyristic acid (at positions 2, 3, 2' and 3') and 2 moles of phosphoryl groups (at positions 1 and 4').

In two in vivo endotoxin assay systems, however, the synthetic as well as the bacterial precursor Ia were of significantly different activity than free lipid A or lipopolysaccharide. The synthetic preparations and the bacterial lipid A precursor were found to be of weak pyrogenicity and to be completely incapable of preparing for or to elicit the local Shwartzman reaction. This suggested that the presence of (R)-3-acyloxyacyl groups which are absent from precursor Ia but present in lipid A (14) is a necessary prerequisite for the (full) expression of these endotoxic reactions.

To test this hypothesis, chemically synthesized complete lipid A was essential. In October 1984 the group of T. Shiba succeded for the first time in the chemical synthesis of E. coli lipid A (15). The structure of synthetic E. coli lipid A corresponds to that shown in Fig. 3 except that GlcN I carries at the glycosidic hydroxyl group a phosphoryl (and not pyrophosphoryl) group and that KDO is lacking (free hydroxyl group in position 6'). The structural identity of the synthetic preparation with natural free E. coli lipid A was demonstrated by ^{1}H-nuclear magnetic resonance. The synthetic lipid A preparation (compound 506) was placed at our disposition to test it for biological activity in comparison with bacterial free E. coli lipid A. These analyses were again performed collaboratively in the Max-Planck-Institut für Immunbiologie (Freiburg) and in our Institute in Borstel and some of the results obtained are summarized in Table 2. We could show that in all test systems including Shwartzman reactivity, pyrogenicity, lethal toxicity, induction of endotoxin cross-tolerance, B-lymphocyte mitogenicity, macrophage activation (prostaglandin E_2 release) and serological cross-reactivity with lipid A antiserum the synthetic preparation 506 was, on a weight basis, of identical activity as its bacterial counterpart (16). Similar results were also obtained by other groups (17, 18).

Synthetic (Prep. 506) and Bacterial Escherichia coli
Lipid A Express Identical Endotoxic Activity (16)

Biological Activity	Dose required to induce endotoxic effect with	
	Synthetic lipid A (Prep. 506)	Bacterial lipid A (E. coli)
	µg	
Lethality (LD_{50})	0.01	0.01
Pyrogenicity (MPD-3)	0.004	0.005
Local Shwartzman Reaction (+ + +)	12.5	12.5
Endotoxin Cross-Tolerance (100 % survival)	30	30
B-Lymphocyte Mitogenicity (Index of 15)	2	2
Macrophage Activation (Release of 10 ng/ml PGE_2)	1	1

These exciting results lead to the following conclusions and considerations:
1) The demonstration of identical NMR spectra and of identical biological activity of synthetic and bacterial lipid A shows that our structural proposal for lipid A is correct, i.e. the structure derived from analytical work has been confirmed by chemical synthesis.
2) For all the biological activities discussed, synthetic lipid A is as active as natural lipid A. This shows that the variable substituents of lipid A such as phosphate, 4-aminoarabinose, the polysaccharide portion of lipopoly-saccharides and contaminating proteins are not required for the expression of endotoxicity. From this it clearly follows that lipid A, with the structure shown (Fig. 3), harbours the endotoxic principle of lipopolysaccharides.
3) Synthetic lipid A as a well defined, homogenous and water soluble material is now available as a reference substance. Therefore, future chemical and biological analyses of lipid A's can be related and compared to synthetic lipid A which should, in the years to come, serve as a standard in endotoxin research.
4) The gate is now open for the construction of lipid A analogues and derivatives at will. Such preparations could be synthesized to perhaps realize the old dream of separating wanted i.e. beneficial endotoxin activities (like induction of tumour necrosis and nonspecific resistance to viral and bacterial infections) from harmful or toxic effects (hypotension, disseminated intravascular coagulation, shock). Such preparations may even find their application in clinical medicine.

We are at present hopeful that by biologically analysing lipid A part structures and analogues a deeper insight into those minimally essential structures and conformations which endow lipid A with its unique and impressive spectrum of biological activities can be obtained. We are, therefore, confident that it is only a matter of time that conclusive correlations between the chemical structure and biological activity of lipid A and endotoxins in general will be established on a molecular level. In the years to come, endotoxin research will certainly be as exciting and rewarding as it has been for several decades.

Acknowledgement

The financial support of the Fonds der Chemischen Industrie is gratefully acknowledged.

References

1. Lüderitz, O., M.A. Freudenberg, C. Galanos, V. Lehmann, E.Th. Rietschel, D.W. Shaw. 1982. In: Microbial Lipids, Vol. 17. C.S. Razin, S. Rottem, eds. Academic Press, London, New York, p. 79-151.

2. Galanos, C., O. Lüderitz, E.Th. Rietschel, O. Westphal. 1977. In: International Review of Biochemistry, Biochemistry of Lipids II, Vol. 14. T.W. Goodwin, ed. University Press, Baltimore, p. 239-335.

3. Zähringer, U., B. Lindner, U. Seydel, E.Th. Rietschel, H. Naoki, F.M. Unger, M. Imoto, S. Kusumoto, T. Shiba. Tetrahedr. Lett. (in press)

4. Brade, H., E.Th. Rietschel. 1984. Eur.J.Biochem. 145, 231-235.

5. Rietschel, E.Th., H.W. Wollenweber, H. Brade, U. Zähringer, B. Lindner, U. Seydel, H. Bradaczek, H. Labischinski, P. Giesbrecht. 1984. In: Handbook of Endotoxins, R.A. Proctor, ed. Vol. 1, Chemistry of Endotoxin. E.Th. Rietschel, ed. Elsevier North Holland, Biomedical Press, p. 187-220.

6. Takayama, K., N. Qureshi, E. Ribi, J.L. Cantrell. 1984. Ref.Infect.Dis. 6, 439-443.

7. Rietschel, E.Th., H. Brade, L. Brade, A. Bünsch, A. Tacken, U. Zähringer. 1985. Forum Mikrobiol. 8, 286-294.

8. Mayer, H., P.V. Salimath, O. Holst, J. Weckesser. 1984. Rev.Infect.Dis. 6, 542-545.

9. Rietschel, E.Th., H.W. Wollenweber, Z. Sidorczyk, U. Zähringer, O. Lüderitz. 1983. In: Bacterial lipopolysaccharides: structure, synthesis, biological activities, L. Anderson, F. Unger, eds. Am.Chem.Soc.Symp.Series, Vol. 231, Washington, D.C. p. 195-218.

10. Imoto, M., H. Yoshimura, M. Yamamoto, T. Shimamoto, S. Kusumoto, T. Shiba. 1984. Tetrahedr. Lett. 25, 2667-2670.

11. Galanos, C., V. Lehmann, O. Lüderitz, E.Th. Rietschel, O. Westphal, H. Brade, L. Brade, M.A. Freudenberg, T. Hansen-Hagge, T. Lüderitz, G. McKenzie, U. Schade, W. Strittmatter, K. Tanamoto, U. Zähringer, M. Imoto, H. Yoshimura, M. Yamamoto, T. Shimamoto, S. Kusumoto, T. Shiba. 1984. Eur.J.Biochem. 140, 221-227.

12. Kotani, S., H. Takada, M. Tsujimoto, T. Ogawa, K. Harada, Y. Mori, A. Kawasaki, A. Tanaka, S. Nagao, S. Tanaka, T. Shiba, S. Kusumoto, M. Imoto, H. Yoshimura, M. Yamamoto, T. Shimamoto. 1984. Infect.Immun. 45, 293-296.

13. Kanegasaki, S., Y. Kojima, M. Matsuura, J.Y. Homma, A. Yamamoto, Y. Kumazawa, K. Tanamoto, T. Yasuda, T. Tsumita, M. Imoto, H. Yoshimura, M. Yamamoto, T. Shimamoto, S. Kusumoto, and T. Shiba. 1984. Eur.J.Biochem. 143, 237-242.

14. Wollenweber, H.W., U. Seydel, B. Lindner, O. Lüderitz, and E.Th. Rietschel. 1984. Eur.J.Biochem. 145, 265-272.

15. Imoto, M., H. Yoshimura, S. Kusumoto, T. Shiba. 1984. Proc.Jap.Acad. 60, 285-288.

16. Galanos, C., O. Lüderitz, E.Th. Rietschel, O. Westphal, H. Brade, L. Brade, M.A. Freudenberg, U. Schade, M. Imoto, H. Yoshimura, S. Kusumoto, and T. Shiba. 1985. Eur.J.Biochem. 148, 1-5.

17. Kotani, S., H. Takada, M. Tsujimoto, T. Ogawa, I. Takahashi, T. Ikeda, K. Otsuka, H. Shimanchi, N. Kasai, J. Mashimo, S. Nagao, A. Tanaka, S. Tanaka, K. Harada, K. Nagaki, H. Kitamura, T. Shiba, S. Kusumoto, M. Imoto, H. Yoshimura. 1985. Infect.Immun. 49, 225-237.

18. Homma, J.Y., M. Matsuura, S. Kanegasaki, Y. Kawakubo, Y. Kojima, N. Shibukawa, Y. Kumazawa, A. Yamamoto, K. Tanamoto, T. Yasuda, M. Imoto, H. Yoshimura, S. Kusumoto, T. Shiba. 1985. J.Biochem. 98, 395-406.

MACROPHAGE ACTIVATION BY MURAMYL PEPTIDES

M. Parant, L. Chedid
Experimental Immunotherapy, Institut Pasteur, 75724 Paris cedex 15, France

Introduction

The mechanisms whereby immunoregulatory agents modify immune responses appear as a multifactorial enhancement of cell function, and such products may alter T cell, B cell, and macrophage functions. Cells of the monocyte/macrophage series have been shown to be critical regulatory cells in immune responses, and many adjuvants exert their primary effects on monocytic cells, which in turn modulate antibody production through the secretion of regulating factors (1). Moreover, these cells are deeply involved in host defense reactions to infections and in nonspecific anti-neoplastic surveillance. Isolation of chemically defined components has been an important factor in investigations to establish cell specific targets of immunoregulators of bacterial origin. Thus, in the course of studies concerned with the biological activities of muramyl peptides, it became apparent that in many cases the effects could be mediated, in some way or another, by the macrophage.

Relation of Muramyl Peptide Structure to Activity

Synthetic muramyl dipeptide or MDP (AcMur-L-Ala-D-iGln) was prepared following a progressive series of biochemical studies attempting to purify active adjuvant components of the bacterial cell wall (reviewed in 2). This monosaccharide dipeptide which represents the active site of larger peptidoglycan fragments was first described at the First International Workshop on the Immunological and Biological Properties of Peptidoglycan which was held in Munich in 1974 (3). Since that time, MDP and a series of analogs and derivatives have been intensively studied and demonstrated to have numer-

Biological Properties of Peptidoglycan

ous potent, diverse, and interesting biological activities (2, 4-6). Though being a copy of peptidoglycan fragment, the small molecular weight (<500) synthetic MDP is non-immunogenic when given alone or in FCA (Freund's complete adjuvant). Antibodies can only be elicited following the immunization with MDP linked to some carrier (7), and such a conjugate was used to obtain monoclonal antibodies to MDP (8). Therefore, the dissociation between immunogenicity and the capacity for displaying immunostimulant effects when administered in saline, has made of MDP and analogs valuable tools to further our understanding of the immune system since several of these agents have been shown to modify both specific and nonspecific immunity.

Chemically different muramyl peptides were comparatively evaluated in a variety of experimental test systems (for review, see 2, 4, 5). It was therefore demonstrated that some of the numerous biological properties exhibited by MDP could be modified or dissociated by chemical substitution (9). Thus, pyrogenicity is not a prerequisite for adjuvant effect or for enhancement of nonspecific resistance to infection, and moreover, adjuvanticity is not a requirement for protection against bacterial infections (4). In the course of these studies, chemical structures responsible for the various activities were recognized in some cases and partially mastered. It is now possible to find MDP analogs which have adjuvant effect and increase nonspecific resistance, yet do not exhibit the pharmacological reactions produced by MDP administration. Up to now, such compounds have mainly been found among the water-soluble MDP analogs (10, 11). Murabutide (MDP(Gln)-OnBu) has thus been selected since it is as effective as MDP in stimulating immune responses but does not retain any side effect (12).

However, hydrophilic muramyl peptides administered in saline do not appear to markedly augment cell-mediated immunity whereas they are very potent in this respect when given as an oily emulsion or in liposomes (reviewed in 13). Because of the major importance of cell-mediated immunity, efforts were made to potentiate these responses in the absence of FIA (Freund's incomplete adjuvant). Progress was made in this direction by synthesizing lipophilic derivatives that may induce delayed-type hypersensitivity to a protein antigen when injected in saline (5, 14, 15). They are often more effective against bacterial challenges (5, 15-17), and an interesting example of anti-viral effect has been reported after intranasal, oral, or intravaginal administration of minute amounts of MTP-PE (muramyl tripeptide-dipalmitoyl phosphatidyl-ethanolamine) in a variety of experimental virus models (18). Repeated intranasal administration of MTP-PE could also induce a high survival rate in the B16/B16 melanoma system in mice (19). Moreover, lipophilic muramyl peptides represent good candidates to be incorporated into liposomes, and they have been shown to prevent and to eradicate lung metastases into B16/B16 melanoma bearing mice (20, 21).

The conjugation procedure with an appropriate carrier -natural or synthetic, immunogenic or not- was also used to increase both the size and the efficiency of MDP and of some analogs. This procedure may reduce the rapid excretion of muramyl peptides and modify the affinity for target cells. The anti-infectious activity of MDP or analogs could be potentiated to a considerable extent after linkage to high molecular weight molecules although the adjuvant effect may not be increased (22). Muramyl peptides have also been shown to elicit antibodies to numerous synthetic antigens after conjugation to a carrier. However, MDP or derivatives were likewise conjugated to natural or synthetic vaccines directly to raise antibodies and to diminish untoward effects linked to carriers (for review, see 13).

Alterations within Monocytes/Macrophages after In Vitro Exposure to Muramyl Peptides

With MDP, like with other bacterial adjuvants, a major mechanism of action appears to be through direct triggering of the macrophage although a direct effect on lymphocytes has also been reported (for review, see 23). Most of the effects on the macrophage/histiocyte series have been demonstrated in vitro, but the same stimulating agents influence immune responses in vivo, so there is good reason to believe that the immunoregulatory role of macrophages operate also in intact animals.

The in vitro assays summarized in Table 1 were performed with MDP, using the inactive stereoisomers MDP(D,D) or MDP(L,L) as negative controls. An increased viability was first observed in cell cultures incubated with MDP, as well as a marked increase in adherence, spreading, and pseudopod formation (24, 25). These effects and the enhancement of chemotactic responses were produced only by the adjuvant-active muramyl peptides (27). In another study concerned with chemotaxis, several lipophilic derivatives were shown to be still more potent than MDP or Murabutide. By contrast, all these compounds were found inactive on the migration of human polymorphonuclear leukocytes (28).

Table 1. Morphological and Biochemical Changes in Macrophages Incubated with MDP

Assay	References
Increased viability	24
Enhancement of adherence and spreading	24, 25
Inhibition of macrophage migration	26
Chemotaxis of human monocytes	27
Increased glucosamine uptake	29
Glucose oxidation	30
Release of superoxide anion	31
Stimulation of chemiluminescence	32
Increased cGMP and cAMP levels	33, 25
Prostaglandin and collagenase synthesis	25
Differential release of hydrolytic enzymes	34
Neutral protease release	35
Inhibition of plasminogen activator secretion	36

Glucosamine uptake by guinea-pig peritoneal macrophages was enhanced by MDP and adjuvant compounds (29). Again in this assay, some 6-O-acyl-MDP derivatives were found to exert a stronger stimulation (28). Direct metabolic activation of macrophages has also been demonstrated by the MDP-induced increase of glucose oxidation (30). After in vitro priming by MDP, the ability of cultured resident mouse macrophages to generate superoxide anion in the presence of PMA (phorbol myristate acetate) was enhanced (31). Macrophages that were activated or elicited in vivo by MDP produced increased amount of O_2^- when triggered with phorbol myristate acetate or PMA (31). A 6-O-acyl derivative of MDP has been reported to be twice as potent as MDP in its capacity to primed cultured macrophages for superoxide release (31). Activation of macrophages in other studies was determined by enhanced levels of intracellular cGMP and cAMP (25,33), collagenase production and prostaglandin synthesis (25). In this work, indomethacin inhibited the production and release of collagenase indicating the prostaglandin involvement in this process. Quantification of enzyme secretion was also used to assess the effect of MDP on macrophages (34, 35). A lipophilic analogue 6-O-stearoyl nor-MDP was more efficacious in inducing secretion of neutral proteases, and a significant increase in the potency was still observed when the drug was presented to the cells in liposome-encapsulated form (35). However, observations regarding the in vitro effects of lipophilic compounds often remained inaccurate in the absence of inactive controls bearing the same lipophilic chain such as those used in some studies (28). As yet, only a few data on the comparative effect of various hydrophilic compounds are available, although they may elicit different reactions. An example has been provided by MDP and Murabutide, both being adjuvant in vivo and in vitro but only MDP was able to induce EP (endogenous pyrogen) production by macrophages (10). However, in a preliminary assay, both water-soluble compounds displayed the same potency in stimulating prostaglandin release (Table 2), (G. Riveau, unpublished data).

Table 2. Prostaglandin Production by Rabbit Macrophages Induced by Pyrogenic MDP and Non-pyrogenic Murabutide

Agent (µg/ml)	Prostaglandin release in cultures¶			
	PGE_2	PGF_2	PGI_2	TxB_2
MDP	++*	++	+++	+
Murabutide	++	++	+++	+

¶elicited peritoneal macrophages incubated for 24h

*evaluated as stimulation index from unstimulated cells: + = ≦ 10 and +++ = ≧ 20.

As in different in vitro assays, enhanced prostaglandin production by macrophages can be observed in response to a variety of stimuli and it usually reflects cell activation. However, it remains evident that PG generation does not necessarily accompany the process of macrophage activation, and that it may be a somewhat incidental consequence of more fundamental processes.

Influence of Muramyl Peptides on Macrophage Functions In Vitro

MDP significantly enhanced both phagocytosis and killing of Listeria monocytogenes when added at the beginning of the macrophage culture. During the 6-h incubation following the pulse exposure to the bacteria, nearly all the MDP-treated quinea-pig macrophages completely killed their intracellular bacteria (37). Incubation with MDP also increased the candidacical activity of mouse adherent peritoneal cells (38), (Table 3).

Table 3. Enhancement of Macrophage Functions Induced by MDP

Assay	References
Greater killing of Listeria	37
Greater killing of Candida	38
Inhibition of tumor cell growth	39
Cytolytic activity of macrophage cell line	40, 41
Tumoricidal activity (MDP + MAF)	42
Tumoricidal activity (MDP in liposomes)	43

Under optimal in vitro conditions, the activation of tumoricidal properties in macrophages by MDP required a 8-h macrophage-glycopeptide interaction (43). In vivo, more than 90% of parenterally administered MDP in saline is cleared from the body within 2-h, and excreted in urine (44). Hence, the administration of water-soluble MDP does not appear a feasible way to stimulate anti-tumor activity in vivo (39). In contrast to free MDP, liposome-encapsulated MDP required less time to stimulate macrophages in vitro (43). A synergistic activation of macrophages by MAF (macrophage-activating factor) and MDP, both free or encapsulated within the same liposome, has also been reported (42), (Table 3).

Stimulation of Factor Production by Monocytes/Macrophages Treated with Muramyl Peptides

The increase in regulatory factor production by macrophages stimulated in vitro by muramyl peptides has been extensively studied, and there is now numerous experimental data demonstrating these activities as summarized in Table 4. Some of these factors are not necessarily related to a specific response to antigenic determinants, but may affect humoral and cellular -natural or acquired- immunity through complex interactions with each other and their target cells.

Table 4. Stimulation of Factor Production Induced by MDP

Assay	References
Production of fibronectin	45
Fibroblast-activating factor	25
Synovial activator activity	46
Inhibitor of plasminogen activator	47
"non-secretable" form of Interferon-α	48
Tumor cytostatic or cytolytic factors	50, 51
Colony-stimulating factor	52
Differentiation-inducing factor	54
B cell-activating factor	56
Lymphocyte-activating factor	58
Endogenous pyrogen production	62
Slow-wave sleep activity	64
Guinea-pig distress factor	66

Macrophages play an important role in attracting fibroblasts and initiating the wound healing process into an inflammatory site by the release of fibronectin (45), and by the production of a fibroblast-activating factor that stimulated the proliferation (25). MDP was capable of inducing human mononuclear cells to produce a synovial activator that elevated plasminogen activator production of synovial fibroblasts (46), and also to secrete a specific inactivator of the urokinase-type plasminogen activator (47).

In an investigation of the molecular basis of monocyte activation, it was shown that cells activated by MDP contained an IFN-α mRNA which appeared to be associated with the presence of an intracellular antiviral reactivity, neutralizable by an antiserum specific for Interferon-α (48). This observation could be related to the enhanced potency of interferon inducers such as LPS or poly I:C when mice had been pretreated with MDP or with a still more efficient 6-O-acyl derivative (49). The tumoricidal action of macrophages could also be attributable to the production of cytotoxic factors though there are several other possible explanations (50, 51). CSF or colony-stimulating factor levels were elevated after incubating macrophages with MDP (52), and these factors defined on the basis of their capacity to stimulate proliferation and differentiation of myeloid colonies, may also be involved in the production of other monokines such as IFN and tumor necrosis factor (TNF), (53). Myeloid leukemic cells in culture that do no longer require CSF for proliferation are blocked at different stages of maturation. In the presence of differentiation-inducing factor (D-factor), they are induced to differentiate and thus to acquire phenotypic and functional markers of mature cells. Conditioned medium from murine (54) or human (55) monocytic cells stimulated with MDP contained D-factor activity which was shown to be distinct from the concomitantly released CSF activity (Parant et al., unpublished data).

Because of the adjuvant effect of muramyl peptides, the induction of B and T lymphocyte-activating factor was evaluated soon after demonstration of enhanced immune responses in vitro. BAF or B cell-activating factor was produced by human monocytes in the presence of MDP. It could in turn stimulate antibody response to a T cell dependent antigen of T cell-depleted murine splenocytes (56). In mice, the adjuvant effect of MDP has mainly been ascribed to the production of lymphocyte-activating factor (LAF), (57), named Interleukin-1 in 1979. IL-1 production was induced by MDP in elicited and resident mouse macrophage (58, 59), guinea-pig and rabbit peritoneal macrophage (60, 61), and in human blood monocyte cultures (58, 61). The response to MDP in mice was found dependent on the strain (59). In no case, freshly harvested cells or unstimulated cells contained IL-1 provided the culture medium has been carefully selected. After incubation with MDP or with another adjuvant analogue, IL-1 was released in the supernatant fluid, but still higher levels were found associated with the cells (Damais et al., unpublished data). When pyrogenic MDP was incubated in cultures, it obviously stimulated monocytic cells to also produce EP (endogenous pyrogen), (62). In vivo, circulating EP induced by MDP administration could be transferred into another recipient rabbit (63). Thus, supernatant fluids from MDP-activated macrophages provoked a fever response and were able to trigger T lymphocytes. In contrast, Murabutide which did not elicit any fever response, was inactive on the release of EP in vivo or in vitro (10). In addition, this analogue did not inhibit the fever responses. As expected, adjuvant Murabutide was as potent as MDP in inducing IL-1 production by various types of monocytic cells (61), clearly indicating that IL-1 can be dissociated from EP in contrast to what had been previously postulated. Therefore, the term Interleukin-1 should be restricted to a mediator potentiating T lymphocyte responses. In the same conditioned medium obtained from Murabutide-treated cells, similar levels of other monokines were obtained (Table 5).

Table 5. Monokine Production Induced by Pyrogenic MDP and by Non-pyrogenic Murabutide

Assay	Incubation of cells* with :	
	MDP	Murabutide
Lymphocyte-activating factor	+	+
Endogenous pyrogen	+	-
Colony-stimulating factor	+	+
Differentiation-inducing factor	+	+
Cytotoxic factor	+	+
Sleep-promoting factor	+	NT

* Rabbit elicited peritoneal macrophages.

Pyrogenic supernatant fluids from monocyte/macrophage cultures incubated with MDP also contained a factor that enhanced duration of slow-wave sleep (64). This activity has not yet been dissociated from the presence of EP. In addition, non-pyrogenic Murabutide was unable to produce sleep modifications (65). The guinea-pig distress factor activity also appeared to be linked to the presence of EP (66).

Role of the Macrophage in Biological Responses of the Host

A number of in vitro studies have demonstrated that the macrophage is an important target cell for the biological effects of MDP. There is also evidence that in vivo treatment with muramyl peptides results in macrophage stimulation. The effect of MDP on the reticulo-endothelial system in mice was demonstrated by an increase in carbon clearance (67). However, the phagocytic index was affected with relatively high dosage levels of MDP, and differences observed in animal species were not always correlated with their capacity to respond to the immunostimulant effects (67, 68). Thus guinea-pigs did not show any stimulation of phagocytosis after MDP treatment, although immune responses were strongly increased and peritoneal cells directly activated by the adjuvant (67).

Several pharmacological reactions inherent to the administration of peptidoglycan can be partially obtained by MDP injection (for review 69), but not by administration of a derivative such as Murabutide (10). Results reported with supernatant fluids from macrophages cultured in the presence of MDP have shown that they can reproduce some of these effects and may be responsible for some in vivo reactions. It was thus assumed that MDP could elicit fever through the release of EP, although a direct effect on the thermoregulatory centers was also suggested by an increase in sensitivity when using the intracerebroventricular route (63). During the fever response induced by EP administration, concomitant changes in plasma metal levels -hypoferraemia and hypercupraemia- were produced in relation with modifications in serum acute-phase protein concentration (70). MDP itself modified iron and copper levels in the same way, whereas Murabutide was inactive. Fluids from macrophage cultures incubated with the non-pyrogenic adjuvant, which contained only

IL-1 activity, were unable to cause any change in blood metal concentrations (71). These results are in accordance with the observation of McAdam that MDP induced serum amyloid A in mice, whereas Murabutide did not (72). More surprisingly, it was shown that macrophage culture fluid containing both EP and IL-1 activities could increase slow-wave sleep (64). This effect was demonstrated after the discovery made by Pappenheimer, Karnovsky and Krueger showing that the slow-wave factor had a muramyl peptide structure (73), and that some MDP analogues increased the slow-wave sleep duration (65).

In specific immune responses, IL-1 and mononuclear phagocytes play vital roles by providing necessary signals for B cell and T cell functions. The correlation between the adjuvant effect in vivo and ability of a compound to induce IL-1 production has raised the possibility that IL-1 is a common mediator of adjuvanticity (57, 58). The adjuvant effect of exogenous IL-1 on the mouse immune response to BSA (bovine serum albumin) is consistent with the hypothesis (57). It has also been shown that mouse strains which respond more strongly to the adjuvant effect of MDP were also higher responders to the stimulation by IL-1. However, both the high and low responding strains responded to MDP with a similar enhancement in carbon clearance (74).

There is evidence that stimulation of the monocyte/macrophage series may be at least partly responsible for enhancement of non-specific resistance to infections. Muramyl peptides have been shown to stimulate natural immunity against various infectious agents (4), and in some cases the resistance is known to be effected mainly by the enhanced microbicidal capacity of activated macrophages, such as against *Listeria monocytogenes*, *Toxoplasma gondii* and *Trypanosoma cruzi* (75-77). By using several MDP analogues, a linear correlation has been observed between change in carbon clearance and protection against *Candida albicans* (78). Cells of the monocyte-macrophage series participate in host defense against viral diseases through a variety of mechanisms, including destruction of virus particles and selective cytotoxicity for virus-infected cells. A protective effect against viral challenges has been obtained with both hydrophilic and lipophilic MDP derivatives. The latter appeared to be more potent when solubilized in saline or incorporated within liposomes (5, 79).

Liposomes have been used to target agents to cells of the reticulo-endothelial system and to prolong their biologic half-life, as has been demonstrated with hydrophilic and lipophilic muramyl peptides (80). Eradication of metastases has been correlated with the activation of tumoricidal macrophages after administration of muramyl peptides entrapped into liposomes (19, 21).

Conclusions

Comparative studies with MDP derivatives have shown that the effects of MDP are less selective than was at first assumed, and that untoward side effects can be avoided by the choice of certain derivatives such as Murabutide or MDP(threonyl), (10, 11). The diversity and the flexibility of synthetic muramyl peptides may lead to the production of agents displaying only the desired activity. The lipophilic compounds have often been shown to exhibit a higher potency particularly as antiviral and antitumoral agents (5, 19-21). They seem to produce stronger side effects than most of the hydrophilic molecules and thereby have not reached the stage of clinical trials yet. However, pharmacological reactions to these agents clearly depend on the route of administration, and for example, are weakened or even abolished when the lipophilic compound is given as an intranasal application which has been demonstrated to be effective (19). Polymerization and conjugation procedures lead to much more potent products (22). Moreover, after linking to some antigenic determinants, such constructs have a potential for developing antibacterial and antiviral agents endowed with both specific and non-specific activities (13). With MDP linked to an antigen, an enhancement of the adjuvant effect has been demonstrated after complexing with monoclonal anti-MDP antibodies (82). This procedure could also enhance the MDP-induced macrophage activation in vitro and this effect is mediated by Fc receptors (83).

The cell that seems to be the most involved -or at least has been the most studied- in muramyl peptide activities is the macrophage, although the neutrophil is likely to represent another effector cell both against infectious challenges and against tumor metastases (19, 81). The monocyte/macrophage is obviously a key cell in immunoregulation. They are required as accessory and antigen-presenting cells in the development of lymphocyte activities, and the principal function of adjuvants may be to induce formation of IL-1 by accessory cells. The central role of activated mononuclear phagocytes in the destruction of intracellular and extracellular parasites, and of certain viruses is well established. Moreover, when appropriately activated, macrophages can lyse neoplastic cells. Thus, the mechanism(s) responsible for the regression of established metastases after the systemic administration of liposomes containing MDP probably involved the activation of macrophages to become tumoricidal as has been confirmed by adoptive transfer experiments (84).In therapeutic studies against cancer metastases, the synergistic action of MAF (macrophage-activating factor) and muramyl peptides (84) may open an exciting field of research by taking advantage of the opportunity offered by modern technology to produce cytokines. Indeed, recombinant IFN-γ and MDP were shown to act synergistically in inducing IL-1 production (85).

Important factors in regulating macrophage development and function are secretory products released from the macrophages themselves. They can act either directly or indirectly via interactions with other cells or molecules, and can initiate complex positive and negative feedback loops (1). Thus, IFN-α may prime macrophages so that following exposure to IFN inducers they produced much more IFN-α (1). IL-1 may play a role as an autostimulating factor in host defense against tumor cells by promoting monocyte-mediated cytotoxicity (86). GM-CSF can also be considered as a positive autoregulatory mediator for mononuclear phagocytes. It can activate them to increased synthesis and secretion of PGE, IFN, IL-1 and TNF (53).

The influence of muramyl peptides in promoting secretion of the various factors by macrophages underlines the central role of the cells in their biological effects (see table 4). Discrepancies that have been reported between in vivo and in vitro activities are likely to be related to the short half-life of hydrophilic molecules. As an example, liposomes represent an efficient targetted delivery system as has been demonstrated in treating primary and/or metastatic cancer. One of the mechanisms responsible for the killing of tumor cells seems to be the production of cytolytic factor(s) by macrophages stimulated with MDP. This anti-tumor substance could be TNF although the characterization has not yet been achieved (51).

Therefore, TNF which was shown to enhance resistance to bacterial infections when used as a crude serum preparation (87) or as recombinant TNF (Parant et al., unpublished results) could be a common factor endowed with both anti-tumor and anti-infectious activities.

References

1. Allison, A.C. 1984. In: Advances in Inflammation Research, Vol.7 (I. Otterness, R. Capetola and S. Wong, eds.). Raven Press, New York. pp.201-221.

2. Lederer, E. 1980. J. Med. Chem. 23, 819-825.

3. Adam, A., Ellouz, F., Ciorbaru, R., Petit, J-F., Lederer, E. 1975. Zeit. Immün-Forsch. 149, 341-348.

4. Parant, M. 1979. Springer Semin. Immunopathol. 2, 101-118.

5. Kotani, S., Takada, H., Tsujimoto, M., Kubo, T., Ogawa, T., Azuma, I., Ogawa, H., Matsumoto, K., Siddiqui, W.A., Tanaka, A., Nagao, S., Kohashi, O., Kanoh, S., Shiba, T., Kusumoto, S. 1982. In: Bacteria and Cancer (J. Jeljaszewicz, G. Pulverer and W. Roszkowski, eds.). Academic Press, New York. pp.67-107.

6. Chedid, L. 1983. Microbiol. Immunol. 27, 723-732.

7. Reichert, C.M., Carelli, C., Jolivet, M., Audibert, F., Lefrancier, P., Chedid, L. 1980. Molec. Immunol. 17, 357-363.

8. Bahr, G.M., Eshhar, Z., Ben-Yitzhak, R., Modabber, F.Z., Arnon, R., Sela, M., Chedid, L. 1983. Molec. Immunol. 20, 745-752.

9. Lefrancier, P., Lederer, E. 1091. In: Progress Chem. Org. Nat. Prod. Vol.40 (W. Herz, H. Grisebach and G.W. Kirby, eds.). Springer-Verlag, New York. pp. 1-47.

10. Chedid, L., Parant, M., Audibert, F., Riveau, G., Parant, F., Lederer, E., Choay, J., Lefrancier, P. 1982. Infect. Immun. 35, 417-424.

11. Byars, N.E. 1984. Infect. Immun. 44, 344-350.

12. Parant, M., Chedid, L. 1984. Infection, 12, 230-234.

13. Audibert, F., Leclerc, C., Chedid, L. 1985. In: Biological Response Modifiers (P.F. Torrence, ed.). Academic Press, New York. (in press).

14. Azuma, I., Sugimura, K., Yamawaki, M., Yemina, M., Kusumoto, S., Okada, S., Shiba, T., Yamamura, Y. 1978. Infect. Immun. 20, 600-607.

15. Parant, M., Audibert, F., Chedid, L., Level, M. Lefrancier, P.,

Choay, J., Lederer, E. 1980. Infect. Immun. 27, 825-831.

16. Dietrich, F.M., Sackmann, W., Mitchell, G.H., Voller, A. 1983. In: Chemotherapy and Immunology in the Control of Malaria, Filariasis and Leishmaniasis (N. Anand and A.B. Sen, eds.). Tata McGraw-Hill Publ., New Delhi. pp. 277-303.

17. Matsumoto, K., Ogawa, H., Kusama, T., Nagase, O., Sawaki, N., Inage, M., Kusumoto, S., Shiba, T., Azuma, I. 1981. Infect. Immun. 32, 748-758.

18. Dietrich, F.M., Lukas, B., and Schmidt-Ruppin, K.H. 1983. 13th Intern. Cong. Chemotherapy, Vienna, P.50.

19. Brownbill, A.F., Braun, D.G., Dukor, P., Schumann, G. 1985. J. Leukocyte Biol. 38, 317-325.

20. Fidler, I.J., Sone, S., Fogler, W.E., Smith, D., Broun, D.G., Tarcsay, L., Gisler, R.H., Schroit, A.J. 1982. J. Biol. Resp. Modif. 1, 43-55.

21. Phillips, N.C., Moras, M.L., Chedid, L., Lefrancier, P., Bernard, J.M. 1985. Cancer Res. 45, 128-134.

22. Chedid, L., Parant, M., Parant, F., Audibert, F., Lefrancier, P., Choay, J., Sela, M. 1979. Proc. Natl. Acad. Sci. USA. 76, 6557-6561.

23. Leclerc, C., Chedid, L. 1982. In: Lymphokines, Vol.7 (E. Pick, ed.). Academic Press, New York, pp.1-21.

24. Tanaka, A., Nagao, S., Imai, K., Mori, R. 1980. Microbiol. Immunol. 24, 547-557.

25. Wahl, S.M., Wahl, L.M., McCarthy, J.B., Chedid, L., Mergenhagen, S.E. 1979. J.Immunol. 122, 2226-2231.

26. Yamamoto, Y., Nagao, S., Tanaka, A., Koga, T., Onoue, K. 1978. Biochem. Biophys. Res. Commun. 80, 923-928.

27. Ogawa, T., Kotani, S., Fukuda, K., Tsukamoto, Y., Mori, M., Kusumoto, S., Shiba, T. 1982. Infect. Immun. 38, 817-824.

28. Kotani, S., Takada, H., Tsujimoto, M., Ogawa, T., Mori, Y., Shiba, T., Kusumoto, S., Inage, M., Kasai, N. 1984. In: Bacterial Endotoxin; Chemical, Biological and Clinical Aspects (J.Y. Homma, S. Kanegasaki, O. Lüderitz, Shiba, T., O. Westphal, eds.). Verlag Chimie, Weinheim, pp.111-146.

29. Takada, H., Tsujimoto, M., Kato, K., Kotani, S., Kusumoto, S., Inage, M., Shiba, T., Yano, I., Kawata, S., Yokogawa, K. 1979. Infect. Immun. 25, 48-53.

30. Imai, K., Tomioka, M., Nagao, S., Kushima, K., Tanaka, A. 1980. Biomed. Res. 1, 300-307.

31. Pabst, M.J., Johnston, R.B. 1980. J. Exp. Med. 151, 101-114.

32. Masihi, K.N., Azuma, I., Brehmer, W., Lange, W. 1983. Infect. Immun. 40, 16-21.

33. Hadden, J.W. 1978. Cancer Treatment Rep. 62, 1981-1985.

34. Imai, K., Tanaka, A. 1981. Microbiol. Immunol. 25, 51-62.

35. Mehta, K., Juliano, R.L., Lopez-Berestein, G. 1984. Immunology, 51, 517-528.

36. Drapier, J.C., Lemaire, G., Petit, J-F. 1982. Int.J.Immunopharmacol. 4, 21-34.

37. Hadden, J.W., Englard, A., Sadlik, J.R., Hadden, E.M. 1979. Int. J. Immunopharmacol. 1, 17-27.

38. Nozawa, R.T., Sekiguchi, R., Yokota, T. 1980. Cell. Immunol. 53, 116-124.

39. Juy, D., Chedid, L. 1975. Proc. Natl. Acad. Sci. USA. 72, 4105-4109.

40. Taniyama, T., Holden, H.T. 1979. Cell. Immunol. 48, 369-374.

41. Matter, A. 1979. Cancer Immunol. Immunother. 6, 201-210.

42. Sone, S., Fidler, I.J. 1980. J. Immunol. 125, 2454-2458.

43. Sone, S., Fidler, I.J. 1981. Cell. Immunol. 57, 42-50.

44. Parant, M., Parant, F., Chedid, L., Yapo, A., Petit, J-F., Lederer, E. 1979. Int.J. Immunopharmacol. 1, 35-41.

45. Tsukamoto, Y., Hesel, W.E., Wahl, S.M. 1981. J. Immunol. 127, 673-678.

46. Hamilton, J.A., Zabriskie, J.B., Lachman, L.B., Chen, Y.S. 1982. J. Exp. Med. 155, 1702-1718.

47. Golder, J.P., Stephens, R.W. 1983. Europ. J. Biochem. 136, 517-522.

48. Stevenson, H.C., Miller, P., Huffer, T., Oldham, R., Kanapa, D., Sen, A. 1985. J. Leukocyte Biol. 37, 585-595.

49. Kotani, S., Takada, H., Tsujimoto, M., Ogawa, T., Mori, Y., Koga, T., Iribe, H., Tanaka, A., Nagao, S., McGhee, J.R., Michalek, S.M., Morisaki, I., Nishimura, C., Ikeda, S., Kohashi, O., Ogawa, H., Ozawa, S., Hamada, S., Kawata, S., Shiba, T., Kusumoto, S. 1983. In: Progress in Immunology, Vol.5 (Y. Yamamura, T. Tada, eds.). Academic Press, Tokyo. pp.1359-1377.

50. Kildhal-Andersen, O., Nissen-Meyer, J. 1985. Cell. Immunol. 93, 375-386.

51. Sone, S., Tachibana, K., Ishii, K., Ogawa, M., Tsubura, E. 1984. Cancer Res. 44, 646-651.

52. Staber, F.G., Gisler, R.H., Schumann, G., Tarcsay, L., Schläfli, E., Dukor, P. 1978. Cell. Immunol. 37, 174-187.

53. Moore, R.N.Hoffeld, J.T., Farrar, J.J., Mergenhagen, S.E., Oppenheim, J.J., Shadduck, R.K. 1981. In: Lymphokines, Vol.3 (E. Pick, ed.). Academic Press, New York. pp.119-148.

54. Akagawa, K., Tokunaga, T. 1980. Microbiol. Immunol. 24, 1005-1011.

55. Parant, M., Vinit, M-A., Damais, C., Riveau, G., Chedid, L. 1985. Exp. Hematol. 13, 221-228.

56. Wood, D.D., Staruch, M.J. 1980. In: Biochemical Characterization of Lymphokines (A.L. DeWeck, F. Kristensen, M. Landy, eds.) Academic Press, New York. pp.423-425.

57. Staruch, M.J., Wood, D.D. 1983. J. Immunol. 130, 2191-2194.

58. Oppenheim, J.J., Togawa, A., Chedid, L., Mizel, S. 1980. Cell. Immunol. 50, 71-81.

59. Abeshira-Amar, O., Damais, C., Parant, M., Chedid, L. 1985. J. Immunol. 134, 365-368.

60. Iribe, H., Koga, T., Onoue, K. 1982. J. Immunol. 129,1029-1032.

61. Damais, C., Riveau, G., Parant, M., Gerota, J., Chedid, L. 1982. Int. J. Immunopharmacol. 4, 451-462.

62. Dinarello, C.A., Elin, R.J., Chedid, L., Wolff, S.M. 1978. J. Infect. Dis. 138, 760-767.

63. Riveau, G., Masek, K., Parant, M., Chedid, L. 1980. J. Exp. Med. 152, 869-877.

64. Krueger, J.M., Walter, J., Dinarello, C.A., Wolff, S.M., Chedid, L. 1984. Amer. J. Physiol. 246, R994-R999.

65. Krueger, J.M., Pappenheimer, J.R., Karnovsky, M.L. 1982. Proc. Natl. Acad. Sci. USA 79, 6102-6106.

66. Byars, N.E. 1984. Infect. Immun. 44, 344-350.

67. Tanaka, A., Nagao, S., Nagao, R., Kotani, S., Shiba, T., Kusumoto, S. 1979. Infect. Immun. 24, 302-307.

68. Waters, R.V., Ferraresi, R.W. 1980. J. Reticuloendothel. Soc. 28, 457-471.

69. Chedid, L., Parant, M., Riveau, G. 1982. In: Immunopharmacology and the Regulation of Leukocyte Function (D.R. Webb, ed.). Marcel Dekker, New York. pp.159-182.

70. Dinarello, C.A. 1984. Rev. Infect. Dis. 6, 51-95.

71. Riveau, G., Parant, M., Damais, C., Parant, F., Parant, F., Chedid, L. 1986. Amer. J. Physiol. (in press).

72. McAdam, K.P.W.J., Foss, N.T., Garcia, C., DeLellis, R., Chedid, L., Rees, R.J.W., Wolff, S.M. 1983. Infect. Immun. 39, 1147-1154.

73. Krueger, J.M., Pappenheimer, J.R., Karnovsky, M.L. 1982. J. Biol. Chem. 257, 1664-1669.

74. Wood, D.D., Staruch, M.J., Durette, P.L., Melvin, W.V., Graham, B.K. 1983. In: Interleukin, Lymphokine, and Cytokine (J.J. Oppenheim, S. Cohen, eds.). Academic Press, New York. pp. 691-696.

75. Fraser-Smith, E.B., Waters, R.V., Matthews, T.R. 1981. Infect. Immun. 34, 676-683.

76. Krahenbuhl, J.L., Sharma, S.D., Ferraresi, R.W., Remington, J.S. 1981. Infect. Immun. 31, 716-722.

77. Kierszenbaum, F., Ferraresi, R.W. 1979. Infect. Immun. 25, 273-278.

78. Fraser-Smith, E.B., Waters, R.V., Matthews, T.R. 1982. Infect. Immun. 35, 105-110.

79. Koff, W.C., Showalter, S.D., Hampar, B., Fidler, I.J. 1985. Science, 228, 495-497.

80. Fogler, W.E., Wade, R., Brundish, D.E., Fidler, I.J. 1985. J. Immunol. 135, 1372-1377.

81. Osada, Y., Otani, T., Une, T., Ogawa, H., Nomoto, K. 1982. J. Gen. Microbiol. 128, 2361-2370.

82. Bahr, G.M., Tello, D.J., Chedid, L. 1985. Infect. Immun. 49, 312-319.

83. Leclerc, C., Bahr, G.M., Chedid, L. 1984. Cell. Immunol. 86, 269-277.

84. Fogler, W.E., Fidler, I.J. 1984. In: Immune Modulation Agents and their Mechanisms (R.L. Fenichel, M.A. Chirigos, eds.). Immunology Series 25, Marcel Dekker, New York. pp.499-512.

85. Damais, C., Parant, M., Chedid, L. 1985. Int. J. Immunopharmacol. 7, 361.

86. Onozaki, K., Matsushima, K., Kleinerman, E.S., Saito, T., Oppenheim, J.J. 1985. J. Immunol. 135, 314-320.

87. Parant, M., Parant, F., Chedid, L. 1980. Infect. Immun. 28, 654-659.

IMMUNOMODULATION BY MURAMYLDIPEPTIDE (MDP)

H.Finger, C.H.Wirsing von Koenig
Institute for Hygiene and Laboratory Medicine,
Municipal Hospital, Krefeld, FRG

The most potent immunological adjuvants are derived from microorganisms, such as Mycobacteria or Bordetella pertussis. The minimal chemical structure capable of effectively replacing mycobacteria in FREUND's complete adjuvant has been identified as a muramyl-dipeptide, generally referred to as MDP. This substance, incorporated in an oil phase or in an aequous solution, was found to potentiate numerous cellular and humoral immune responses (1,2). It was active when administered orally or parenterally (1,3) and did not show the various side effects of FREUND's complete adjuvant (2,4). Thus, MDP appeared to fulfil every requirement for a human application. Furthermore, MDP can easily by altered by or substituted with other amino acids , sugars or lipids. Several hundred different preparations of MDP and its analogues have been tested for their immunomodulating activities during the recent years. On the basis of this large amount of experimental data (cf.5), the immunomodulating activities of MDP and other immunological adjuvants, such as killed B.pertussis organisms (PO), will be compared.

Table 1 summarizes the main characteristics of PO, a potent immunological adjuvant. PO increases the immune response to soluble protein antigens and to heterologous erythrocytes.

Table 1 : Main In-Vivo Characteristics of Potent Immunological Adjuvants such as killed Cells from Bordetella pertussis

1. Accelerated, increased and prolonged primary antibody response to soluble protein antigens and heterologous erythrocytes.
2. Capacity to increase the primary immune response to "saturated" antigen doses. No measurable effect with subimmunogenic antigen doses.

Biological Properties of Peptidoglycan

3. Increased development of IgG and IgM memory cells after effective priming.
4. Increased systemic anaphylactic sensitivity in mice and rats.
5. Induction of delayed type hypersensitivity reactions to protein antigens.
6. Altered non-specific resistance to viral and bacterial infections.
7. Prompt induction of experimental "allergic" diseases, when incorporated in an oily phase.
8. Immunosuppressive activity, when administered under special experimental conditions.

The additional injection of 2×10^9 PO together with the primary immunization of mice with a suboptimal dose of 5×10^6 sheep erythrocytes (SE) results in an accelerated, increased and prolonged development of direct (DPFC) and indirect plaque-forming cells (IPFC). Moreover, PO are also capable of increasing the immune response to maximal antigen doses. The PO-mediated multiplication of antibody-forming cells is not merely due to a shifting of plasma cells or their precursors from outside of the spleen, and this could be proven by the significantly increased serum antibody titers (6). However, PO are evidently not capable of initiating an immune response unless the dose of SE is immunogenic by itself. Furthermore, PO significantly increases the process of priming for the secondary response, due to an increased formation of 19S and 7S memory cells (cf.6).

In order to test for the adjuvant activity of MDP, NMRI mice were immunized subcutaneously with a suboptimal dose of 500 µg bovine serum albumin (BSA) either alone or together with MDP. Mice immunized with BSA only showed no detectable antibody formation on days 14 or 30 after primary immunization and only minimal BSA-antibody titers (mean titer 1:10) 5 days after the booster injection. In contrast, animals immunized with BSA and 100 µg MDP showed higher serum BSA-antibody titers (mean titer 1:100) 5 days after secondary immunization. No mouse of the group primarily immunized with BSA only died spontaneously during the period of observation. However, 5 of 12 mice in the group primarily immunized with MDP and BSA succumbed to systemic anaphylaxis shortly after the intravenous challenge with BSA. The surviving animals also exhibited transient signs of anaphylaxis (7). Similar data were obtained by a further experiment, displayed in table 2.

Table 2 : Development of susceptibility to systemic anaphylaxis after primary intraperitoneal immunization of NMRI mice with 4.0 mg of BSA and simultaneous injection of different adjuvants. Challenge was performed on day 20 by an intravenous injection of 100 µg BSA.
20 mice were tested in each group.

Adjuvant added	% of mice with lethal anaphylaxis after challenge
None	0 %
2 x $1o^9$ PO	100 %
100 µg MDP	40 %
100 µg PG	25 %
50 µg ET 1	0 %
50 µg ET 2	0 %

Abbreviations:
PO : killed B.pertussis organisms
MDP: Muramyl-dipeptide
PG : purified peptidoglycan from B.pertussis
ET 1: Lipopolysaccharide from B.pertussis
ET 2: Lipopolysaccharide from Salmonella typhi

Furthermore, the sera of mice primed with a mixture of BSA and MDP contained significantly higher levels of IgG_1 five days after the booster injection as compared to the corresponding controls (Fig.1). Since IgG_1 antibodies are known to mediate anaphylactic reactions in mice, MDP, by increasing the production of such antibodies, may produce anaphylactic hypersensitivity. Conflicting results were obtained regarding the influence of MDP on the production of IgE antibodies. Whilst some authors found increased IgE responses to ovalbumin in the murine host, other reports indicated that the conjugation of MDP to the same antigen would selectively suppress the IgE response, and this could be employed as a possible therapy in human allergic diseases (cf.5).

Fig. 1

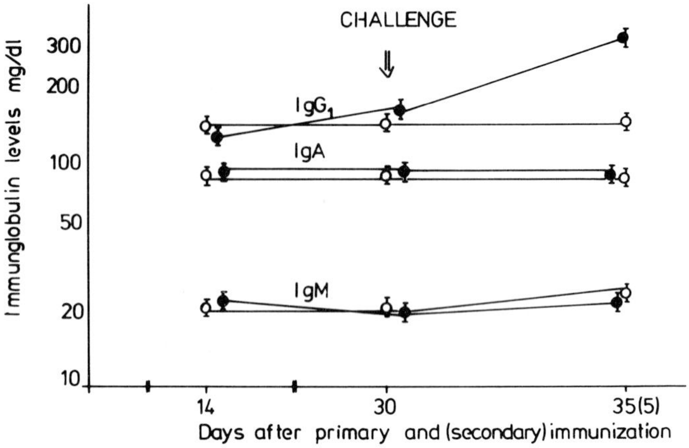

Effect of MDP on the serum immunoglobulin concentration during primary and secondary immune response of NMRI mice to BSA.

o = primary immunization with 500 µg BSA subcutaneously

● = primary immunization with 500 µg BSA and 100 µg MDP injected together subcutaneously.

The animals were challenged on day 30 by an i.v.injection of 100 µg BSA. 6 mice were used per point.

The effect of MDP on the immune response of NMRI mice to SE is displayed in figure 2. No significant differences between both groups were found during the primary response. After the booster injection, the antibody producing spleen cells of the group primarily immunized with MDP and SE significantly exceeded those of the control group only on day 4 after the secondary immunization.

Fig.2

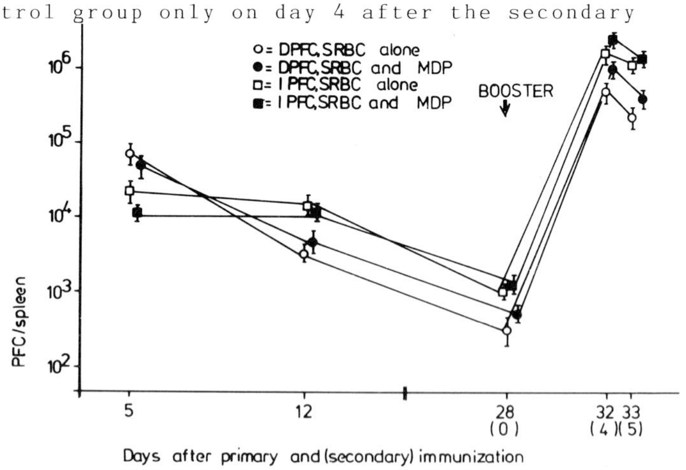

Effect of MDP on the development of DPFC and IPFC in primary and secondary immune response of NMRI mice to SE. 6 mice per point.

Using guinea pigs as test animals, it has been repeatedly reported that MDP injected in a water-in-oil emulsion is able to produce delayed type hypersensitivity recations. The same holds true for the production of experimental "allergic" diseases, such as encephalomyelitis, experimental "allergic" aspermatogenic orchiepididymitis and experimental "allergic" uveo-retinitis. In these experimental models, however, MDP was found to be significantly less active than mycobacteria or a water-soluble extract of mycobacteria (cf.5).

In analogy to the action of other bacterial adjuvants, MDP and several of its structural analogues were tested for their ability to increase resistance to different infections. MDP was found to be effective in mediating a non-specific resistance to Klebsiella pneumoniae infection in mice. Contrasting all other immunological adjuvants, however, MDP displayed its activity only when administered with or very shortly before the infection. Numerous experiments employing a variety of infectious agents have been performed and positive and negative effects have been described. In some experiments, MDP was less effective than its analogues, whereas in other experiments no differences could be found (cf.5). It is generally accepted, however, that MDP and its analogues are unable of increasing the resistance to an infection with Listeria monocytogenes (8). A comparison between the effects of MDP and PO on a listeric infection is shown in figure 3.

Fig. 3

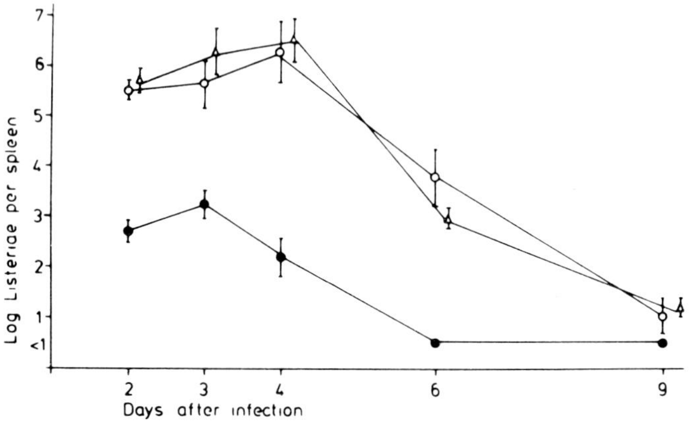

Legend to fig.3: Influence of $3x10^9$PO (●) or 500 µg MDP (△) on listeric infection, when given i.p. 5 days before the i.v. infection with $3.6x10^3$ viable Listeriae.(○) = Controls. Five mice were used per point

---

While the injection of PO five days before the infection has an impressive effect, the application of 500 µg MDP was completely ineffective. Similarly, a blockage of the mononuclear system by dextran sulfate 500 could not be overcome by pretreatment with MDP as again in contrast to the action of PO (cf.5).

There is now increasing evidence that all immunologic adjuvants produce immunosuppressive effects under special experimental conditions. Thus, it is not surprising that a limited suppression of the antibody response to different antigens was also observed after treatment with MDP (cf.5).

Evaluating the numerous reports regarding the adjuvant activity of MDP and its analogues, it could be suggested that the stimulatory effect of this substance on the immune reactions is considerably less pronounced than the responses observed after treatment with established immunoadjuvants, such as mycobacteria or PO.

Acknowledgement

This work was supported by the Schutzkommission beim Bundesminister des Innern.

References

1. Chedid,L.,F.Audibert,P.Lefrancier,J.Choay,E.Lederer.1976 Proc.Nat.Acad.Sci.USA 73: 2472

2. Chedid,L.,F.Audibert.1977. in: Immunopathology (P.A.Miescher,Ed.). Schwabe,Basel-Stuttgart, p.382

3. Audibert,F.,L.Chedid,P.Lefrancier,J.Choay,E.Lederer.1977. Ann.Immunol.(Inst.Pasteur) 128 C: 653

4. Audibert,F.,L.Chedid,P.Lefrancier,J.Choay.1976. Cell.Immunol. 21: 243

5. Finger,H.,C.H.Wirsing von König, in: Modulation of the immune response by microorganisms (J.Zschiesche,Ed.). VEB G.Fischer, Leipzig, in press

6. Finger,H.,1975 in:The immune system and infectious diseases (E.Neter and F.Milgrom,Eds.).S.Karger,Basel, p.132

7. Heymer,B.,H.Finger,C.H.Wirsing.1978. Z.Immunitätsforsch.Klin.Immunol. 155: 87

8. Finger,H.,C.H.Wirsing von König. 1980. Infection and Immunity 27: 288

DISACCHARIDE MDP ANALOGS. PREPARATION AND SOME BIOLOGICAL PROPERTIES

M. Zaoral, J. Farkaš, M. Ledvina, J. Ježek
Institute of Organic Chemistry and Biochemistry, Czechoslovak Academy of Sciences, 166 10 Prague 6, Czechoslovakia

J. Rotta, M. Rýc
Institute of Hygiene and Epidemiology, 100 42 Prague 4, Czechoslovakia

Introduction

Three reasons focused our attention on MDP analogs with an oligosaccharide instead of monosaccharide moiety and modified also in the peptide part: 1) MDP has strong immunoadjuvant, pyrogenic and distinct cancerostatic activity (1); 2) MDP analog containing normuramic instead of muramic acid and α-aminobutyric acid instead of alanine in position 1 of the peptide part has strong immunoadjuvant, decreased pyrogenic and strong cancerostatic activity (2); 3) MDP analogs containing an oligosaccharide instead of monosaccharide residue have, as a rule, stronger immunoadjuvant and higher cancerostatic activity than their monosaccharide counterparts (3). Often they have decreased pyrogenic activity. To achieve an improvement of the parameters of the biological effects we decided to incorporate the structural changes mentioned sub 1 - 3) into a single molecule.

Results

For the preparation of the desired compounds we adapted (Fig. 1) the synthetic scheme of Durette, Meitzner and Shen (4). The original

Biological Properties of Peptidoglycan

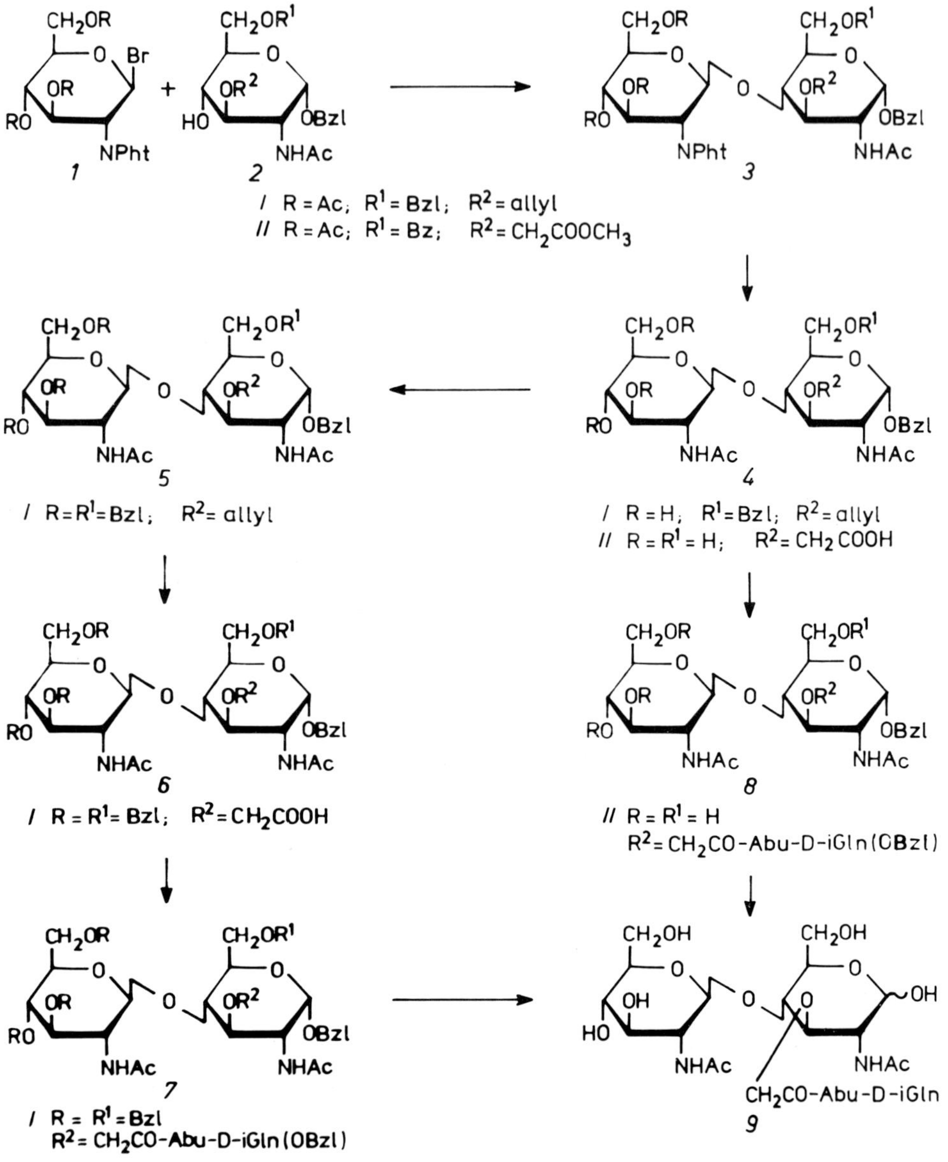

Fig. 1
Scheme of the synthesis of N-acetylglucosaminyl-(ß1-4)-N-acetyl-normuramyl-α-aminobutyryl-D-isoglutamine benzylglycoside benzyl-ester 8 and of N-acetylglucosaminyl-(ß1-4)-N-acetylnormuramyl-α-aminobutyryl-D-isoglutamine 9.

scheme has two evident weak points. One is just in the "silver-assisted" glycosylation, in the preparation of the disaccharide derivative 3 from the corresponding monomeric units 1 and 2. The coupling reaction gives low yield (38%) and it decreases considerably the preparative value of the scheme. The second weak point is the benzylation (5) of compound 4 to yield 5 (route I). The reaction is slow, it gives rise to a considerable quantity of side products, the purification of the reaction mixture is complicated and the yield is unsatisfactory. We succeeded in removing the weak points. The first one by carrying out the coupling reaction in the absence of base (6), the second by irradiating the reaction mixture by ultrasound. Thus we increased the yield of the coupling reaction up to 90% and of the benzylation reaction up to 71% (pure product). However, the synthetic scheme has still a weak point. It is the removal of the allyl protecting group from compound 5 which requires two-stage reaction and a relatively expensive reagent [tris(triphenylphosphine)rhodium(I)chloride]. For that reason we have worked out an alternative synthesis (route II) using as a passive component a normuramic acid instead of glucosamine derivative in the glycosylation reaction. The yield of the glycosylation drops to about 60% but this is well counterbalanced by considerable simplification of the synthesis. So far we have prepared, and investigated in some detail, two compounds. One is the disaccharide-dipeptide 9, the second its partially protected precursor, the benzylglycoside benzyl ester 8. The benzylglycoside benzyl ester has very strong immunoadjuvant and no pyrogenic effect (Table 1). The free disaccharide-dipeptide 9 has also a very strong immunoadjuvant effect. In contrast to compound 8 it has a clean-cut pyrogenic effect but which is considerably lower than that of MDP.

Table 1

Biological Activities of Disaccharide-dipeptides

Compound	Activity	
	Immunoadjuvant	Pyrogenic
Benzylglycoside benzyl ester 8	++++	0
Disaccharide-dipeptide 9	+++	+
MDP	++	++

References

1. Adam, A., E. Lederer. 1984. Medicinal Research Reviews. 4, 111.

2. McLaughlin, Ch.A., S.M. Schwartzman, B.L. Horner, G.H. Jones, J.G. Moffatt, J.J. Nestor, Jr., D. Tegg. 1980. Science 208, 415.

3. Rostovceva, L.I., T.M. Andropova, V.P. Malkova, I.B. Sorokina, V.T. Ivanov. 1981. Bioorgan. Chimija 7, 1843.

4. Durette, P.L., E. Meitzner, T.Y. Shen. 1979. Carbohyd. Res. 77, C1.

5. Harrison, R., H.G. Fletcher. 1965. J. Org. Chem. 30, 2317.

6. Garegg, P.J., T. Norberg. 1979. Acta Chem. Scand. 33B, 116.

BIOLOGICAL ACTIVITY OF SYNTHETIC DISACCHARIDE-PEPTIDE ANALOGUES OF PEPTIDOGLYCAN

M. Rýc, J. Rotta, R. Straka
Institute of Hygiene and Epidemiology, Prague, Czechoslovakia

M. Zaoral, J. Farkaš, M. Ledvina, J. Ježek
Institute of Organic Chemistry and Biochemistry, Prague, Czechoslovakia

J. Pekárek
Institute of Sera and Vaccines, Prague, Czechoslovakia

Introduction

Chemically well-defined synthetic compounds able to amplify both specific and nonspecific immune response and not exhibiting toxic properties have been under investigation for many years. Since the discovery of muramyldipeptide (MDP) as the minimal peptidoglycan subunit with pronounced immunostimulating activity (1,2), hundreds of compounds analogous to or derived from peptidoglycan subunit structure have been prepared and tested. MDP exhibits some toxic activities known to be exerted by the whole, natural peptidoglycan molecule (3). This refers for example to its ability to elicit fever in rabbits and to damage their blood platelets (4). Apyrogenic, adjuvantly active analogues of MDP have been prepared by the replacement of the D-isoglutaminyl residue by D-glutamine alkyl esters (5). Some investigators have found that a chemically synthetized disaccharide dipeptide (GMDP) exhibits a higher immunoadjuvant activity than MDP (6). In the present paper, two distinct toxic properties, namely pyrogenicity in rabbits and platelet-damaging activity, were correlated with the immunostimulating potency of several synthetic analogues of MDP that differ from MDP in sugar moiety. The immunoadjuvant activity of these materials was assayed in terms of delayed-type hypersensitivity to ovalbumin and to antigen extracts from an M-positive strain of

Biological Properties of Peptidoglycan

Group A streptococci. Experimental allergic encephalitis of guinea pigs was used as another criterion of the immunostimulating activity of the compounds tested.

Results

Pyrogenicities were measured in rabbits injected intravenously with 40, 200 or 1000 nM of synthetic material in 1 ml of saline (Figs. 1 and 2). Each response curve in the Figures is an average for 3 rabbits measured rectally at the indicated times. Glycodipeptide (MDP) elicited a distinct febrile response (Fig. 1). Dipeptidyldisaccharide was less, but also definitely pyrogenic, whereas dipeptidylglucosamine and disaccharide dipeptide were non-pyrogenic. The chemical formulae are also indicated in the Fig. 1. It should be pointed out that in dipeptidyldisaccharide and dipeptidylglucosamine, the glucosamine was linked to D-isoglutamine and that the compound called disaccharide dipeptide contained - as compared to MDP - GlcNAc, L-aminobutyric acid instead of L-alanine, and two benzyl groups.

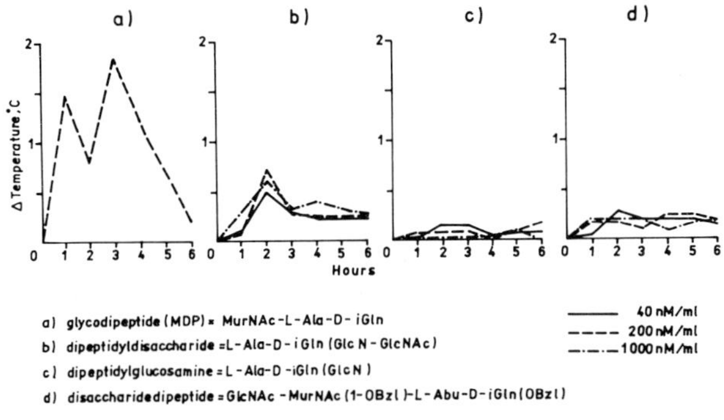

Fig. 1. Febrile response of rabbits to (a) glycodipeptide, (b) dipeptidyldisaccharide, (c) dipeptidylglucosamine, (d) disaccharide-dipeptide. Each curve shows average values of 3 animals.

Pyrogenic activities of some other MDP analogues with different changes in the sugar moiety, namely (a) material in which the lactyl group in N-acetylmuramic acid was replaced by an glycolyl group and the L-alanine by L-aminobutyric acid, (b) a compound with L-diamino-butyric acid instead of L-alanine, (c) a galacto-MDP with galactosamine instead of glucosamine, are demonstrated in Fig. 2. A fourth compound (d) was N-acetylchitotetrose composed of four subunits of N-acetylglucosamine; this material was weakly pyrogenic, as were also N-acetylchitotriose and N-acetylchitobiose (not shown in the Fig. 2).

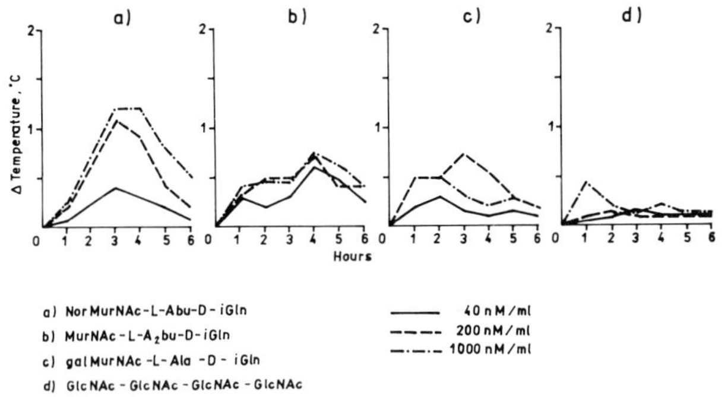

Fig. 2. Febrile response of rabbits to MDP analogues with different changes in the sugar moiety.

The platelet-damaging effect of these materials was studied after in vitro incubation of isolated rabbit platelets with 100-500 μg of analogue per 1 ml. The study was done on ultrathin sections in the electron microscope. Three different ultrastructural patterns were observed: a) no change, b) a degranulation of platelets consisting in the loss of granules containing biogenic amines, and c) a complete lysis. The results are summarized in the Table 1. MDP caused complete lysis; the lytic reaction was dose-dependent, but the doses of MDP able to cause complete lysis of platelets were, in comparison with natural peptidoglycans, very high (up to 500 μg/ml). Only the disaccharide dipeptide and the nor-MDP derivative showed no effect on rabbit platelets; however, the latter substance was pyrogenic.

The influence of MDP-analogues on the delayed-type hypersensitivity reaction to ovalbumin was tested in guinea pigs. A commercial product of ovalbumin as antigen, in a dose of 200 μg, was mixed with 400 nM of test material in a volume of 0.2 ml for each animal. In this solution an equal volume of 0.2 ml of Freund's incomplete adjuvant was emulsified; a volume of 0.4 ml of the total mixture was administered per animal, 0.2 ml being injected into each hind

Table 1. Platelet Damaging Effect of Some Synthetic MDP Analogues

Material	Effect
MurNAc-L-Ala-D-iGln (= MDP)	Lysis
L-Ala-D-iGln(GlcN-GlcNAc)	Degranulation
L-Ala-D-iGln(GlcN)	Degranulation
GlcNAc-MurNAc(1-OBzl)-L-Abu-D-iGln(OBzl)	No effect
norMurNAc-L-Abu-D-iGln	No effect
galMurNAc-L-Ala-D-iGln	Degranulation

footpad. Fourteen days after this sensibilization, the animals were injected intradermally with two concentrations (50 and 250 /ug) of ovalbumin at two sites of the back. The skin indurations were measured with a calliper after 24 hrs. Of the materials tested, only disaccharide dipeptide exhibited a higher immunostimulation than MDP; dipeptidyldisaccharide and dipeptidylglucosamine were less active (Fig. 3).

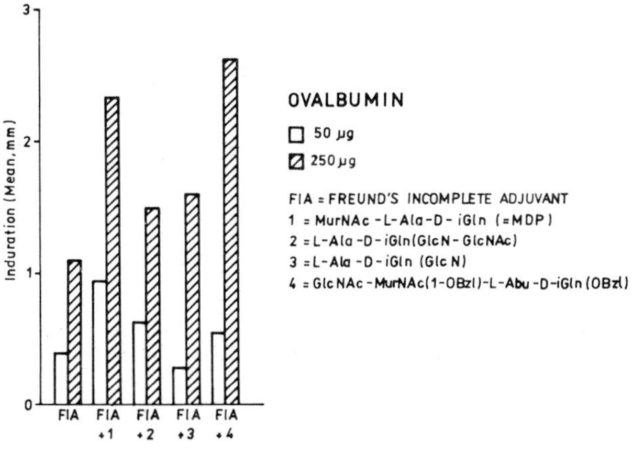

Fig. 3. Delayed-type hypersensitivity reaction to ovalbumin induced by some synthetic MDP analogues.

Still more pronounced results were obtained if a weaker antigen was used. This antigen was hydrochloric acid extract made at pH 2.5 by 10 min extraction at 100° C of an M-positive Group A streptococcus type 24 strain. For the sensibilization, 500 /ug of this antigen was used, and the skin reactions were measured after the administration of 100 μg and 400 μg of the antigen. Again, only disaccharide dipeptide exhibited an immunostimulating activity comparable to that of MDP (Fig. 4).

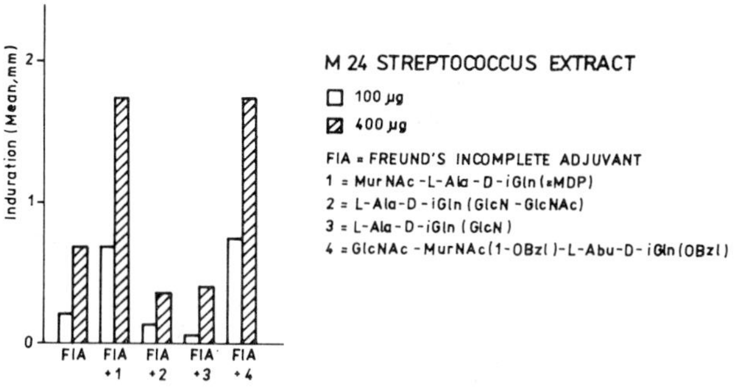

Fig. 4. Delayed-type hypersensitivity reaction to M 24 streptococcus extract induced by some synthetic MDP analogues.

In the last series of experiments, the immunomodulatory activity of selected analogues was tested in experimental allergic encephalitis. Guinea pigs were injected into footpads with rabbit encephalitogen (0.4 mg) mixed with one of the analogues or Freund's complete adjuvant (FCA). The Table 2 shows that - as compared with MDP or FCA - only disaccharide dipeptide and the nor-MDP derivative were active. However, the latter compound was also pyrogenic. All the other materials tested were ineffective, including the apyrogenic N-acetylchitotetrose.

Table 2. Experimental Allergic Encephalitis

Adjuvant Material	No. of Animals In Experiment	No. of Animals With Paresis
Freund's complete adjuvant	10	8
MurNAc-L-Ala-D-iGln (=MDP)	10	6
L-Ala-D-iGln(GlcN-GlcNAc)	10	0
L-Ala-D-iGln(GlcN)	10	0
GlcNAc-MurNAc(1-OBzl)-L-Abu-D-iGln(OBzl)	10	6
norMurNAc-L-Abu-D-iGln	10	8
galMurNAc-L-Ala-D-iGln	10	0

Our results are summarized in Table 3. It is evident that of all the materials examined only disaccharide dipeptide exhibited, in the tests performed, "desirable" features, i.e. being nontoxic it produced high immunostimulation. This material thus seems to be promising for further study as a nontoxic immunostimulating compound. It will be necessary, of course, to subject it to further immunological investigations before a final evaluation can be made.

Table 3. Comparison of Toxic and Immunostimulating Effects of Some Synthetic MDP Analogues

Material	Pyrogenicity	Platelets Damage	DTH[a)]	EAE[b)]
MurNAc-L-Ala-D-iGln (=MDP)	+++	++	++	++
L-Ala-D-iGln(GlcN-GlcNAc)	+	+	±	-
L-Ala-D-iGln(GlcN)	-	+	±	-
GlcNAc-MurNAc(1-OBzl)--L-Abu-D-iGln(OBzl)	-	-	++	++
norMurNAc-L-Abu-D-iGln	++	-	++	++

a) delayed-type hypersensitivity
b) experimental allergic encephalitis

References

1. Ellouz, F., A. Adam, R. Ciorbaru, E. Lederer. 1974. Biochem. Biophys. Res. Commun. 59, 1317.

2. Kotani, S., Y. Watanabe, F. Kinoshita, T. Shimono, I. Morizati, T. Shiba, S. Kusumoto, Y. Tarumi, K. Ikenaka. 1975. Biken J. 18, 105.

3. Kotani, S., Y. Watanabe, T. Shimono, K. Harada, T. Shiba, S. Kusumoto, K. Yokogawa, M. Taniguchi. 1976. Biken J. 19, 9.

4. Rotta, J., M. Rýc, K. Mašek, M. Zaoral. 1979. Exp. Cell Biol. 47, 258.

5. Lefrancier, P., M. Derrien, X. Jamet, J. Choay, E. Lederer, F. Audibert, M. Parant, F. Parant, L. Chedid. 1982. J. Med. Chem. 25, 87.

6. Tsujimoto, M., F. Kinoshita, T. Okunnaga, S. Kotani, S. Kusumoto, K. Yamamoto, J. Shiba. 1979. Microbiol. Immunol. 23, 933.

ENZYMATIC OBTENTION OF BIOLOGICALLY ACTIVE GLYCOPEPTIDES FROM *ACTINOMADURA* R 39.

M. Guinand, P. Morel, M.J. Vacheron, G. Michel

Laboratoire de Biochimie Microbienne, Université Claude Bernard, Lyon I, 69622 Villeurbanne, Cedex, France.

P. Dupassieu, D. Yavordios

Laboratoires de Thérapeutique Moderne
01400 Châtillon-sur-Chalaronne, France.

Introduction

The disaccharide-dipeptide, *N*-acetyl-β-D-glucosaminyl-(1→4)-*N*-acetylmuramyl-L-alanyl-D-isoglutamine, or GMDP, has been chemically synthesized (1,2). It was found to possess a higher immunoadjuvant activity in comparison to that of MDP (*N*-acetylmuramyl-L-alanyl-D-isoglutamine) (3). Recently GMDP has been obtained by an enzymatic degradation of the peptidoglycan of *Actinomadura* R 39 (4). The structure of the peptidoglycan was demonstrated in (5) and is given in Fig. 1. The peptidoglycan was successively hydrolyzed in [1] by the lysozyme, in [2] by the DD-carboxypeptidase G from *Streptomyces albus* G and in [3] by the γ-D-glutamyl-*meso*-diaminopimelate endopeptidase I from *Bacillus sphaericus* 9602. The enzymes were used in their soluble form.

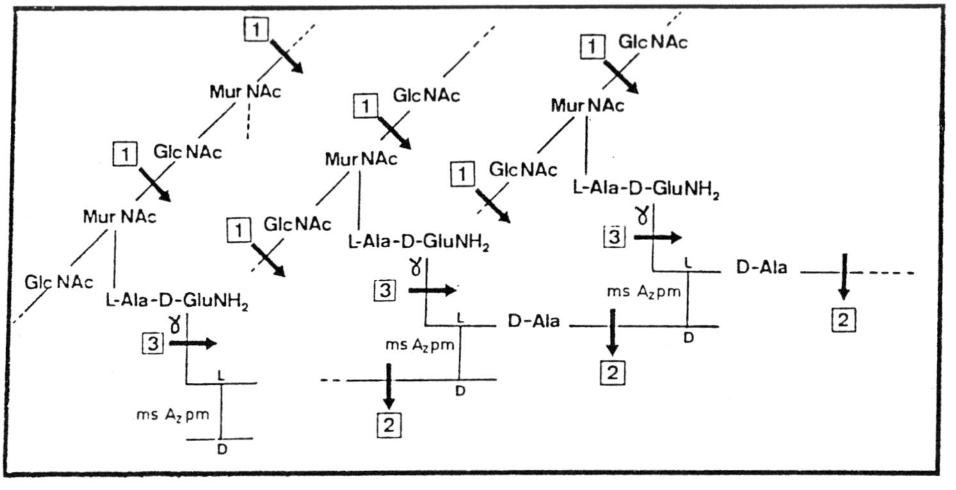

Fig. 1 : Peptidoglycan of *Actinomadura* R 39 (from M. Guinand *et al.* 4).

Biological Properties of Peptidoglycan

We managed to improve the method by immobilization of two of the enzymes, DD-carboxypeptidase G and endopeptidase I, and we produced substantial amounts of GMDP and of related glycopeptides whose chemical synthesis is difficult if not impossible (submitted for publication).
A survey of the method is now reported as well as a preliminary study of the biological properties of the glycopeptides.

Immobilization of DD-carboxypeptidase G and properties of the bioreactor.

The DD-carboxypeptidase G (EC 3.4.17.8) was purified as described in (6) and 12 units of enzyme were obtained from 25 liters of *Streptomyces albus* G. They were immobilized on Ultrogel AcA 22 treated with glutaraldehyde as recommended in (7); two reactors were prepared : the first one was saturated in enzyme with 0.9 unit of DD-carboxypeptidase G linked to 1 ml of gel and it was used to study the properties of the immobilized enzyme. The second reactor was made with 11.1 units of DD-carboxypeptidase G immobilized on 60 ml of gel ; it was packed in a column (2.6 x 12 cm) and it was used for the digestion of the peptidoglycan.
The properties of the saturated gel were studied with Ac_2-L-Lys-D-Ala-D-Ala as substrate. The heat stability of immobilized DD-carboxypeptidase G was examined in comparison with that of the free enzyme. At the inactivation temperature, where the remaining activity is half of the original activity, the immobilized enzyme is more stable than the free enzyme by 10°C. It was also found that the immobilized enzyme retained roughly its initial activity even after one year.
The kinetic constant, K'_m, of the immobilized enzyme was studied at 37°C in a small packed bed reactor of 0.3 ml. Various S_o concentrations of Ac_2-L-Lys-D-Ala-D-Ala were pulsed through the column at various flow rates and samples were taken out for enzymatic assays of released D-Ala (8) until the maximum substrate conversion rate was obtained. K'_m has been determined according to the Michaelis-Menten equation : $P \cdot S_o - K'_m \cdot \mathrm{Log}(1-P) = V_m / V_s$ (9). S_o is the initial concentration of substrate and P the maximum substrate conversion rate. The different $P \cdot S_o$ datas were plotted versus $-\mathrm{Log}(1-P)$. Straight and parallel lines were obtained at the different space velocities and a K'_m of 2 mM was calculated from the slopes. It is only 6 times greater than that for the free enzyme (10). Thus the properties of the immobilized DD-carboxypeptidase G (stability towards heat and time as well as kinetic constants) are all suitable to its permanent use as a bioreactor, at any flow rates experimented.

Utilization of the immobilized DD-carboxypeptidase G

Over a culture of 70 l of *Actinomadura* R 39, 7 g of cell walls were obtained as described in (4). They were solubilized by the lysozyme by portion of 1 g and the solutions were separately pulsed and recycled through the reactor thermostated at 37°C at a flow rate of 6 $cm.h^{-1}$. Samples have been analysed for amino terminal groups (11) until the completion of the reaction. The products were then eluted and separated through Sephadex G50-G25 columns according to a classical procedure. A disaccharide-peptide monomer (K_{av} 0.60) and a disaccharide-peptide dimer (K_{av} 0.42) were subsequently desalted on Trisacryl GF 05. From the 7 g of walls the yield was 750 mg of monomer and 250 mg of dimer. The monomer or GMTP : N-acetyl-β-D-glucosaminyl-(1→4)-N-acetylmuramyl-L-alanyl-D-isoglutaminyl-(L) *meso*-diaminopimelyl(L)-(D-alanine) was shown to be a mixture of disaccharide-tetrapeptide (87%) and disaccharide-tripeptide (13%) ; the dimer, di-GMTP, possess an osidic linkage (MurNAc→GlcNac) between two moles of GMTP.

Immobilization of endopeptidase I and properties of the immobilized enzyme

The endopeptidase I was obtained from 100 liters of *Bacillus sphaericus*. The enzyme was extracted from the sporulation medium and from the spores and was purified as described in (12). Five units of enzyme were obtained. We previously demonstrated that endopeptidase I was firmly bound to phenyl-Sepharose by hydrophobic interactions (12) ; the purified enzyme solution was then simply loaded at pH 8 on a phenyl-Sepharose column (2.6 x 12 cm) and the properties of the gel were studied with GMTP as substrate. The conversion of GMTP into GMDP was analysed by quantitative reversed phase HPLC in an ammonium phosphate buffer at pH 5.5. The inactivation temperature which is already very high, 80°C, for the free enzyme is maintained but not enhanced by immobilization. The time stability is constant over several months. A K'_m of 2.5 mM was determined as described above in a small packed bed reactor of 0.3 ml. It is smaller than that (4.34 mM) for the free enzyme. This can be due to the fact that endopeptidase I is naturally particulate and acts better in its immobilized form than in its soluble form. So, the immobilization of endopeptidase I is particularly compatible for its permanent use.

Production of glyco-dipeptides with immobilized endopeptidase I

The compounds GMTP (256 mg) or di-GMTP (151 mg) were pulsed and recycled through the reactor of endopeptidase I at a flow rate of 6 $cm.h^{-1}$. Aliquots were removed and analysed by HPLC until the completion of the reaction. The purification of the resulting compound GMDP or di-GMDP was achieved by chromatography in water on a Dowex-50 (H^+) column as described previously (4). The yield was 136 mg of GMDP and 95 mg of di-GMDP.

Biological activities

In the present paper two assays are only reported : the phagocytic activity of macrophages (Table 1) and the antibody production in mice (Table 2).

Table 1 Effect of Glycopeptides on Carbon Clearance by Reticuloendothelial System of Mice*.

Material tested	Amount mg / kg	Phagocytic index(.10^{-4})	T95 SEM (.10^{-4})	Statistical significance
Control		272	38	
MDP	200	378	115	NS
MDP	2.000	579	109	S
Control		293	117	
MDP	200	410	77	NS
MDP	2.000	715	88	S
Control		272	38	
GMDP	200	405	59	S
GMDP	2.000	459	92	S
Control		293	117	
GMDP	200	443	67	S
GMDP	2.000	505	100	S
Control		297	51	
GMDP	200	213	74	NS
GMDP	2.000	436	110	S
Control		360	80	
GMTP	200	291	90	NS
GMTP	2.000	436	120	NS
Control		293	117	
di-GMDP	200	531	90	S
di-GMDP	2.000	357	99	NS
Control		297	51	
di-GMDP	200	297	51	NS
di-GMDP	2.000	353	74	NS
Control		360	80	
di-GMTP	200	216	40	S
di-GMTP	2.000	279	40	NS

* - The amount of injected carbon is 80 mg/kg
- Each assay was performed on 10 male mice (24 to 26 g).

The phagocytic activity of macrophages was assayed by measuring the rate of clearance of colloïdal carbon by the reticuloendothelial system according to the method of Biozzi *et al.* (13). The phagocytic index K was calculated and K was statistically analysed using Student's test. The compound GMDP seemed to cause a significant increase in the phagocytic activity of macrophages as compared with the injection of MDP.

The antibody production was studied with the plaque-forming-cell (PFC) antibody response of mice and the results were statistically analysed using Student's test. They are presented in Table 2. Only the compound di-GMDP seemed to be an activator of the antibody production.

Table 2 Plaque-Forming-Cell Antibody Response of Mice to the Glycopeptides*

Compound	Amount µg/kg	Number of plaques/spleen (mean)	795 SEM	Statistical significance
Control		71.404	17.762	
MDP	500	120.301	21.697	S
MDP	5.000	61.545	13.639	NS
Control		89.036	36.139	
MDP	50	95.129	28.218	NS
MDP	500	93.563	36.860	NS
MDP	5.000	99.439	38.068	NS
Control		73.131	22.759	
GMDP	50	125.089	32.689	S
GMDP	500	89.684	37.213	NS
GMDP	5.000	101.877	29.581	NS
Control		71.404	17.762	
GMDP	500	120.745	19.657	S
GMDP	5.000	74.890	14.111	NS
Control		64.611	16.205	
GMTP	50	101.853	26.960	S
GMTP	500	94.451	35.514	NS
GMTP	5.000	80.473	38.208	NS
Control		66.562	24.406	
di-GMDP	50	123.878	41.838	S
di-GMDP	500	127.228	37.310	S
di-GMDP	5.000	151.911	34.950	S
Control		64.611	16.205	
di-GMTP	50	70.478	32.737	NS
di-GMTP	500	107.678	33.158	S
di-GMTP	5.000	122.592	45.113	NS

*Each assay was performed on 10 female mice (20 to 22 g)

In conclusion, we managed to perfect a system of bioreactors. It can be used for the production of large amounts of various glycopeptides from bacterial cell-walls. Preliminary assays have shown the interest of those glycopeptides which possess various biological activities.

Acknowledgement

Support for research grants from CNRS (UA 528) is gratefully acknowledged.

References

1. Kusumoto, S., Yamamoto, K., Shiba, T. (1978) Tetrahedron Lett.45,4407-4410.

2. Durette, P.L., Meitzner, E.P., Shen, T.Y. (1979) Carbohydr.Res.77, C1-C4.

3. Tsujimoto, M. Kinoshita, F. Okunnaga, T., Kotani, S., Kusumoto, S., Yamamoto, K., Shiba, J. (1979) Microbiol. Immunol. 23, 933-936.

4. Guinand, M., Françon, A.,Vacheron, M.J., Michel G., (1984), Eur. J.Biochem., 143, 359-362.

5. Ghuysen, J.M., Leyh-Bouille, M., Campbell, J.M., Moreno, R., Frère, J.M., Duez, C., Nieto, M., Perkins, H.R. (1973) Biochemistry, 12, 1243-1251.

6. Duez, C., Frère, J.M., Geurts, F., Ghuysen, J.M., Dierickx, L., Delcambe, L., (1978), Biochem. J., 175, 793-800.

7. Weston, P.D., Avrameas, S., (1971), Biochem., Biophys. Res. Commun. 45, 1574-1580.

8. De Cohen, J., Lamotte-Brasseur, J., Ghuysen, J.M., Frère, J.M., Perkins, H.R., (1981), Eur. J. Biochem., 121, 221-232.

9. Tosa,T., Mori,T.,Chibata, I., (1971), J. Ferment, Technol., 49, 522-528.

10. Leyh-Bouille, M., Ghuysen, J.M., Bonaly, R., Nieto, M., Perkins, H.R., Schleifer, K.H., Kandler, O., (1970), Biochemistry, 9, 2961-2970.

11. Ghuysen, J.M., Tipper, D.J., Strominger, J.L., (1966). In Methods in Enzymology, (Neufeld, E.T. and Ginsburg V. ed.) Vol. 8, pp.685-699, Academic Press, New-York.

12. Garnier, M., Vacheron, M.J., Guinand, M., Michel, G., (1985), Eur. J. Biochem., 148, 539-543.

13. Biozzi, G., Bennacerraf, B., Halpern, B.N. (1953) Brit. J. Exp. Pathol., 34, 441-457.

ADJUVANT ACTIVE PEPTIDOGLYCANS INDUCE THE SECRETION OF A CYTOTOXIC FACTOR BY MACROPHAGES

F. Vacheron, M. Guenounou, C. Nauciel

Laboratoire de Microbiologie, Faculté de Médecine de Paris-Ouest, 92380 Garches, France

Introduction.

Previous studies have shown that peptidoglycans (PG) exert an adjuvant effect on immune responses (1) and can stimulate nonspecific resistance to tumor growth in rat (2) and mouse (3). It has also been observed that there is a relationship between the structure of PG and their immunomodulating properties (3, 4).The aim of the present study was to analyse the role of macrophages in the PG-induced resistance to tumor growth.

Materials and Methods

Activating agents.

PG from Bacillus megaterium, Staphylococcus aureus, Micrococcus lysodeikticus and Corynebacterium poinsettiae were purified as previously described (4). Peptide subunit structures are reported in Table 1. Lipopolysaccharide (LPS) from Salmonella typhimurium was purchased from Difco Laboratories.

Macrophage cytotoxic activity.

Peritoneal exsudate cells (PEC) were harvested from 7 to 10 weeks old DBA/2 mice, 3 days after the intra-peritoneal injection of 2 ml of thioglycollate broth. PEC were washed with Hanks balanced solution (HBSS) adjusted at 1.25 x 10^6/ml in tissue culture medium (TCM) consisting of RPMI 1640 supplemented with 2 mM glutamine, antibiotics and 10 % foetal calf serum (FCS). A volume of 0.2 ml of cell suspension was distributed in 96-well flat-bottomed microtiter plates. After a 2 h incubation at 37°C, nonadherent cells were removed by 3 washes with HBSS. P815 mastocytoma cells, maintained by serial passages in the peritoneal cavity of DBA/2 mice, were washed in HBSS and adjusted at 1.25 x 10^5/ml in TCM. Then 0.2 ml of this cell suspension was added to the macrophage monolayers. Cultures were incubated at 37°C in a 5 % CO_2 atmosphere for 24 h. Tritiated thymidine (^{3}H-thymidine, 0.2 µCi per well) was added for the final 4 h of incubation and P815 cells were collected with an automatic harvester. Incorporated radioactivity was measured by liquid scintillation spectrophotometry. Each determination was carried out in triplicate.

Biological Properties of Peptidoglycan

Table 1. Peptide Subunit Structure of PG

Bacterium	Peptide sequence
B. megaterium	L-Ala-D-Glu-m-A_2pm-D-Ala
S. aureus	L-Ala-D-Glu(NH_2)-L-Lys-D-Ala
M. lysodeikticus	L-Ala-D-Glu(Gly)-L-Lys-D-Ala
C. poinsettiae	Gly-D-Glu-L-Hsr-D-Ala

Cytotoxic factor production.
PEC adjusted at 0.5 x 10^6/ml in TCM were plated in 16-mm wells (0.5 ml per well) or in 25 cm^2 flasks (5 ml per flask) and incubated for 2 h at 37°C. Non-adherent cells were removed by washing. Adherent cells were cultured in serum-free RPMI 1640 medium containing 25 mM HEPES. Various concentrations of activating agents were added and culture supernatants were harvested 24 h later.
The cytotoxic activity of supernatants was assayed on L-929 cells. Target cells in TCM were seeded in microtiter plates (40,000 cells/well) in the presence of actinomycin D (1 µg/ml) and various dilutions of supernatants in a final volume of 0.2 ml. In control wells target cells were incubated with TCM instead of macrophage supernatant. After 18 h of incubation at 37°C, the nonadherent (dead) cells were removed by washing and the remaining adherent cells were stained with crystal violet (5). Absorbance was read at 545 nm . Each determination was carried out in triplicate.

Cytotoxic factor characterization.
A sample of 100 ml of supernatants from macrophage cultures stimulated with 100 µg/ml of PG from B. megaterium was concentrated by ammonium sulfate precipitation (at a saturation of 80 %) dialysed against 0.1 M Tris-HCl buffer (pH 7.3) applied to a column of Ultrogel AcA 54 (Pharmacia) (2.6 x 90 cm) and eluted with the same buffer. Fractions of 7.5 ml were collected, sterilized by filtration, diluted 1:50 in TCM and tested for cytotoxic activity on L-929 cells. Another sample of concentrated supernatants was dialysed against 0.025 M imidazole-HCl buffer (pH 7.4) and fractionated by chromatofocusing on a PBE 94 column (Pharmacia) (1 x 35 cm) with 300 ml of polybuffer 74 pH 7.4. Fractions of 7.5 ml were collected, dialysed against RPMI 1640, sterilized by filtration, diluted 1:4 in TCM and assayed for cytotoxic activity.

Results

Activation of macrophages for cytotoxicity by PG.

Thioglycollate induced macrophages were not cytotoxic for P815 cells. In the presence of PG they inhibit the growth of the target cells. PG from B. megaterium and S. aureus were more active than PG from M. lysodeikticus and C. poinsettiae (Table 2).

Table 2 Growth Inhibition of P815 Mastocytoma Cells by PG-Stimulated Macrophages

Macrophages	PG	µg/ml	^{3}H-thymidine incorporation	(% growth inhibition)
-	-	-	19,088 ± 2,724	
+	-	-	19,606 ± 3,505	
-	B. megaterium	100	18,339 ± 1,115	(4)
+	B. megaterium	10	5,433 ± 735	(71)
+	B. megaterium	100	2,654 ± 345	(86)
+	S. aureus	10	3,485 ± 472	(82)
+	S. aureus	100	461 ± 135	(97)
+	M. lysodeikticus	10	19,539 ± 2,023	(0)
+	M. lysodeikticus	100	12,550 ± 1,577	(34)
+	C. poinsettiae	10	12,707 ± 2,650	(33)
+	C. poinsettiae	100	7,121 ± 1,628	(62)

Release of a cytotoxic factor by macrophages stimulated by PG.
Macrophages incubated with 1 to 10 µg/ml of PG from B. megaterium or S. aureus released cytotoxic activity in supernatants. PG from M. lysodeikticus and C.poinsettiae were active only at higher concentration (100 µg/ml) and the level of activity released was low (Table 3).

Properties of the cytotoxic factor.
The factor was not inactivated by heating at 56°C for 30 min, but destroyed at 80°C. Cytotoxic activity was not inhibited by serum or trypsin inhibitors such as soja bean trypsin inhibitor. The cytotoxic factor was however not stable in the presence of serum.
After gel filtration a major peak of cytotoxic activity was observed in the range of 60-70,000 daltons (Fig. 1). After chromatofocusing a single peak of activity was present at pH 4.8.

Discussion.

The data presented here show that PG activate the cytotoxic activity of

Table 3. Cytotoxic Activity on L-929 Cells of Supernatants from Macrophage Cultures Stimulated by PG or LPS.

Stimulant	μg/ml	% cytotoxicity*
-	-	0
B. megaterium PG	1	74
	10	92
	100	89
S. aureus PG	1	0
	10	84
	100	93
M. lysodeikticus PG	1	0
	10	2
	100	25
C. poinsettiae PG	1	0
	10	2
	100	32
LPS	0.1	77
	1	88

* Supernatants were assayed at 1:10 dilution

macrophages and induce the release of a cytotoxic factor. PG from B. megaterium and S. aureus were very active. Previous studies have shown that these preparations exerted immunoenhancing activities, including the ability to inhibit tumor growth (3, 4). Adjuvant inactive PG from M. lysodeikticus or C. poinsettiae have only a slight effect on macrophage cytotoxicity.
Several studies have shown that macrophages can release cytotoxic factors after stimulation by various inducers such as LPS. Different factors have been described. Some of them are low molecular weight components : oxygen metabolites (6), thymidine (7) or C3a (8). Arginase is of a molecular weight higher than the factor described in the present study (9). Factors with protease activity have also been reported (10). Unlike these factors the cytotoxic factor we have found is not inhibited by serum or trypsin inhibitors. It shares several properties with a factor released by macrophages after induction by calcium ionophore (11) or BCG and LPS (12). Its relationship with tumor necrosis factor has been suggested (12).

Figure 1. Ultrogel AcA 54 Chromatography of Cytotoxic Factor

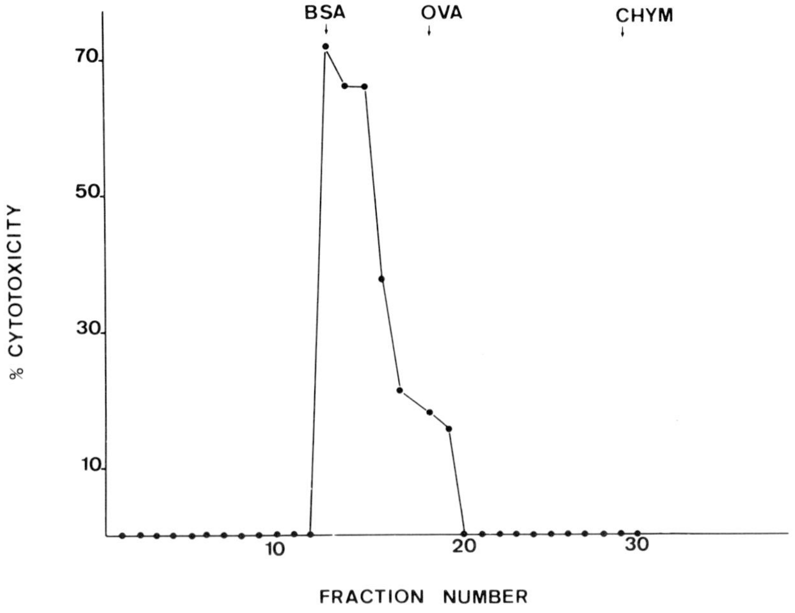

Molecular weight markers were bovine serum albumin (BSA), ovalbumin (OVA) and chymotrypsinogen (CHYM)

The ability of PG from some bacterial species to induce the secretion of a cytotoxic factor by macrophages may explain, at least in part, their activity against experimental tumors. Furthermore our results confirm that macrophages are an important target of PG (13).

References

1. Nauciel, C., J. Fleck, J.P. Martin, M. Mock, H. Nguyen-Huy. 1974. Eur.J. Immunol. 4, 352.

2. Nauciel, C., A.F. Goguel. 1977. J.Natl.Cancer Inst. 59, 1723.

3. Goguel, A.F., G. Lespinats, C. Nauciel. 1982. J.Natl.Cancer Inst. 68, 657.

4. Guenounou, M., A.F. Goguel, C. Nauciel. 1982. Ann.Immunol.(Paris). 133D, 3.

5. Fish, H., G.E. Gifford. 1980. J.Immunol.Methods. 57, 311.

6. Nathan, C.F., S.C. Silverstein, L.H. Brühner, Z.A. Cohn. 1979. J.Exp.Med. 149, 100.

7. Opitz, H.G., D. Niethammer, R.C. Jackson, H. Lemke, R. Huguet, H.D. Flad. 1975. Cell.Immunol. 18, 70.

8. Ferluga, J., H.U. Schorlemmer, L.C. Baptista, A.C. Allison. 1978. Clin.Exp. Immunol. 31, 512.

9. Currie, G.A. 1978. Nature. 273, 758.

10. Reidarson, T.H., G.A. Granger, J. Klostergaard. 1982. J.Natl.Cancer Inst. 69, 889.

11. Drysdale, B., C.M. Zacharchuck, H.S. Shin. 1983. J.Immunol. 131, 2362.

12. Männel, D.N., R.N. Moore, S.E. Mergenhagen. 1980. Infect.Immun. 30, 523.

13. Vacheron, F., M. Guenounou, C. Nauciel, 1983. Infect. Immun. 42, 1049.

STIMULATION OF NONSPECIFIC RESISTANCE AGAINST AEROGENIC VIRAL AND BACTERIAL INFECTIONS BY MURAMYL DIPEPTIDE COMBINED WITH TREHALOSE DIMYCOLATE

K.N. Masihi, W. Brehmer, W. Lange

Robert Koch Institute, Federal Health Office, 1000 Berlin 65, F.R.G.

Introduction

There are a multitude of infections which affect man and domestic animals. Many infectious diseases like smallpox, yellow fever, rubella, rabies, mumps, measles, and hepatitis B can be prevented by the use of appropriate vaccines, but despite major advances, infections still constitute an important source of morbidity and mortality. It is difficult to envisage that specific vaccines or acceptable chemotherapy will be developed in the near future for many infections like the rhinoviruses with over 100 serotypes or the coxsackie B viruses with 23 types. In addition to this, certain of the current viral vaccines are not completely satisfactory. For instance, the nature of the influenza viruses present inherent problems. The frequent antigenic drift of the surface antigens of the influenza viruses necessitates an annual review of the vaccine composition. The efficacy of influenza vaccines is particularly limited in a shift situation when viruses possessing completely different antigens appear. A recent WHO report (No. 698, 1983) stated that "effective control of influenza epidemics or individual infections has not yet been achieved... Particular emphasis should be placed on the feasibility of developing vaccines with a wide specificity of protection against different influenza A viruses or variants". In the light of this background, the concept of enhancing nonspecific host defence mechanisms against infections offers an alternative and an attractive avenue for research.

Biological response modifiers of bacterial origin are widely used as immunostimulants to enhance a variety of host immune

Biological Properties of Peptidoglycan

responses. Detailed investigations on adjuvant-active mycobacteria, corynebacteria, and nocardia have led to the identification of their cell wall structure as a mycolic acid-arabinogalactan-mucopeptide complex which contains immunopotentiating moieties. The minimum structure essential for the bacterial adjuvanticity is the peptidoglycan monomer MDP. In addition to this glycopeptide, other immunopotentiating components present on cell walls are glycolipids like 6,6-diesters of trehalose. We investigated the ability of these biological response modifiers to stimulate nonspecific resistance against aerogenic influenza virus and Mycobacterium tuberculosis infections.

Results and Discussion

NMRI mice pretreated with synthetic MDP, natural TDM, or the combination of MDP+TDM were infected by an aerosol of mouse-virulent influenza virus or pathogenic M.tuberculosis organisms. There was a relatively long interval of 3-4 weeks between pretreatment and aerosol infection. Regardless of the MDP analog employed, no detectable resistance to influenza virus infection was observed (Table 1). None of the MDP analogs that were tested, nor TDM, could restrict the growth of M.tuberculosis; on the contrary, they slightly increased the multiplication of bacteria. Aqueous suspensions of MDP, TDM, or a mixture of the two were also found to be ineffective against both aerosol infections (data not shown). Significant protection against influenza virus-induced mortality and resistance to the multiplication of M.tuberculosis, as indicated by the reduced number of colonies in the lung, were obtained only when MDP or analogs were combined with TDM in 1 % squalane-in-water preparations (Table 1). Similar results were obtained with 6-O- acyl analogs and ubiquinone derivative of MDP. A greater degree of protection was induced by the combination of TDM with lipophilic derivatives like the seryl analog of desmethyl-MDP.

Table 1. Effect of MDP and Analogs Alone and in Combination with TDM Against Influenza Virus and M. tuberculosis Infections in Mice.

MDP analog	TDM	% surviving influenza infection	P	Viable median count of M. tuberculosis per lung (log 10)	P
MDP	-	0		7.43	
	+	60	0.001	5.74	0.05
MDP-seryl	-	0		7.20	
	+	55	0.001	5.38	0.01
MDP-des-seryl	-	0		7.17	
	+	80	0.001	5.19	0.01
MDP-Abu	-	0		7.30	
	+	58	0.001	6.50	0.05
-	+	15		6.86	
vehicle control (1 % squalane)	-	0		6.61	

Groups of 10-20 mice were given an aerosol of A/PR/8/34 influenza virus or M. tuberculosis, strain H37Rv, respectively 3 or 4 weeks after intravenous pretreatment with squalane-in-water preparations containing MDP or analog (300 µg), TDM (75 µg for influenza, 150 µg for M. tuberculosis) or MDP+TDM combination (150+75 µg for influenza, 300+150 µg for M. tuberculosis). MDP groups were compared with respective MDP+TDM group for the statistical analysis.

The efficacy of MDP+TDM combination was further checked by the measurement of the virus concentration in the infected lungs by hemadsorption with guinea-pig erythrocytes on BHK-21 tissue cultures. Peak virus concentrations were attained on day 3 but the MDP+TDM pretreated mice showed markedly decreased levels of virus and the clearance of detectable infectious virus occured earlier in this group (Figure 1). Also, the hemagglutination-inhibiting antibodies could be detected already on day 5 in the MDP+TDM pretreated mice but only on day 7 in the control group. It should be mentioned that all survivors of the pretreated groups showed a long-lasting resistance against lethal rechallenge 6 months later, even if a heterologous influenza A strain was used for rechallenge.

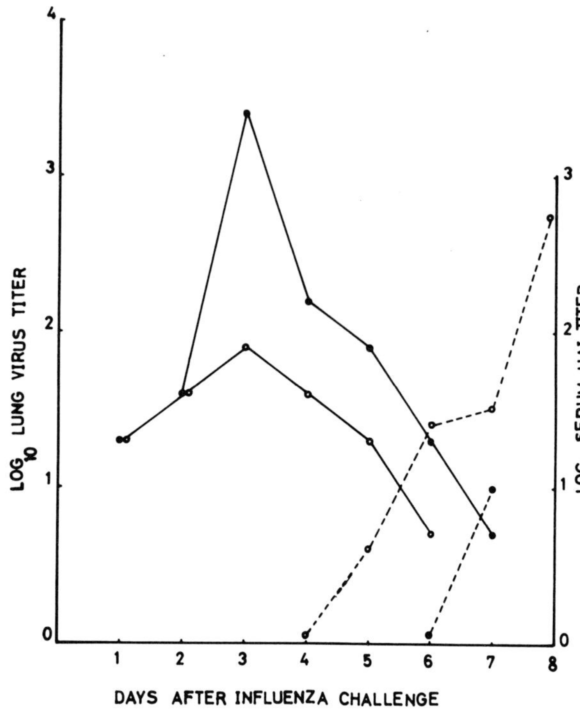

Figure 1. Comparison of lung virus titers and serum hemagglutination-inhibiting antibodies of mice pretreated with 1 % squalane-in-water preparation alone (•) or containing MDP+TDM combination (o).

The mechanisms involved in the protective activity of MDP+TDM combination were investigated in further experiments. Natural killer cell activity of 3-4 week pretreated animals was not significantly different from controls at the time of infection. In contrast to natural killer cells, macrophages appear to be among the target cells for the immunopotentiating activity of MDP+TDM combination 2-4 weeks after pretreatment. To elucidate the role of the macrophage in the resistance induced by MDP+TDM combination, animals that were pretreated 21 days earlier with the immunostimulants were given silica, dextran sulfate, and carrageenan, and 24 hours later, an aerosol of influenza virus.

The results presented in Table 2 show that the resistance to influenza virus that was induced by MDP+TDM combination was abrogated by treatment with all the three agents known to selectively inhibit or impair the macrophage function in vivo.

Table 2. Effects of Macrophage Inhibitory Agents on the In Vivo Resistance of MDP+TDM Combination Against Influenza Virus Infection

Pretreatment (150+75 μg)	Compound	Dose (mg)	% survival	P
MDP+TDM	-	-	90	
MDP+TDM	silica	3	10	0.01
MDP+TDM	carrageenan	1	10	0.01
MDP+TDM	dextran sulfate	1	20	0.01

Groups of 10 mice were pretreated with 1 % squalane-in-water preparations 3 weeks before aerosol infection with A/PR/8/34 influenza virus.
Anti-macrophage compounds were administered intravenously 24 h before infection.

Additional experiments were performed with the aim of detecting functional alterations in the macrophages from MDP+TDM pretreated animals. Phagocytosis stimulates oxidative metabolism of the phagocytic cells which results in increased hexose monophosphate shunt activity and the generation of activated oxygen metabolites like singlet oxygen, superoxide anion, hydrogen peroxide, and hydroxyl radical with the emission of photons. The release of energy in the form of light can be detected by luminol-dependent chemiluminescence. Spleen cells from animals pretreated with the combination of MDP+TDM exhibited markedly enhanced chemiluminescence activity in response to stimulation by Staphylococcus aureus (Figure 2) and similar results were obtained with zymosan.

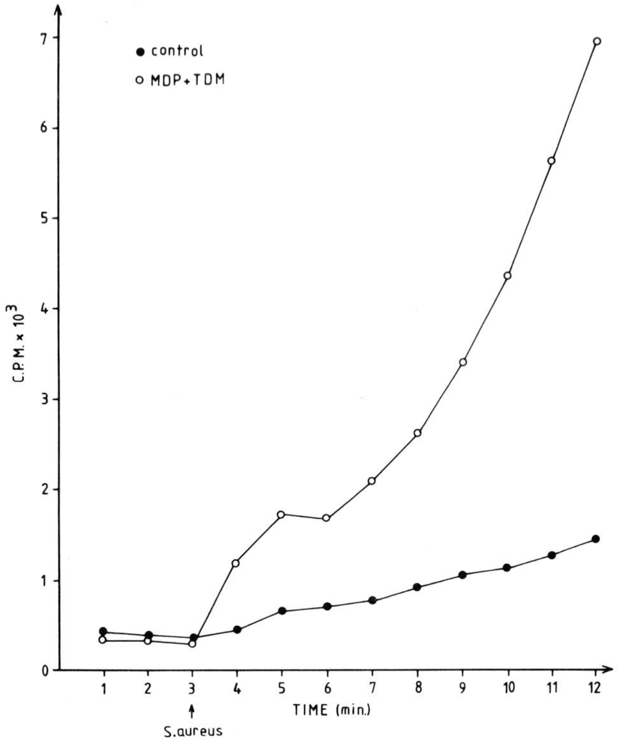

Figure 2. Chemiluminescence response of spleen cells from mice pretreated with 1 % squalane-in-water preparation alone or containing MDP+TDM combination.

In summary, the nonspecific host defence mechanisms against viral and intracellular bacterial infections can be enhanced by potent biological response modifiers and warrants further investigation.

INCREASED ADJUVANT ACTIVITY OF MDP BY DIRECT COUPLING OF MDP TO THE IMMUNOGEN

M. Jolivet, F. Audibert, L. Chedid

Immunothérapie Expérimentale, Institut Pasteur, Paris, France

E.R. Clough

International Minerals and Chemical Corporation, Northbrook, Illinois 60062

Introduction

It is generally assumed that synthetic antigens and vaccines require the addition of an adjuvant. Fifteen years ago several investigators identified the active cell wall structure that can substitute the Mycobacteria in Freund's complete adjuvant (1,2). This molecule is N-acetylmuramyl-L-Ala-D-isoGln, designated muramyl dipeptide or MDP (3). This glycopeptide was effective in stimulating the immune response _in vivo_ even when administered in saline (4). However, after intravenous injection of labelled MDP solution about 65 % of the injected dose was recovered in the urine within 30 min (5). In order to delay its elimination, MDP can be conjugated to the immunogen (6).

In our study, we coupled MDP derivatives to different immunogens including a natural vaccine (diphtheria toxoid) and two synthetic vaccines (carboxy terminal peptide of HCG beta subunit and a synthetic malarial peptide copying a part of the circumsporozoite protein coupled to tetanus protein).

Results

1. Diphtheria toxoid

Toxin was treated with formalin or with glutaraldehyde in presence of L-lysine or a derivative of MDP, the MDP-L-lysine (MDP-Lys) which has a reactive amino group and the same activities as MDP.

Biological Properties of Peptidoglycan

The conjugates contained 8 to 11 molecules of MDP per molecule of diphtheria toxoid and presented a single band by SDS PAGE (data not shown).

Table 1. Antibody Response to Toxoids Conjugated to MDP-L-Lysine.

Immunogen	Coupling agent	Primary response	Secondary response
A	formalin	2.5	1,357
B	formalin	4.5	2,833
C	formalin	15	7,071
D	glutaraldehyde	6.5	2,218
E	glutaraldehyde	12	3,500
F	glutaraldehyde	17	8,722

Toxoid was formed according to the method of Relyveld (7) with the addition of MDP-Lys to the reaction mixture.
(A) Toxin treated by formalin (3Lf = 7.5 μg protein), (B) Toxoid (3Lf) mixed with 0.5 μg of MDP, (C)Toxoid (3Lf) linked to 0.5 μg of MDP, (D) Toxin treated by glutaraldehyde (3Lf), (E)Toxoid (3Lf) mixed with 0.8 μg of MDP, (F)Toxoid (3Lf) linked to 0.8 μg of MDP. Mice (eight per group) received subcutaneously 3 Lf toxoids in PBS. On day 30, animals were boosted with the same amount of toxoid without adjuvant. Mice were bled at days 21 and 36. Sera were titrated by passive hemagglutination and the results meaned. Protein content in each conjugate was measured by Folin reaction and MDP content by a colorimetric method (8).

As can be seen on Table 1, in secondary response 0.5 μg of MDP-Lys derivative mixed with the toxoid enhanced the antibody titer, but a more marked increase was observed in the case of MDP-Lys conjugated to the protein with formalin or glutaraldehyde. In this case, MDP was not immunogenic because anti-MDP antibody was not detected using ELISA.

2. Synthetic vaccines

Synthetic vaccines are usually constructed with a synthetic antigen coupled to a protein carrier and administered with adjuvants. In laboratory models, tetanus toxoid is often used as a carrier and Freund's complete adjuvant as an adjuvant. However, FCA is unacceptable for human use because of its unwanted side effects. Thus, we have constructed two hapten-carriers with or without MDP-Lys coupled to the conjugate.

The first example is constructed from synthetic carboxy terminal peptide of HCG beta subunit (residues 109-145). This peptide is lacking in the human luteinizing hormone (hLH) and other pituitary derived glycoprotein hormone (9).

Table 2. Antibody Response of Rabbits Injected with Synthetic Beta HCG subunit peptide-tetanus toxoid.

Rabbit	Immunogen[1]	Adjuvant	Primary response[2] anti-βhCG	Primary response[2] anti-TT	Secondary response[2] anti-βhCG	Secondary response[2] anti-TT
1	pep-TT	MDP	37	1	6	4
2	pep-TT-MDP		82	4	50	32
3	pep-TT	FCA	53	52	100	256
4	pep-TT	FIA + MDP	25	8	32	128
5	pep-TT-MDP		94	26	97	64

1. The 109-145 synthetic peptide was conjugated to tetanus toxoid (TT) with glutaraldehyde used as the coupling agent, according to a previously described method (10). Fourteen μg of MDP are present in 200 μg of conjugate.
   Five rabbits received 200 μg of conjugate (multiple intradermal sites) in PBS without or with adjuvants (MDP 100 μg, FCA or FIA). On day 75, animals were boosted with the same amount of conjugates in PBS. Rabbits were bled on days 40 and 97 and sera tested.
2. Anti-βhCG antibodies were measured by a radioimmunoassay (RIA) and results were expressed in percentage of binding. Anti-tetanus antibodies titers are expressed as the inverse of the maximal dilution of serum agglutaning tetanus-sensitized sheep erythrocytes x 10^{-3}.

As can be seen on Table 2, a strong primary response was observed against HCG beta subunit when MDP was coupled to the conjugate. In aqueous medium, the percentage of binding with the β-subunit is of 82 % when MDP is coupled compared to 37 % when MDP is only mixed with the conjugate. The same results were obtained in oil medium (rabbits 4 and 5).

Analysis of the secondary response show that a strong antibody response is obtained against the carrier and all the rabbits. Nevertheless, the carrier antibody reponses of rabbits treated with the MDP conjugates were only slightly elevated as compared to rabbits treated with pep-TT mixed with adjuvant.

The second example of a synthetic vaccine is constituted by a synthetic 24 amino acid peptide representing the immunodominant epitope of the sequence of the circumsporozoite protein of P.knowlesi (12). This peptide contains two additional amino acids : tyrosine at the N-terminus and cystein at the C-terminus. Moreover, Gysin (13) has shown that antibodies to the peptide react with the surface membrane of sporozoites and that they neutralize the infectivity of the parasite.

In our study, we have made conjugates containing different substitution ratios of the synthetic malarial peptide and MDP-Lys on tetanus toxoid.

Figure 1.

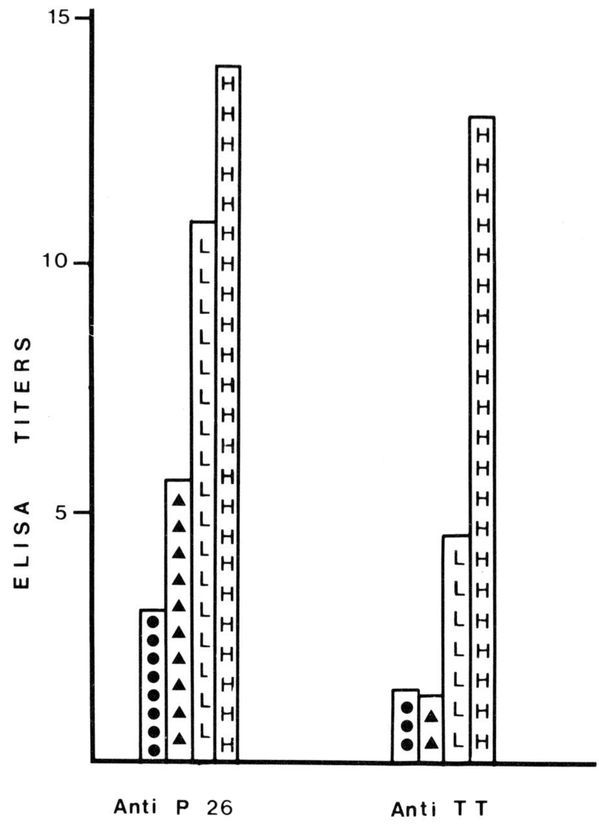

This peptide (P26) represents 24 amino acids present in the natural sequence of the P.knowlesi circumsporozoite protein and two additional amino acids.
Pep-TT was made according to a previously described method (10). For conjugates L (low), H (high) varying amounts of the peptide and MDP-Lys were cross-linked to tetanus toxin using glutaraldehyde. Amino equivalents peptide/MDP-Lys ratio are the following : L 1/0.5, H 1/2.
Six adult female BALB/c mice received pep-TT (50 μg in PBS) ( ) or in presence of MDP (100 μg) ( ) or conjugate L(L) or conjugate H(H). On day 21, animals were boosted with the same amount of conjugate in PBS. Mice were bled on day 28. Sera from animals in each group were pooled and tested for total anti-peptide and anti-tetanus toxoid antibodies by ELISA as previously described (11). Titers

are expressed as inverse of the dilution of serum giving an O.D. in ELISA of two times background x 10^{-3}.

The results in Figure 1 show that the mice treated with conjugates containing MDP-Lys had higher antibody responses to both the peptide and the carrier. Moreover, the secondary antibody response of the conjugate containing 0.3 µg of MDP-Lys is about the same as when the same quantity of antigen is mixed with 100 µg of this adjuvant.
Antibodies were tested in circumsporozoite protein assay and were shown to be effective.

Discussion

These experiments demonstrate that the covalent linkage of MDP-Lys to an immunogen molecule increases the adjuvant activity of MDP. In addition, it does not seem that MDP-Lys became a hapten since no anti-MDP antibody was detected by ELISA.
Pharmacokinetic studies showed that 5 % of MDP was recovered in urine 30 minutes after the injection of MDP linked to the macromolecule in comparison with 65 % of free MDP.
If synthetic peptides are to be used for the production of future vaccines, methods of increasing the immunogenicity of conjugates must be found. Our results demonstrated that very low doses of adjuvant can stimulate the production of protective antibodies. Finally, we have constructed a synthetic vaccine with three parts which can provoke protective antibodies : an antigenic (synthetic peptide copying part of protein natural sequence), an immunogenic (protein carrier) and adjuvant (MDP derivative). This type of vaccine excludes the parts of conventional vaccines which provoke side effects.

Acknowledgements

We thank Dr. Jean Choay and Dr. Pierre Lefrancier for the samples of synthetic compounds. We are grateful to M. Hattab for his technical assistance, and to C. de Champs for typing the manuscript.

References

1. Lederer, E. 1971. J.Med.Chem. 23, 819.
2. Azuma, I., S. Kishimoto, Y. Yamamura, J.F. Petit. 1971. Jap.J. Microbiol. 15, 193.
3. Ellouz, F., A. Adam, R. Ciorbaru, E. Lederer. 1974. Biochem. Biophys.Res.Com. 59, 1317.
4. Audibert, F., L. Chedid, P. Lefrancier, J. Choay. 1976. Cell. Immunol. 21, 243.
5. Parant, M., F. Parant, L. Chedid, A. Yapo, J.F. Petit, E. Lederer. 1979. Int.J.Immunopharmac. 1, 35.
6. Arnon, R., M. Sela, M. Parant, L. Chedid. 1980. Proc.Natl.Acad. Sci. USA 77, 6769.
7. Relyveld, E.H. 1973. C.R.Acad.Sc.Paris 277, 613.
8. Reissig, J.L., J.L. Strominger, L.F. Leloir. 1956. J.Biol.Chem. 217, 959.
9. Morgan, F.J., S. Birken, R.E. Canfield. 1975. J.Biol.Chem. 250, 5247.
10. Audibert, F., M. Jolivet, L. Chedid, R. Arnon, M. Sela. 1982. Proc.Natl.Acad.Sci. USA 79, 5042.
11. Jolivet, M., F. Audibert, E.H. Beachey, A. Tartar, H. Gras-Ma Masse, L. Chedid. 1983. Biochem.Biophys.Res.Com. 117, 359.
12. Godson, G.N., J. Ellis, P. Svec, D.H. Schlesinger, V. Nussenzweig. 1983. Nature 305, 29.
13. Gysin, J., J. Barnwell, D.H. Schlesinger, V. Nussenzweig, R.S. Nussenzweig. 1984. J.Exp.Med. 160, 935.

CONTRIBUTORS AND PARTICIPANTS

R. Barot-Ciorbaru
Université Paris Sud,
Institut de Biochimie,
Bât 432, 91405 Orsay,
France

G. Barratt
Université Paris Sud,
Institut de Biochimie,
Bât 432, 91405 Orsay,
France

G. Baschang
CIBA-GEIGY Limited,
K-122.4.39,
Postfach, CH-4002 Basel,
Switzerland

H. Bauer
Institut für Kristallographie, Takustr. 6,
D-1000 Berlin 33
Germany

W.G. Bessler
Arbeitsbereich Mikrobiologie
und Immunologie
Universität Tübingen,
Auf der Morgenstelle 28
D-7400 Tübingen
Germany

U. Bläsi
Institut für Genetik und
Mikrobiologie der
Universität München,
Maria-Wardstr. 1a
D-8000 München 19.
Germany

H. Brade and L. Brade
Forschungsinstitut Borstel,
Parkallee 4b
D-2061 Borstel,
Germany

E. Bräutigam
Institut für Biologie II,
Mikrobiologie, Albert-Ludwigs-Universität,
Schänzlestr. 1,
Germany

H. Brunner
Bayer AG, Pharma-Forschungszentrum, Institut für Chemotherapie,
D-5600 Wuppertal,
Germany

B. Christensson
Department of Infectious
Diseases, University
Hospital,
S-221 85 Lund,
Sweden

R. Dziarski
Indiana University School of
Medicine, Northwest Center
for Medical Education,
3400 Broadway, Gary, IN 46408,
U.S.A.

J. Endl
Boehringer Mannheim GmbH,
Bahnhofstr. 9-15,
D-8132 Tutzing,
Germany

H.B. Evans
Department of Immunology,
Royal Liverpool Hospital,
P.O. Box 147,
Liverpool, L69 3BX, U.K.

H. Feucht
Institut für Immunologie
der Universität München,
Schillerstr. 42,
D-8000 München 2, Germany

F. Fiedler
Institut für Mikrobiologie
und Genetik der
Universität München
Maria-Ward-Str. 1a
D-8000 München 19
Germany

H. Finger
Städtische Krankenanstalten
Krefeld, Institut für Hygiene und Laboratoriumsmedizin, Lutherplatz 40,
D-4150 Krefeld 1
Germany

A. Fleer
Department of Clin.
Bacteriology, University
Hospital, Utrecht,
Netherlands

H. Formanek
Botanisches Institut der
Universität München
Menzingerstr. 67
D-8000 München 19, Germany

S. Formanek
Botanisches Institut der
Universität München
Menzingerstr. 67
D-8000 München 19, Germany

A. Fox
Department of Microbiology
and Immunology, USC School
of Medicine, Columbia
SC 29208, U.S.A.

N. Franken
Boehringer Mannheim GmbH
Bahnhofstr. 9-15
D-8132 Tutzing, Germany

K. Gadilhe
Lehrstuhl für Mikrobiologie
Technische Universität
München, Arcisstr. 21,
D-8000 München 2, Germany

I. Ginsburg
Department of Oral Biology
Hebrew University, Hadassah
School of Medicine, Ein Kerem,
Jerusalem, Israel

J. Gmeiner
Institut für Mikrobiologie
Technische Hochschule
Darmstadt
Schnittspahnstr. 9
D-6100 Darmstadt, Germany

P. Goroncy-Bermes
Institut für Medizinische
Mikrobiologie und Immunologie
Ruhr-Universität
Universitätsstr. 150
D-4630 Bochum 1, Germany

M. Guinand
Université Cl. Bernard
Laboratoire de Biochimie
microbienne, Lyon I
43 Bd du 11 novembre 1918
69622 Villeurbanne, France

F. Hagen
Lehrstuhl für Mikrobiologie
Technische Universität
München, Arcisstr. 21
D-8000 München 2, Germany

R. Hakenbeck
Max-Planck-Institut
für Molekulare Genetik
Ihnestr. 63-73
D-1000 Berlin 33, Germany

A. Hartinger
Städtisches Krankenhaus
München-Bogenhausen
Institut für Medizinische
Mikrobiologie, Immunologie
und Krankenhaushygiene
Englschalkingerstr. 77
D-8000 München 81
Germany

M. Hasmann
Lehrstuhl für Mikrobiologie
Technische Universität München
Arcisstraße 21
D-8000 München 2
Germany

J. van Heijenoort
Biochimie-Bâtiment 432
Université Paris Sud
Orsay 91405
France

B. Henkel
Lehrstuhl für Mikrobiologie
Technische Universität München
Arcisstraße 21, D-8000 München 2
Germany

C. Herzog
Department für Innere
Medizin
Universitätskliniken
CH-4031 Basel
Switzerland

B. Heymer
Universität Ulm
Abteilung für Pathologie
Oberer Eselsberg
D-7900 Ulm, Germany

J.-V. Höltje
Max-Planck-Institut für
Entwicklungsbiologie
Abteilung Biochemie
Spemannstr. 35
D-7400 Tübingen, Germany

K. Huber
Fachbereich Biologie (10)
Mikrobiologie, Technische
Hochschule Darmstadt
Schnittspahnstr. 9
D-6100 Darmstadt, Germany

L. Johannsen
Robert-Koch-Institut
Nordufer 20
D-1000 Berlin 65
Germany

M. Jolivet
Immunothérapie Expérimentale
Institut Pasteur, 28, rue du
Dr. Roux, 75724 Paris cedex 15
France

U.J. Jürgens
Institut für Biologie II
Mikrobiologie, Albert-
Ludwigs-Universität
Schänzlestr. 1
D-7800 Freiburg, Germany

O. Kandler
Botanisches Institut der
Universität München
Menzingerstr. 67
D-8000 München 19
Germany

H.J. Kolb
Städtisches Krankenhaus
München-Harlaching
Klinisch-chemisches
Institut, Sanatoriumsplatz 2
D-8000 München 90
Germany

T. Komuro
Max-Planck-Institut für
Immunbiologie
Stübeweg 51
D-7800 Freiburg
Germany

S. Kotani
Department of Microbiology
and Oral Microbiology
Osaka University
Dental School
1-8 Yamadoakoa, Suita
Osaka 565, Japan

H.P. Kroll
Bayer AG
Pharma-Forschungszentrum
Institut für Chemotherapie
D-5600 Wuppertal
Germany

J.M. Krueger
Department of Physiology
The Chicago Medical School
3333 Green Bay Road
North Chicago, IL 60064
U.S.A.

T. Kuchenbauer
Kinderklinik Josefinum
Kapellenstr. 30
D-8900 Augsburg
Germany

E. Kwa
Robert-Koch-Institut
Nordufer 20
D-1000 Berlin 65
Germany

H. Labischinski
Robert-Koch-Institut
Nordufer 20
D-1000 Berlin 65
Germany

H. Leying
Institut für Medizinische
Mikrobiologie und Immunologie
Ruhr-Universität
Universitätsstr. 150
D-4630 Bochum 1, Germany

M. Loos
Institut für Medizinische
Mikrobiologie
Universität Mainz
Obere Zahlbacherstr. 67
D-6500 Mainz, Germany

W. Lubitz
Institut für Genetik
und Mikrobiologie
Universität München
Maria-Ward-Str. 1a
D-8000 München 19, Germany

W. Marget
Universitäts-Kinderklinik
Lindwurmstr. 4
D-8000 München 2
Germany

H.H. Martin
Fachbereich Biologie (10)
Mikrobiologie
Technische Hochschule Darmstadt
Schnittspahnstr. 9
D-6100 Darmstadt, Germany

S.A. Martin
Massachusetts Institute
of Technology
Building 56-035
Cambridge, MA 02139
U.S.A.

K.N. Masihi
Robert-Koch-Institut
Nordufer 20
D-1000 Berlin 65, Germany

J.v. Mayenburg
Technische Universität
München
Fakultät für Medizin
Ismaningerstr. 22
D-8000 München 80, Germany

C. Nauciel
Immunochimie des Protéines
Institut Pasteur
75724 Paris Cedex 15
France

H. Nowack
Luitpold-Werk
Chemisch-pharmazeutische
Fabrik
Zielstattstr. 9-11
D-8000 München 70
Germany

M. Parant
Immunothérapie Expérimentale
Institut Pasteur
28 rue du Dr. Roux
75724 Paris Cedex 15
France

V. Paulicks
Lehrstuhl für Mikrobiologie
Technische Universität
München
Arcisstr. 21
D-8000 München 2, Germany

J.-F. Petit
Institut de Biochimie, Bat 432
Université Paris Sud
91405 Orsay Cedex, France

D. Rehbinder
Boehringer Ingelheim
Diagnostica
Gutenbergstr. 3
D-8046 Garching, Germany

S. Reißenweber
Lehrstuhl für Mikrobiologie
Technische Universität München
D-8000 München 2, Germany

B. Rieth
Institut für Biologie II
Mikrobiologie
Albert-Ludwigs-Universität
Schänzlestr. 1
D-7800 Freiburg, Germany

E.Th. Rietschel
Forschungsinstitut Borstel
Institut für Experimentelle
Biologie und Medizin
Parkallee 1-42
D-2061 Borstel, Germany

R.S. Rosenthal
Department of Microbiology
and Immunology
Indiana University
School of Medicine
Indianapolis, IN 46223, USA

M. Rýc
Institute of Hygiene
Epidemiology
Srobarova 48
Praha 10, CSSR

K.H. Schleifer
Lehrstuhl für Mikrobiologie
Technische Universität
München
Arcisstr. 21
D-8000 München 2, Germany

K. Schorn
Lehrstuhl für Mikrobiologie
Technische Universität München
Arcisstr. 21
D-8000 München 2, Germany

R. Schubert
Hygiene-Institut
Paul-Ehrlich-Str. 40
D-6000 Frankfurt, Germany

F. Schumacher-Perdreau
Hygiene-Institut
Goldenfelsstr. 21
D-5000 Köln 41, Germany

G. Schumann
Ciba-Geigy Ltd.
R-1056.321
P.O. Box
CH-4002 Basel, Switzerland

J.H. Schwab
Department of Microbiology
and Immunology
804 FLOB 231-H
University of North Carolina
Chapel Hill, N.C. 27514, USA

P.H. Seidl
Lehrstuhl für Mikrobiologie
Technische Universität
München
Arcisstr. 21
D-8000 München 2, Germany

J. Seidl
Lehrstuhl für Mikrobiologie
Technische Universität München
Arcisstr. 21
D-8000 München 2, Germany

J. Smalley
Department of Immunology
Royal Liverpool Hospital
P.O. Box 147
Liverpool, L69 3BX, U.K.

H. Takada
Max-Planck-Institut für
Immunbiologie
Stübeweg 51
D-7800 Freiburg, Germany

J.P. Tenu
Institut de Biochimie, Bât
432, Université de Paris Sud
91405 Orsay, France

J. Tomašić
Institute of Immunology
Department of Radioimmunology
Rockefellerova 10
41000 Zagreb, Yugoslavia

T. Torsvik
Institute of Microbiology
and Plant Physiology
Allegaten 70
University of Bergen
5000 Bergen, Norway

K.-D. Tympner
Städtisches Krankenhaus
München-Harlaching
Kinderklinik
Sanatoriumsplatz 2
D-8000 München 90, Germany

J. Verhoef
Department of Clin. Bacte-
riology, University Hospital
Utrecht, The Netherlands

E.D. Wachsmuth
CIBA-GEIGY Ltd.
CH-4002 Basel, Switzerland

M. Wagner
Zentralinstitut für Mikro-
biologie und Experimentelle
Therapie
Beutenbergstr. 11
DDR-Jena, GDR

J. Wecke
Robert-Koch-Institut
Nordufer 20
D-1000 Berlin 65, Germany

L. Weiss
Institut für Klinische Chemie
Städtisches Krankenhaus
München-Harlaching
Sanatoriumsplatz 2
D-8000 München 90, Germany

H.I. Wergeland
Department of Microbiology
and Immunology
The Gade Institute
MFH-Bygget
5016 Haukeland Sykehus
Norway

C.H. Wirsing von Koenig
Städtische Krankenanstalten
Krefeld
Institut für Hygiene und
Laboratoriumsmedizin
Lutherplatz 40
D-4150 Krefeld 1, Germany

M. Zaoral
Institute of Organic Chemistry
and Biochemistry
Czechoslovak Academy of Sciences
166 10 Prague 6, Flemingovo nâm 2,
CSSR

E. Zauner
Lehrstuhl für Mikrobiologie
Technische Universität
München
Arcisstr. 21
D-8000 München 2, Germany

A.R. Zeiger
Jefferson Medical College of
Thomas Jefferson University
Department of Biochemistry
Philadelphia 19107, USA

S. Zeller
Lehrstuhl für Mikrobiologie
Technische Universität
München
Arcisstr. 21
D-8000 München 2, Germany

R. Ziegelmaier
Behringwerke AG, Mikrobiologie
Diagnostika
D-3550 Marburg, Germany

P. Zwerenz
Lehrstuhl für Mikrobiologie
Technische Universität
München
Arcisstr. 21
D-8000 München 2, Germany

AUTHOR INDEX

SUBJECT INDEX

The abbreviation of Peptidoglycan is PG.

M